CHARGE SENSITIVITY APPROACH TO ELECTRONIC STRUCTURE AND CHEMICAL REACTIVITY

Advanced Series in Physical Chemistry

Editor-in-Charge

Cheuk-Yiu Ng, *Ames Laboratory USDOE, and Department of Chemistry, Iowa State University, USA*

Associate Editors

Paul F. Barbara, *Department of Chemistry, University of Minnesota, USA*

Sylvia T. Ceyer, *Department of Chemistry, Massachusetts Institute of Technology, USA*

Hai-Lung Dai, *Department of Chemistry, University of Pennsylvania, USA*

Benny Gerber, *The Fritz Haber Research Center and Department of Chemistry, The Hebrew University of Jerusalem, Israel, and Department of Chemistry, University of California at Irvine, USA*

James J. Valentini, *Department of Chemistry, Columbia University, USA*

Vol. 1: Physical Chemistry of Solids: Basic Principles of Symmetry and Stability of Crystalline Solids
H. F. Franzen

Vol. 2: Modern Electronic Structure Theory
ed. D. R. Yarkony

Vol. 3: Progress and Problems in Atmospheric Chemistry
ed. J. R. Barker

Vol. 4: Molecular Dynamics and Spectroscopy by Stimulated Emission Pumping
eds. H.-L. Dai and R. W. Field

Vol. 5: Laser Spectroscopy and Photochemistry on Metal Surfaces
eds. H.-L. Dai and W. Ho

Vol. 6: The Chemical Dynamics and Kinetics of Small Radicals
eds. K. Liu and A. Wagner

Vol. 7: Recent Developments in Theoretical Studies of Proteins
ed. R. Elber

Advanced Series in Physical Chemistry – Vol. 8

CHARGE SENSITIVITY APPROACH TO ELECTRONIC STRUCTURE AND CHEMICAL REACTIVITY

Roman F Nalewajski
Jacek Korchowiec

K.Gumiński Dept. of Theoretical Chemistry, Jagiellonian University, Cracow, Poland

World Scientific
Singapore • New Jersey • London • Hong Kong

Published by

World Scientific Publishing Co. Pte. Ltd.

P O Box 128, Farrer Road, Singapore 912805

USA office: Suite 1B, 1060 Main Street, River Edge, NJ 07661

UK office: 57 Shelton Street, Covent Garden, London WC2H 9HE

British Library Cataloguing-in-Publication Data
A catalogue record for this book is available from the British Library.

CHARGE SENSITIVITY APPROACH TO ELECTRONIC STRUCTURE AND CHEMICAL REACTIVITY

ISBN 981-02-2245-9

Printed in Singapore.

PREFACE

When investigating a variety of chemical phenomena, chemists have always tried to formulate adequate concepts and criteria for understanding the geometrical/electronic structure and reactivity of molecules; trends in the site or path selectivities of chemical processes, at least in a family of related species, could then be predicted using such information on isolated reactants. Even before the advent of quantum chemistry some important electronic structure concepts have been formulated, e.g., the Lewis electron pair model, and classical electrostatic considerations have been employed in an attempt to rationalize selected topics in molecular structure and chemical reactivity.

Very successful static reactivity indices have been formulated already within the Hückel molecular orbital theory, to be used to diagnose relative local preferences of systems with conjugated π-bonds towards a localized attack by a small electrophilic, nucleophilic, or radical agents. Such criteria, including the π-electron densities, polarizabilities, free valences (Coulson, Longuet Higgins, Mc Weeney, Thiele, Werner), densities of frontier electrons (Fukui), hyperconjugation indices (Fukui, Yonezawa, Fueno), and stabilization energy (Nakajima), have substantially increased our understanding of elementary reactivity trends of π-electron systems. The index measuring the localization energy in the transition-state of the electrophilic substitution in aromatic systems (σ-complex) has also been proposed (Wheland and Dewar) to reflect trends in activation energy. The *valence state electron pair repulsion* (VSEPR) model (Gillespie), a natural extension of the Lewis idea, has also become one of the most successful and widely used qualitative tool for predicting the geometrical arrangement of ligands in the closed-shell molecules. Finally, the electrostatic potential distributions around molecules (Scrocco, Tomasi) have also been advocated as reliable guiding tools for diagnosing chemical reactivity trends of an initial stage of the reactant approach toward each other.

These early, one-reactant indices have already identified the major aspects of the static (electronic) theory of chemical reactivity. More specifically, they have emphasized the open character of reactants, capable of accepting (donating) electrons from (to) the reaction partner (inter-reactant charge transfer), and the crucial role of the reactant responses (generalized polarizabilities) to the external potential perturbation due to the presence of the other reactant. The dominant, frontier orbital channels of such general charge polarizabilities (sensitivities) have also been properly identified by these early orbital theories.

Some of these early concepts have also been employed as two-reactant criteria, which take into account the electronic properties of both, be it non-polarized reactants. For example, many important reactivity preferences can be rationalized by examining the overall matching of the electrostatic potentials or the interaction between frontier orbitals of large reactants (Fukui, Fujimoto, Yonezawa), e.g., the Woodward-Hoffmann rules, etc. A modern extension of the VSEPR model, the principle of the spatial complementarity of distributions of the Laplacian of the reactant electron densities (Bader et al.), which combines the regions of the electron depletion of one reactant with the regions of the electron concentration of the other reactant, represents yet another successful two-molecule approach to the electronic structure and chemical reactivity.

In the past two decades an alternative approach has been developed, using the electronegativity (of Pauling and Mulliken), hardness/softness (of Pearson and Parr) and related concepts. These quantities have great intuitive appeal and, as demonstrated by Parr at al., can be rigorously defined in the density functional theory of Kohn, Hohenberg and Sham. Some important rules of chemistry, e.g., the Sanderson *electronegativity equalization* (EE) principle, the *hard/soft acids and bases* (HSAB) rules of Pearson, and the structural rules of Gutmann, have all been rationalized in terms of these fundamental properties. The Fukui reactivity indices can also be formulated in terms of the relevant charge responses of open systems (so called Fukui function quantities), introduced by Parr and Yang. These concepts have been supplemented by the corresponding non-local, second-order properties, e.g., the hardness/softness kernels and matrices, which form the basis of a systematic treatment of molecular responses to external perturbations; they also define the relevant quadratic Taylor expansions of the energy in reactive systems. This so called *charge sensitivity analysis* (CSA), or *electronegativity equalization method* (EEM), offers a thermodynamic-like description of the system equilibrium charge redistributions, in response to an electron population and/or external potential displacements (perturbations), basic independent "coordinates" of chemical processes. This novel approach can be formulated at alternative levels of resolution, i.e. of discretizing the fundamental Euler equation of density functional theory; they range from the local description, in terms of the electron density distribution, to a global one, using the number of electrons as the system populational variable.

Although the conceptual and computational apparatus of the CSA is still being developed, it has already been shown to provide an attractive and efficient tool for predicting reactivity trends in a variety of chemical reactions, including processes in the heterogeneous catalysis. Various environmental effects in reactive systems can be directly described in terms of charge sensitivities. In particular, there has been a remarkable progress in treating the donor-acceptor (charge transfer) interactions at both conceptual and applicative levels. The new approaches to chemical reactivity problems in an open system of reactants have also been suggested and alternative reference frames in the electron population space of constituent atoms have been examined. This fast progress calls for a monographic survey presenting the physical basis of this novel approach, its density functional roots, and summarizing a variety of concepts and typical applications. The present monograph is an exposition of the charge sensitivity approach to problems in the theory of molecular structure and chemical reactivity.

After an introductory outlook we present the charge sensitivity analysis and its illustrative applications in the four following chapters, starting from atomic development (Chapter 2). In Chapter 3 the basic concepts, relations, computational schemes and alternative collective populational coordinate systems for general molecular systems are presented. Specific concepts for the donor-acceptor reactive systems and mapping relations between molecular populational and geometrical degrees of freedom are the subject of Chapter 4. We illustrate the applications of the CSA to model catalytic systems in Chapter 5, focusing mainly on the two-reactant treatment of the *in situ* (isoelectronic) processes in chemisorption systems. All illustrative examples correspond to the atomic resolution, in which atomic net charges (electron populations) and the corresponding external potential variables of constituent atoms represent the independent degrees of freedom. The scope of these four chapters has been mainly limited to the research carried out in our own group in Cracow. In Chapter 6 we survey the local charge sensitivities in Kohn-Sham theory, starting with the previous development by Parr, Yang, Berkowitz, Cohen et al.; we next supplement it with the constrained equilibrium local theory required to describe interacting, mutually closed reactants. Finally, in Chapter 7 an overall exposition of the elements of the CSA in the orbital resolution is given.

Throughout the book the density functional roots of charge sensitivity analysis are emphasized, sources of the charge sensitivity data are outlined, and the relevant energy expressions in terms of charge sensitivities are given. Although we have intended to provide a fairly complete and rigorous exposition of the background material, this monograph can not be considered as a systematic review of all recent developments in the field. In writing this book we hope to alert a general chemical community to the growing potential, variety of concepts, and future prospects of the CSA. A special emphasis has been placed upon concepts and quantities of already demonstrated applicability to systems involving large reactants. We have also stressed the method versatility and flexibility in describing various molecular fragments, e.g., atoms-in-molecule, functional groups, etc., under different experimental or hypothetical conditions.

The intended reading audience for this monograph is the broad community of chemists, and particularly physical chemists. We would hope that this book will offer to the reader a new outlook on molecular systems, close to an intuitive chemical thinking, and a deeper understanding of responses of reactants to typical perturbations. We would also expect that it will help all those interested in studying and modelling elementary chemical reactions, and particularly, catalytic processes. This monograph can also be used as supplementary reading for graduate students, who are interested in the physical justification of mostly intuitive concepts and rules of chemistry.

The senior author would like to thank Professor Robert G. Parr for a constant inspiration over a number of years. Members of the Quantum Chemistry Group of the Jagiellonian University, past and present, have contributed to the results reported in this book. We are also greatly indebted to Professors Wilfried J. Mortier and Karl Jug, and to members of their research groups, for many stimulating discussions on the CSA. We have also enjoyed useful comments on catalytic systems made by Professors Robert A. Schoonheydt and Małgorzata Witko, and their associates. Mr. Artur Michalak has made a great effort in preparing some of the book figures. Last but not least, the authors gratefully acknowledge a financial support from the State Committee for Scientific Research in Poland and from the Commission of European Communities (COST D3 and D5 Actions).

Cracow R. F. N.
September 1996 J. K.

This book is dedicated to the memory of my parents
R. F. N.

ADVANCED SERIES IN PHYSICAL CHEMISTRY

INTRODUCTION

Many of us who are involved in teaching a special-topic graduate course may have the experience that it is difficult to find suitable references, especially reference materials put together in a suitable text format. Presently, several excellent book series exist and they have served the scientific community well in reviewing new developments in physical chemistry and chemical physics. However, these existing series publish mostly monographs consisting of review chapters of unrelated subjects. The modern development of theoretical and experimental research has become highly specialized. Even in a small subfield, experimental or theoretical, few reviewers are capable of giving an in-depth review with good balance in various new developments. A thorough and more useful review should consist of chapters written by specialists covering all aspects of the field. This book series is established with these needs in mind. That is, the goal of this series is to publish selected graduate texts and stand-alone review monographs with specific themes, focusing on modern topics and new developments in experimental and theoretical physical chemistry. In review chapters, the authors are encouraged to provide a section on future developments and needs. We hope that the texts and review monographs of this series will be more useful to new researchers about to enter the field. In order to serve a wider graduate student body, the publisher is committed to making available the monographs of the series in a paperbound version as well as the normal hardcover copy.

Cheuk-Yiu Ng

CONTENTS

Preface v
Introduction viii
List of Acronyms xiii

1. INTRODUCTORY SURVEY 1
   1. Global and Regional Equilibria and the Associated Charge Sensitivities 2
   2. The Chemical Potential/Electronegativity Equalization 4
   3. Alternative State Variables and Principal Derivatives 5
   4. Modelling the AIM Hardness Matrix 8
   5. Direct and Indirect Modelling of Charge Responses 10

2. ATOMIC CHARGE SENSITIVITIES 11
   1. Alternative Representations of Atomic States, Potentials, Derivatives and Their Reduction 13
   2. Qualitative Discussion of Atomic Sensitivities 20
   3. Reduction of Derivatives: Equipotential Sensitivities 24
   4. Diatomic Donor-Acceptor Systems 28
      4.1. Harpoon Mechanism 28
      4.2. Interaction Energy 30
      4.3. Chemical Potential Equalization Perspective on Harpoon Mechanism 34
   5. Combination Rules 37
   6. Illustrative Qualitative Applications 40
      6.1. Electronegativity Equalization Rules 40
      6.2. Hard-Soft-Acids-and-Bases Principles 40
      6.3. Directing Influence of Ligands 43
   7. General Legendre Transformed Representations 45
   8. More about the Interaction Energy in Donor-Acceptor Systems 48

3. CONCEPTS AND RELATIONS OF MOLECULAR CHARGE SENSITIVITY ANALYSIS 51
   1. Hardness and Softness Quantities in the AIM-Resolution 52
   2. Charge Sensitivities in the Local Description 57
   3. Relative Internal Development 60
   4. Determining Charge Sensitivities from Canonical AIM Hardnesses 64
   5. *In Situ* Quantities in the Donor-Acceptor Systems 67
   6. Collective Charge Displacements 69
      6.1. Populational Normal Modes 70
      6.2. Minimum Energy Coordinates 79
      6.3. Collective Modes of Molecular Fragments 87

4. CONCEPTS FOR CHEMICAL REACTIVITY 91
   1. Partitioning of the Fukui Function Indices in Reactive Systems 93
   2. Internal Hardness Decoupling Modes 96
   3. Externally Decoupled Modes 105
   4. Inter-Reactant Modes 124

5. Mapping Relations 131
6. Stability Considerations 141
7. Partitioning of the Interaction Energy 149
8. Mutual Polarization of Reactants 152
9. Intersecting-State Model of Charge Transfer Processes 157

5. ILLUSTRATIVE APPLICATIONS TO MODEL CATALYTIC SYSTEMS 161
1. Rutile Surface Clusters and the Water-Rutile System 163
1.1. PNM Resolution of the AIM FF Indices 163
1.2. Use of the MEC in Diagnosing Surface Clusters 163
1.3. Water-Rutile Molecular Chemisorption 166
1.4. The Transition-State for the Water Dissociation 170
1.5. Polarizational Patterns of the Closed Adsorbate and Substrate 173
2. Toluene - [V_2O_5] System 175
2.1. Surface Cluster and Adsorption Arrangements 175
2.2. Energetical Considerations 186
2.3. Relative Role of the Diagonal and off-Diagonal Isoelectronic Responses 187
2.4. Direction of the Inter-Reactant Charge Transfer 188
2.5. Relative Role of the P and CT Components in the Overall Isoelectronic Responses 189
2.6. Mutual Polarization of Reactants 189
2.7. Charge Transfer Induced Displacements 191
2.8. Toluene Activation and Surface Reconstruction Trends from the Overall Diagrams 194
2.9. External Charge Transfer Effects 196
2.10. Concluding Remarks 197
3. Allyl-[MoO_3] System 197
3.1. Charge Reconstruction Patterns 212
3.2. IRM Representation of the Inter-Reactant CT Effects 214
3.3. Density Difference Diagram 215
3.4. Bond-Order Analysis 215
3.5. Implications for the Mechanism of Catalytic Allyl Oxidation to Acrolein 218
4. Conclusion 219

6. CHARGE SENSITIVITIES IN KOHN-SHAM THEORY 220
1. Conventional and Density Functional Quantum Mechanics 220
2. Electronic Softness Quantities in Kohn-Sham Theory 222
3. Nuclear Reactivity Indices 225
4. Softness and Hardness Kernels and Their Spectral Analysis 226
5. Isoelectronic Kernels 227
6. Charge Sensitivities in the Constrained Equilibrium State 229
6.1. Electronegativity/Chemical Potential Equalization Equations 229

6.2. Hardness (Interaction) and Softness (Response) Quantities 230
6.3. Equalization of the Local Hardnesses in the Constrained Equilibrium State of $M^+ = (A^+|B^+)$ 235
6.4. AIM Discretization 236
6.5. Kohn-Sham-Type Scheme for Subsystems 239
7. Conclusion 241

7. ELEMENTS OF THE ORBITALLY-RESOLVED CSA 243
1. Populational Variables and Energy Gradients 243
2. Hardness Derivatives 247
3. Nature of the Orbital Discretization 248
3.1. Local and Global State Variables in the Orbital Description 248
3.2. Equilibrium Criteria and Charge Sensitivities in the Local SO Description 250
3.3. Equilibrium Criteria and Charge Sensitivities in the Global Description 253
4. Frontier Orbital Considerations 255
5. Global Frontier Orbital Properties of Open Reactive Systems 257
6. Constrained Energy Functions and Lagrange Multipliers 261
7. Occupational Derivatives of the Constrained Energy Functions 264

APPENDIX A: Variational Principle for the Fukui Function and the Minimum Hardness Principle 267

APPENDIX B: Relaxed Fragment Transformation of the Hardness Matrix in $M = (A|B)$ Systems 270

APPENDIX C: The AIM/PNM-Resolved Intersecting-State Model 272

APPENDIX D: Bond-Multiplicities From One-Determinantal, Difference Approach 274

APPENDIX E: GUHF and UHF Energy Expressions and Their Derivatives with Respect to the Spin-Resolved Density Matrices 277

REFERENCES 279

SUBJECT INDEX 285

LIST OF ACRONYMS

A	Acid, Acidic (Reactant)
AIM	Atoms-in-Molecules (Level of Resolution)
AO	Atomic Orbitals
as	Active Site
B	Base, Basic (Reactant)
CBO	Charge-and-Bond-Order (Matrix, Data)
CI	Configuration Interaction (Method)
CNDO	Complete Neglect of Differential Overlap (Approximation, Method)
CS	Charge Sensitivities
CSA	Charge Sensitivity Analysis
CT	Charge Transfer (External or Internal)
DFT	Density Functional Theory
E	External (Stability/Instability)
EDM	Externally Decoupled Modes
EDM (→IDM)	EDM Determined to Resemble the IDM (MOC)
EDM (→MEC)	EDM Determined to Resemble the MEC (MOC)
EE	Electronegativity Equalization
EEM	Electronegativity Equalization Method
EPS	Electron Population Space
ES	Electrostatic Interactions (Contribution to the Energy)
f	Forward (Direction of Charge Transfer)
FF	Fukui Function
G	Group (Level of Resolution)
g	Global (Level of Resolution)
g.s.	Ground-State
GUHF	Generalized UHF (Approximation, Method)
H	Hard (Acid, Base)
HF	Hartree-Fock (Theory)
HK	Hohenberg-Kohn (DFT)
HOMO	Highest Occupied Molecular Orbital (Frontier MO)
HSAB	Hard-Soft-Acids-and-Bases (Principle)
I	Internal (Stability/Instability)
IDM	Internally Decoupled Modes
IEE	Internal Electronegativity Equalization
INDO	Intermediate Neglect of Differential Overlap (Approximation, Method)
INM	Internal Normal Modes
IRM	Inter-Reactant Modes
ISM	Intersecting-State Model
KS	Kohn-Sham (DFT)
L	Local (Level of Resolution)

ℓ	Localized (Modes)
LCAO	Linear Combinations of Atomic Orbitals
LCMO	Linear Combinations of (Reactant) MO
LDA	Local Density Approximation (KS DFT)
LSDA	Local Spin-Density Approximation (KS DFT)
LTR	Legendre Transformed Representations
LUMO	Lowest Unoccupied Molecular Orbital (Frontier MO)
M	Molecular System
m-	*Meta*- Position in the Substituted Benzene Ring
MEC	Minimum Energy Coordinates
MINDO	Modified INDO (Approximation, Method)
MNDO	Modified Neglect of Differential Overlap [NDDO (Neglect of Diatomic Differential Overlap) Approximation, Method]
MO	Molecular Orbital (Theory, Level of Resolution)
MOC	Maximum Overlap Criterion
MOE	Minimum Orbital Energy (Principle)
NPS	Nuclear Position Space
o-	*Ortho*- Position in the Substituted Benzene Ring
P	Polarization/Polarizational
p-	*Para*- Position in the Substituted Benzene Ring
PNM	Populational Normal Modes
r	Reservoir (External, Macroscopic, of Electrons)
REC	Relaxational Coordinates
REM	Relaxational Modes
RHF	Restricted HF (Approximation, Method)
S	Soft (Acid, Base)
SAL	Separated Atoms/Ions Limit
SCCC	Self-Consistent Charge and Configuration (Method)
SCF	Self Consistent Field (Method, Iterative Procedure)
Scaled-INDO	INDO Method with Scaled Parameters for the AIM Charge and Configuration
SINDO1	Symmetrically Orthogonalized INDO (Method)
SO	Spin-Orbitals
SRL	Separated Reactants Limit
TS	Transition State
UHF	Unrestricted HF (Approximation, Method)
VNM	Vibrational Normal Modes
VSEPR	Valence State Electron Pair Repulsion (Model)
VSIE	Valence State Ionization Energies
ZDO	Zero Differential Overlap
ZINDO	Zerner's INDO (Method)

CHAPTER 1

INTRODUCTORY SURVEY

In recent years much attention has been directed toward the understanding of the electronic structure and chemical reactivity of molecular systems in terms of their charge distributions. Historically, the Hilbert space of atomic orbitals has first been used to define a variety of quantities of *atoms-in- molecules* (AIM), e.g., the classical populational analysis schemes [e.g., 1-3], designed to define the AIM effective oxidation state (net charge), and more recent quantum-mechanical chemical valence/bond multiplicity indices [e.g., 4-6]. Later, the rigorous partitioning scheme in the physical space has been proposed by Bader et al. [7], which generates alternative structural information in the atomic resolution [7, 8], which can eventually be used to diagnose reactivity. The latter, predicting the site and path selectivities of chemical reactions from properties of reactants, is the main goal of the present book. We shall describe a perturbative approach of predicting the open reactants behaviour, in the presence of each other, from their charge sensitivities and the simplified, but realistic, representation of their mutual interactions.

The CSA (EEM) [9, 10] has drawn a considerable attention recently. It is based upon various generalized charge "polarizabilities" defined within the *density functional theory* (DFT) or related physical models in the AIM resolution. Various charge sensitivities, including hardness, softness, and the Fukui function quantities, have extraordinary potential for quantifying chemical concepts and providing them a theoretical basis [11-13]. They have also helped to justify some intuitive chemical concepts, rules, and the observed reactivity preferences. These sensitivities are of great interest in the theory of chemical reactivity, and catalysis in particular, because they provide a basis for both qualitative interpretation and quantitative description of regional acidic/basic properties, as well as the reactant polarization and charge transfer effects due to the molecular environment and/or structural changes. The charge sensitivities also reflect direct and indirect couplings between larger parts of a given molecular, reactive or catalytic system, e.g., the functional groups, adsorbate and substrate, etc.

The CSA represents the linear response treatment of molecular or reactive systems, based upon the system canonical hardness tensor or kernel, defining the underlying quadratic Taylor expansion of the energy. Selected responses to various electron population and external/chemical potential displacements, under alternative closure constraints (constrained equilibrium states), can be determined from the relevant EE equations. Some of the constrained equilibrium responses are related to the familiar intermediate stages of a reaction mechanism, e.g., the intra-reactant *polarization* (P) and the inter-reactant *charge transfer* (CT) components. The relative role of functional groups, the *in situ* characteristics of reactants, and the site-selectivity trends in a reactive system under consideration, are all related to such constrained equilibrium states, in which some barriers preventing a spontaneous flow of electrons are present.

The canonical input parameters of the CSA in the AIM resolution (Born-Oppenheimer approximation) are the initial chemical/external potential data and the hardness tensor reflecting the charge couplings between the system constituent atoms, for the actual AIM valence states; they all can be modelled semiempirically using available finite-difference expressions in terms of the atomic electron affinities and ionization potentials. In this chapter we shall introduce some of these general concepts and modelling techniques.

Throughout the book the atomic units are used, unless specified otherwise.

1. Global and Regional Equilibria and the Associated Charge Sensitivities

Charge sensitivities are represented by the corresponding derivatives, e.g., $P^{R}_{Y,X} \equiv (\partial p_X / \partial t_Y)_R$, which measures a response of the parameter p characterizing the fragment X of a molecule M, per unit displacement of another parameter t of the equilibrium state of the fragment Y of M; this response is measured under specific constraints imposed on the two fragments in question and the molecular reminder, Z′ = (R|Z), consisting in general case of the mutually closed freely relaxing part R and the rest, Z, the charge distribution of which is assumed to remain "frozen" (Fig. 1). Such a sensitivity may correspond to a specific, mental division (d) of M into mutually closed fragments $M^{(d)}$ = (X|Y|R|Z) when the numbers of electrons in each part are held fixed. Moreover, depending upon the nature of the initial displacement (perturbation) of Y and the constraints imposed, these molecular fragments may be either externally closed or opened (coupled to independent, external electron reservoirs), as shown in Fig. 1. One could call the Y, X, R, Z, the displacement, response, relaxing, and frozen molecular fragments, respectively. When R vanishes (an empty set of AIM), Z = Z′, the corresponding derivative, $P^{R}_{Y,X} = P_{Y,X}$, represents the so-called rigid sensitivity; similarly, when R = Z′ (Z vanishes), the corresponding sensitivity for the assumed partitioning M = (X|Y|R) measures the quantity P coupling X and Y under condition of the fully relaxed system remainder, i.e., when R is free to adjust its electron distribution to the displaced state of X and the perturbed state of Y. Finally, when Z′ vanishes and X = Y = M one deals with the global sensitivity of the system as a whole, when no dividing walls for the flow of electrons inside M are present. Clearly, there is a multitude of additional, partially relaxed sensitivities, when some barriers for the internal flows between AIM in $M^{(d)}$ have been partially removed. Each system division d defines a particular constrained (regional) equilibrium state of $M^{(d)}$, as schematically shown in Fig. 1b.

One can therefore define a wide range of sensitivities for each chemically interesting partitioning of M, in terms of which one can probe reactivity trends or monitor the progress of a given process (chemical reaction, changes induced by the presence of the reaction partner or a catalyst, change in the system geometrical structure, etc.). They offer alternative reactivity criteria and provide a quantitative framework for many intuitive and qualitative considerations in chemistry. Such generalized "polarizabilities" can be classified in accordance with the general type of equilibrium to which they correspond. We call the global (or resultant) sensitivities those related to the global equilibrium in M (Fig. 1a), when all constituent AIM are mutually opened. This state is marked by the constant chemical potential throughout the system, $\mu^{M}_{X} = \mu^{M}_{Y} = \mu^{M}_{R} = \mu^{M}_{Z} = \mu$; this can be envisaged as resulting from coupling all molecular fragments (mutually open) to a common electron reservoir r, characterized by its chemical potential $\mu_r = \mu$. Similarly, the sensitivities corresponding to the general regional (constrained) equilibrium of Fig. 1b are called the regional(group) sensitivities; the chemical potential now reaches different values within each freely relaxing subsystem, $\mu_X^M \neq \mu_Y^M \neq \mu_R^M$, due to the hypothetical walls preventing a flow of electrons between subsystems. The regional equilibrium can be realized by coupling relaxing fragments to their respective (mutually closed) reservoirs. Clearly, the frozen fragment, Z (closed AIM), is not in an internal equilibrium and will exhibit different chemical potentials on its constituent atoms.

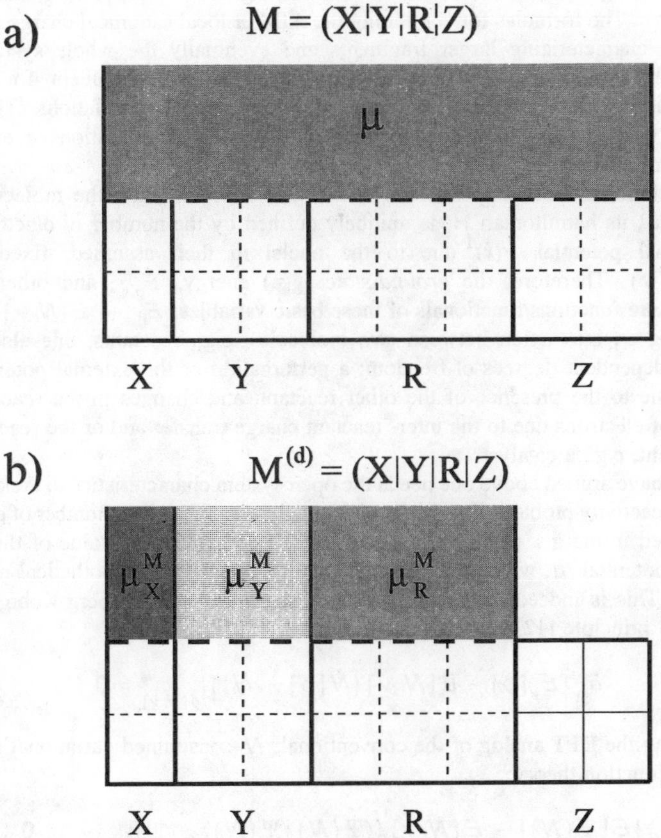

Figure 1. Graphical representation of the global equilibrium (a) and constrained equilibrium (b) states, corresponding to the resultant and regional charge sensitivities, respectively, of molecular fragments in the general partitioning of a molecular system M: M = (X¦Y¦R¦Z) and $M^{(d)}$ = (X|Y|R|Z), respectively.

2. The Chemical Potential/Electronegativity Equalization

All charge sensitivities can be expressed in terms of the canonical (basic) quantities relevant for the assumed level of resolution (local, AIM, molecular fragment, global), using the standard Jacobian and chain-rule transformations of derivatives. Moreover, the derivatives at a less resolved level can always be expressed in terms of derivatives defined at a more resolved level. The local resolution, in which one is interested in local responses to local perturbations, e.g., $P(\mathbf{r},\mathbf{r}') \equiv \delta p(\mathbf{r}')/\delta t(\mathbf{r})$, will require the relevant basic kernels to calculate general sensitivities and the associated energy changes. In the AIM resolution the corresponding basic matrices will be needed, reflecting a response on one atom to a test perturbation on another atom, e.g., $P^{c}_{i,j} \equiv (\partial p_j/\partial t_i)_c$, under relevant constraints c. The formulas for combining the AIM or local canonical charge sensitivities into those characterizing larger fragments and eventually the whole system, can be formulated in a direct or recursive forms; the first is convenient for qualitative considerations, while the latter is preferred for numerical calculations. They can be developed using the appropriate chemical potential equalization equations or, equivalently, the standard chain-rule transformations of derivatives.

In the familiar Born-Oppenheimer (adiabatic) approximation the molecular system identity, i.e., its hamiltonian H, is uniquely defined by the number of electrons N and the external potential $v(\mathbf{r})$ due to the nuclei in their assumed, fixed positions: $\mathrm{H} = \mathrm{H}(N, v)$. Therefore, the *ground-state* (g.s.) energy, $E_{\mathrm{g.s.}}$, and other molecular properties are functions/functionals of these basic variables: $E_{\mathrm{g.s.}} = E[N, v]$, etc. When considering an interaction between two molecules, e.g., reactants, one also considers similar independent degrees-of-freedom: a perturbation of the external potential of one reactant due to the presence of the other reactant, and changes in the reactant overall numbers of electrons due to the inter- reactant charge transfer and/or the reactive system environment, e.g., a catalyst.

As we have argued above one needs the open system characteristics to tackle a general chemical reactivity problem. For the fixed external potential a given number of electrons N can be fixed in the g.s. of an open system by the corresponding value of the electronic chemical potential, μ, which may be attributed to an external, hypothetical macroscopic reservoir. This is indeed rigorously stated by the general Hohenberg-Kohn (HK) DFT variational principle [12, 14-17]:

$$\delta_{\rho}\{E_v[\rho] - \mu[N, v](N[\rho] - N)\}\big|_{\rho[N, v]} = 0\,, \tag{1}$$

representing the DFT analog of the conventional, N-constrained variational principle of the wave-function theory:

$$\delta_{\Psi}\{E[\Psi(N)] - E[N, v](\langle\Psi(N)\,|\,\Psi(N)\rangle - 1)\}\big|_{\Psi[N, v]} = 0\,; \tag{2}$$

here $N[\rho] = \int \rho\, d\mathbf{r}$, and the g.s. wave-function $\Psi_{\mathrm{g.s.}} = \Psi[N, v]$ and density $\rho_{\mathrm{g.s.}} = \rho[N, v]$ define the g.s. energy:

$$E[N, v] = \langle\Psi[N, v]|\mathrm{H}(N, v)|\Psi[N, v]\rangle = E\big[N[\rho_{\mathrm{g.s}}], v[\rho_{\mathrm{g.s}}]\big] \equiv E_v[\rho_{\mathrm{g.s}}]\,. \tag{3}$$

In the above equation we have indicated that also the external potential, and thus $\Psi_{g.s.}$ and $E_{g.s.}$, are the unique functionals of the $\rho_{g.s.}$, as demonstrated by Hohenberg and Kohn [14]. The energy functional can be written in the familiar form

$$E_v[\rho] = \int \rho(\mathbf{r})\, v(\mathbf{r})\, d\mathbf{r} + F[\rho] , \tag{4}$$

separating the external potential dependent electron-nuclei attraction energy from the remaining, universal part, generating the sum of the electron kinetic and repulsion energies, represented by the operators T_e and V_{ee}, respectively:

$$F[\rho_{g.s.}] = \langle \Psi[N, v] | T_e + V_{ee} | \Psi[N, v] \rangle . \tag{5}$$

Equation 1 shows that the closed system (fixed N) can be alternatively viewed as the open system, in contact with an appropriate electron reservoir, the chemical potential of which fixes the system correct number of electrons.

The system global chemical potential, the Lagrangian multiplier associated with the N-constraint in Eq. 1, can be alternatively interpreted as the following global or local derivatives of the system g.s. energy [11, 12]:

$$\mu \equiv \partial E[N, v] / \partial N = \delta E_v[\rho] / \delta \rho(\mathbf{r}) \big|_{\rho[N, v]} \equiv \mu(\mathbf{r}) . \tag{6}$$

The first, partial derivative is the negative of the system electronegativity, $\mu = -\chi$, while the second, functional derivative with respect to the density, generates the local value of the chemical potential (negative of the local electronegativity), $\mu(\mathbf{r}) = -\chi(\mathbf{r})$. The above equation implies the equalization of the electronegativity throughout the whole globally relaxed electron distribution, in accordance with the familiar intuitive principle of Sanderson [18]. Similarly, in the regional equilibrium case the chemical potential (electronegativity) is equalized throughout each molecular fragment, electron population (density) of which is allowed to freely relax (fragments X, Y, and R in Fig. 1b) [19].

The chemical potential/electronegativity equalization of Eq. 6 has a number of other important implications. For example, as we shall demonstrate latter in the book, also the so called local hardness becomes equalized in the global equilibrium state at the global hardness level [20].

3. Alternative State Variables and Principal Derivatives

Equation 6 also shows that the system chemical potential, populational gradient, is the state variable conjugate to N. The familiar Hellmann-Feynman theorem [11, 21],

$$dE[N, v] = \mu[N^0, v^0]\, dN + \int \rho_{g.s.}(\mathbf{r})\, dv(\mathbf{r})\, d\mathbf{r} , \tag{7}$$

where $\rho_{g.s.} = \rho[N^0, v^0]$, identifies the g.s. density as the conjugate of v:

$$\rho_{g.s}(\mathbf{r}) = \delta E[N, v] / \delta v(\mathbf{r}) . \tag{8}$$

Clearly, any pair of the above global (N, μ) and local (v, ρ) parameters may be used to uniquely specify the system g.s.: $\{N, v\}$, $\{\mu, v\}$, $\{N, \rho\}$, and $\{\mu, \rho\}$. This implies that the resultant charge sensitivities corresponding to these alternative g.s. specifications, represented by the derivatives of the relevant Legendre transforms of the system g.s. energy, can also be reduced to basic second-order derivatives of the energy [21]:

$$\eta[N, v] = \partial^2 E[N, v]/\partial N^2 = (\partial\mu[N, v]/\partial N)_v \qquad \textit{global hardness}, \qquad (9a)$$

$$f(\mathbf{r}) = [\partial\rho(\mathbf{r})/\partial N]_v = [\delta\mu[N, v]/\delta v(\mathbf{r})]_N, \qquad \textit{Fukui function}\ \text{(FF)}, \qquad (9b)$$

$$\beta(\mathbf{r}, \mathbf{r}') = \delta^2 E[N, v]/\delta v(\mathbf{r})\,\delta v(\mathbf{r}') = [\delta\rho_{g.s.}(\mathbf{r}')/\delta v(\mathbf{r})]_N, \ \textit{polarization kernel.} \qquad (9c)$$

They define the second-order change of $E_{g.s.}$,

$$d^2 E[N, v] = \frac{1}{2}\Big[\eta(dN)^2 + 2dN\int f(\mathbf{r})\,dv(\mathbf{r})\,d\mathbf{r} + \iint dv(\mathbf{r})\,\beta(\mathbf{r}, \mathbf{r}')\,dv(\mathbf{r}')\,d\mathbf{r}\,d\mathbf{r}'\Big], \qquad (10)$$

which, together with Eq. (7), give the relevant quadratic Taylor expansion of the system electronic energy:

$$\Delta^{1+2} E_{g.s.} = dE[N, v] + d^2 E[N, v]. \qquad (11)$$

For a general constrained-equilibrium density one needs a more detailed specification, involving both external potential and electron density: $\{\rho, v\}$. The corresponding state $\Psi[\rho]$ can be formally defined via the Levy constrained search construction [22],

$$Q[\rho] = \min_{\Psi \to \rho} \langle \Psi | \mathrm{T_e} + \mathrm{V_{ee}} | \Psi \rangle, \qquad \Psi_{opt} = \Psi[\rho], \qquad (12)$$

where the search involves all the pure and mixed state wavefunctions Ψ giving the current density ρ [23, 24]. The above construction also formally defines the constrained-equilibrium energy $E[\rho, v] = \int \rho\, v\, d\mathbf{r} + Q[\rho]$. Clearly, for the globally relaxed charge distribution $Q[\rho_{g.s.}] = F[\rho_{g.s.}]$.

The quadratic Taylor expansion of $E[\rho, v]$ reads:

$$\Delta^{1+2} E[\rho, v] = \{dE[\rho, v]\} + \{d^2 E[\rho, v]\}$$

$$= \Big\{\int\Big[\frac{\delta E}{\delta\rho(\mathbf{r})}\,d\rho(\mathbf{r}) + \frac{\delta E}{\delta v(\mathbf{r})}\,dv(\mathbf{r})\Big]d\mathbf{r}\Big\}$$

$$+\left\{\frac{1}{2}\iint\left[d\rho(\mathbf{r})\,\frac{\delta^2 E}{\delta\rho(\mathbf{r})\,\delta\rho(\mathbf{r}')}\,d\rho(\mathbf{r}') + 2\,d\rho(\mathbf{r})\,\frac{\delta^2 E}{\delta\rho(\mathbf{r})\,\delta v(\mathbf{r}')}\,dv(\mathbf{r}') + \right.\right.$$

$$\left.\left. dv(\mathbf{r})\,\frac{\delta^2 E}{\delta v(\mathbf{r})\,\delta v(\mathbf{r}')}\,dv(\mathbf{r}')\right]d\mathbf{r}\,d\mathbf{r}'\right\}$$

$$\equiv \{\textstyle\int[\mu(\mathbf{r})\,d\rho(\mathbf{r}) + \rho(\mathbf{r})\,dv(\mathbf{r})]\,d\mathbf{r}\} + \{\frac{1}{2}\iint[d\rho(\mathbf{r})\,\eta(\mathbf{r},\mathbf{r}')\,d\rho(\mathbf{r}')$$

$$+ 2\,d\rho(\mathbf{r})\,\delta(\mathbf{r}-\mathbf{r}')\,dv(\mathbf{r}') + dv(\mathbf{r})\,\beta(\mathbf{r},\mathbf{r}')\,dv(\mathbf{r}')]\,d\mathbf{r}\,d\mathbf{r}'\}\,. \tag{13}$$

The hardness kernel [12, 25-27],

$$\eta(\mathbf{r},\mathbf{r}') = \frac{\delta\mu(\mathbf{r}')}{\delta\rho(\mathbf{r})} = \frac{\delta^2 Q[\rho]}{\delta\rho(\mathbf{r})\,\delta\rho(\mathbf{r}')}\,, \tag{14}$$

is the principal second-order quantity since, as shown by Berkowitz and Parr [27], the polarizational kernel $\beta(\mathbf{r},\mathbf{r}')$ can be expressed in terms of quantities derivable solely from $\eta(\mathbf{r},\mathbf{r}')$.

As shown in Appendix A the FF minimizes the system quadratic energy $\Delta^{1+2}E[\rho, v]$ for constant v, and is responsible for the local hardness equalization in the global equilibrium charge distribution [20, 28, 29].

In the above *"fine-grained"* local formulation the state parameters ρ and v are defined for each infinitesimal volume element. In many chemical problems a less resolved, *"coarse-grained"* description is sufficient. For example, in an approximate, phenomenological AIM-resolution one would attribute the effective electron population and external potential variables, $\boldsymbol{N} = (N_1, N_2, ..., N_m)$ and $\boldsymbol{v} = (v_1, v_2, ..., v_m)$, to each constituent atom, thus expressing the system energy as function of these two vectors of the AIM state variables: $E^{AIM}(\boldsymbol{N}, \boldsymbol{v})$. Clearly, a similar discretization involving molecular orbitals (*MO-resolution* [30, 31]) or larger molecular fragments (*group-resolution*) can also be envisaged.

As we shall demonstrate later in the book, the principal second-order quantity in the AIM-resolution is the AIM hardness tensor:

$$\boldsymbol{\eta} = \partial^2 E^{AIM}/\partial\boldsymbol{N}\,\partial\boldsymbol{N} = \partial\boldsymbol{\mu}/\partial\boldsymbol{N} = \{\partial\mu_i/\partial N_j\}\,, \tag{15}$$

where the populational gradient $\boldsymbol{\mu} = (\mu_1, \mu_2, ..., \mu_m) = \partial E^{AIM}/\partial\boldsymbol{N}$ is the vector of the AIM chemical potentials. Again the AIM polarization (linear response) matrix,

$$\boldsymbol{\beta} = (\partial^2 E^{AIM}/\partial\boldsymbol{v}\,\partial\boldsymbol{v})_N = \{(\partial N_i/\partial v_j)_N\}\,, \tag{16}$$

and other charge sensitivities in this resolution can be expressed in terms of the canonical AIM hardness tensor alone; here $N = \Sigma_i^m N_i$ is the global number of electrons. The inverse of $\boldsymbol{\eta}$, $\boldsymbol{\sigma} = \boldsymbol{\eta}^{-1} = \{\partial N_i / \partial \mu_j\}$, is the system softness matrix, representing the electron population compliant matrix [9, 32].

4. Modelling the AIM Hardness Matrix

Let us consider first the asymptotic behaviour of the off-diagonal, coupling AIM hardness $\eta_{i,j}$, at large internuclear distance, $R_{i,j}$, where the point-charge interaction between AIM, $E_{i,j} \approx (Z_i - N_i)(Z_j - N_j)/R_{i,j}$, may be adopted. In this limit one may approximate [10] the system energy as the sum of atomic energies, $E_i(N_i, \boldsymbol{v})$, depending solely on the atom net charge, and the above pair-interaction energies, $\{E_{i,j}\}$. Such a crude, though realistic approximation immediately gives at large distances

$$\eta_{i,j} \approx 1/R_{i,j}, \tag{17}$$

independent of the AIM chemical environment in a molecule.

A more adequate modelling is suggested by the standard (closed-shell) SCF MO theory [33-35], in which the charge distribution in the Hilbert (function) space of *atomic orbitals* (AO) $|\boldsymbol{\lambda}\rangle = \{|\lambda_p\rangle\}$ is defined by the familiar *charge-and-bond-order* (CBO) matrix, $\mathbf{P} = \mathbf{C}\,\mathbf{o}\,\mathbf{C}^\dagger$, elements of which may be considered as independent AIM populational variables; here $\mathbf{C}$ is the matrix of the LCAO MO coefficients defining *molecular orbitals* (MO) in terms of AO, $|\boldsymbol{\psi}\rangle = |\boldsymbol{\lambda}\rangle\,\mathbf{C}$, and $\mathbf{o} = \{n_i\,\delta_{i,j}\}$ is the diagonal matrix of the MO electron occupations.

For the closed-shell one-determinant function in the usual LCAO MO approach the energy function can be expressed in terms of the $\mathbf{P}$ and Fock matrices, $\mathbf{F(P)} = \mathbf{h} + \mathbf{G(P)}$, where

$$\mathrm{G}_{p,q}(\mathbf{P}) = \sum_{r,s}^{AO} \mathrm{P}_{r,s}\,[(pq|rs) - \frac{1}{2}(ps|rq)] \equiv \sum_{r,s}^{AO} \mathrm{P}_{r,s}\,(pq||rs), \tag{18}$$

$(pq|rs)$ stands for the two-electron repulsion integral between the $\Omega^{\lambda}_{p,q}(1) = \lambda_p^*(1)\,\lambda_q(1)$ and $\Omega^{\lambda}_{r,s}(2) = \lambda_r^*(2)\,\lambda_s(2)$ distributions of two electrons, and $\mathrm{h}_{p,q} = \langle p|h|q\rangle$ is the matrix element of the one-electron operator h. The relevant energy expression is given by the trace:

$$E(\mathbf{P}) = \frac{1}{2}\,\mathrm{tr}\{[\mathbf{h} + \mathbf{F(P)}]\,\mathbf{P}\}; \tag{19}$$

hence

$$\partial E(\mathbf{P})/\partial \mathrm{P}_{p,q} = \mathrm{F}_{q,p} \quad \text{and} \quad \partial^2 E(\mathbf{P})/\partial \mathrm{P}_{p,q}\,\partial \mathrm{P}_{r,s} = (pq||rs). \tag{20}$$

For the orthogonal AO basis set or the assumed unit metric of the AO overlap integrals, $\langle p|q\rangle = \delta_{p,q}$, common in the semi-empirical SCF MO approaches, the electron population of atom i involves only the AO charges (diagonal CBO elements): $N_i = \Sigma_p^i P_{p,p} \equiv \mathrm{tr}_i\, \mathbf{P}$. The actual changes in the electron populations, accompanying typical chemical transformations, are predominantly due to variations in populations of the valence-shell AO. Therefore, to a very good approximation one may adopt the *frozen-core* assumption and thus limit the tr_i to the valence-shell AO only. This is the usual practice in the semiempirical calculations. Equation 20 then indicates that the hardness matrix elements measure the coulomb-exchange repulsion of the valence-shell electrons of the atom(s) involved. One could therefore identify $\eta_{i,j}$ as the coulomb-exchange interaction between the valence shell *s*-electrons:

$$\eta_{i,j} \approx (s_i s_i || s_j s_j) = (s_i s_i | s_j s_j) - \frac{1}{2}(s_i s_j | s_j s_i) \equiv \gamma_{i,j} - \frac{1}{2} K_{i,j}, \quad (21)$$

as required by the invariance requirement with respect to the AIM hybridization. A similar averaged quantities are introduced in the *Zero Differential Overlap* (ZDO) type theories (CNDO, INDO, MINDO, etc.). The consistent ZDO (CNDO) approximation neglects the exchange integrals $K_{p,q}$, so at this level of modelling the AIM hardness matrix is defined by the respective valence-shell, coulomb electron repulsion integrals:

$$\eta_{i,j} \approx \gamma_{i,j}\,. \quad (22)$$

This approximation has the intuitively correct asymptotic behaviour at both large internuclear distances (Eq. 17) and reproduces the finite-difference diagonal atomic hardness [13] through the familiar Pariser formula [36]:

$$\eta_{i,i} \approx I_i - A_i \approx \gamma_{i,i}, \quad (23)$$

where I_i and A_i stand for the ionization potential and electron affinity of the neutral atom i .

The off-diagonal coulomb integral can be interpolated from the two diagonal integrals via the known formulas of the semi-empirical theories, e.g., the Mataga-Nishimoto [37],

$$\gamma_{i,j} \approx (d_{i,j} + R_{i,j})^{-1}, \quad (24)$$

or Ohno [38],

$$\gamma_{i,j} \approx (d_{i,j}^2 + R_{i,j}^2)^{-1/2}, \quad (25)$$

formulas, where $d_{i,j} = 2/(\eta_{i,i} + \eta_{j,j})$. The latter will be used in all illustrative applications of the AIM-resolved CSA presented in this book.

5. Direct and Indirect Modelling of Charge Responses

The main objectives of the CSA in the AIM resolution are molecular responses to the dN ($d\boldsymbol{N}$) and $d\boldsymbol{v}$ perturbations, which molecules undergo in chemical reactions. The *direct* modelling of the hardness matrix [9], representing the charge couplings (electron interactions) between constituent AIM, can be used to *indirectly* determine such responses, via inverting the hardness tensor into the softness matrix, which in turn generates the global (S), AIM (S_i) softnesses, and the associated FF indices,

$$f_i = \partial N_i / \partial N = (\partial N_i / \partial \mu) / (\partial N / \partial \mu) \equiv S_i / S, \tag{26}$$

via the straightforward, additive formulas:

$$S_i = \sum_{j=1}^{m} \sigma_{i,j}, \qquad S = \sum_{i=1}^{m} S_i. \tag{27}$$

The AIM FF indices represent important single-reactant CT reactivity criterion [39, 40], which can be directly generated from the canonical hardness matrix [41].

The alternative way of modelling the response (softness/FF) quantities is to directly model the FF [12, 39, 40, 42] and/or the softness matrix. This in turn indirectly determines the system hardness data. Consider the familiar case of the SCF LCAO MO theory of the previous section. The dependence of N_i on N is through $\mathbf{P} = \mathbf{P}(N, v)$:

$$\partial \mathbf{P} / \partial N = \mathbf{C}\,\mathbf{o}\,(\partial \mathbf{C}^{\dagger} / \partial N) + (\partial \mathbf{C} / \partial N)\,\mathbf{o}\,\mathbf{C}^{\dagger} + \mathbf{C}\,(\partial \mathbf{o} / \partial N)\,\mathbf{C}^{\dagger}, \tag{28}$$

where the first two terms represent the MO relaxation contribution, for the *"frozen"* MO occupations, and the third term account for the effect of changing MO occupations for the *"frozen"* shapes of MO. All these derivatives can be estimated via finite differences, by performing separate SCF calculations for the neutral molecule M, cation M^+, and anion M^- [12, 39], or by the Kohn-Sham DFT calculations [42].

In the crude one-determinantal approximation neglecting the orbital relaxation, as in the familiar Koopmans theorem, only the third term contributes, and the changes in the MO occupations are limited to the frontier MO (f): the *highest occupied* (HOMO) and *lowest unoccupied* (LUMO). For an electrophilic attack (f = HOMO) $dN = do_{f,f} = -1$ and for a nucleophilic attack (f = LUMO) $dN = do_{f,f} = 1$, with the unchanged occupations of the remaining MO. Hence the Koopmans approach gives $(\partial P_{p,q} / \partial N)^f = C_{p,f}\, C^*_{q,f}$. In the radical attack one usually takes the average of the nucleophilic and electrophilic estimates.

The local FF can be similarly estimated from the corresponding differences involving $\rho^{M}_{g.s.}$, $\rho^{M^+}_{g.s.}$, and $\rho^{M^-}_{g.s.}$, while the AIM electron populations $\boldsymbol{N}^{M}$, $\boldsymbol{N}^{M^+}$, and $\boldsymbol{N}^{M^-}$, generate the finite-difference AIM FF indices [12, 39]. A similar procedure within Kohn-Sham theory [15], requiring one SCF calculations, has been suggested by Gázquez *et al.* [42].

CHAPTER 2

ATOMIC CHARGE SENSITIVITIES

Chemists usually view molecular processes from the atomic perspective. This approach calls for an appropriate definition of AIM, which is not unique; however, a very elegant quantum mechanical definition based upon the topology of the charge density is available [7]. In a qualitative thinking about changes the constituent atoms undergo in a molecule a useful perturbative description [25, 43-45] involves spherical AIM, each characterized by its electron population, $N = N^0 + \Delta N$ (or net charge, $q = Z - N$) and an *effective nuclear charge,* $Z = Z^0 + \Delta Z$; here N^0 is the initial electron population and Z^0 stands for the nuclear charge of the reference state of the atom in question. Therefore, the displaced state of the AIM, perturbed relative to the initial state of the atom/ion, is characterized by the populational displacement ΔN, due to charge redistribution in a molecule, and the external potential displacement phenomenologically represented by a change in the effective nuclear charge, ΔZ; the latter models the effects due to the field of the remaining AIM or the presence of the other reactant. This very way of describing variations in the external potential, be it very approximate, automatically preserves a spherical character of the perturbed atom.

Let us consider as an illustrative example a formation of a chemical bond between two atoms. Due to the electronegativity difference there will take place the charge transfer between the two AIM, described by their populational displacements. Moreover, each atom experiences some contraction of its electron cloud due to the presence of the other atom; this effect can be described by appropriate displacements in the atomic effective charges.

The equilibrium energy change of an atom, due to the perturbations ΔN and ΔZ, can be estimated using the following Taylor expansion in powers of ΔN and ΔZ:

$$\Delta E(\Delta N, \Delta Z) \equiv E(N, Z) - E(N^0, Z^0) = \left(\frac{\partial E}{\partial N}\Delta N + \frac{\partial E}{\partial Z}\Delta Z\right) + \frac{1}{2}\left(\frac{\partial^2 E}{\partial N^2}(\Delta N)^2 + 2\frac{\partial^2 E}{\partial N\,\partial Z}\Delta N\,\Delta Z + \frac{\partial^2 E}{\partial Z^2}(\Delta Z)^2\right) + \ldots \quad (29)$$

The derivatives appearing in the quadratic expansion of Eq. 29 can be extracted from the atomic data. We consider a continuous smooth surface $E(N, Z)$, shown in Fig. 2, containing all physically meaningful points $\{E(N^0, Z^0)\}$ for integral values of these two state variables, which uniquely determine the AIM hamiltonian $\mathrm{H}(N^0, Z^0)$ and the ground-state energy $E(N^0, Z^0) = \langle\Psi_{g.s}|\mathrm{H}|\Psi_{g.s}\rangle$. It is the main purpose of this chapter to explore these derivatives and to use them in generating the charge sensitivity derivatives $(\partial y/\partial x)_c$, with the response (y) and perturbation (x) quantities involving N, Z and their conjugates given by the corresponding partials of the energy; such a derivative measures the response in y per unit displacement in x under constraints c. These general derivatives correspond to alternative specifications of the atomic states, which we consider in the next section; each representation has its own "thermodynamic" potential given by the relevant Legendre transform of the energy [21, 45]. All charge sensitivities can be reduced to expressions in terms of basic second-order derivatives of energy in Eq. 29.

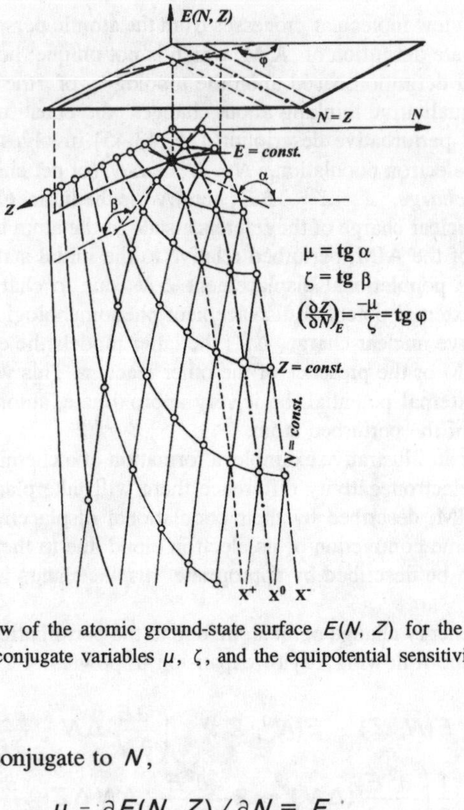

Figure 2. Perspective view of the atomic ground-state surface $E(N, Z)$ for the physically meaningful region $N \leq Z + 1$. The conjugate variables μ, ζ, and the equipotential sensitivity $(\partial Z/\partial N)_E$ are also interpreted.

The state variable conjugate to N,

$$\mu = \partial E(N, Z) / \partial N \equiv E_N, \tag{30}$$

is the atomic chemical potential which can be approximated by the corresponding finite difference expression [11, 12]. From the Hellmann-Feynman theorem one obtains the conjugate of Z,

$$\zeta = \partial E(N,Z) / \partial Z = \langle \Psi_{g.s} | \partial H / \partial Z | \Psi_{g.s} \rangle = - \langle r^{-1} \rangle_{g.s.} \equiv V_{ne} / Z \equiv E_Z, \tag{31}$$

measuring the electron-nuclear attraction energy per unit nuclear charge.

The canonical second-order derivatives of the atomic energy (see Eq. 29 and Table 1) include atomic hardness $\eta = \partial\mu / \partial N = E_{NN}$ of Parr and Pearson [12, 13], $\alpha = \partial\zeta / \partial N = \partial\mu / \partial Z \equiv E_{ZN} = E_{NZ}$, and $\beta = \partial\zeta / \partial Z \equiv E_{ZZ}$. The diagonal derivatives η and β couple N and Z with their respective conjugates, while the mixed derivative α monitors the interaction between N and Z. Among the four state variables it suffices to specify one of the population related parameters (N or μ) and one of the external potential related parameters (Z or ζ) to define atomic state uniquely. Therefore, there are four alternative *Legendre transformed representations* (LTR) [21, 45, 46] of such spherical atoms. Clearly, the variables $N = \int \rho(\mathbf{r})\, d\mathbf{r}$ and $\zeta = -\int \rho(\mathbf{r})\, r^{-1}\, d\mathbf{r}$, both linear functionals of electron density ρ, can be classified as "extensive": $\boldsymbol{X} = (N, \zeta)$; thus, by the same "thermodynamic" analogy, their conjugates μ and Z will be termed as "intensive"[1]: $\boldsymbol{P} = (\mu, -Z)$. The negative sign of Z in $\boldsymbol{P}$ comes from the differentiation of the principal potential $F(\boldsymbol{X})$ (Table 1): $\boldsymbol{P} = \partial F(\boldsymbol{X})/\partial \boldsymbol{X}$.

1. Alternative Representations of Atomic States, Potentials, Derivatives and Their Reduction

Figures 2 and 3 show the concavity of the atomic energy function with respect Z and its convexity with respect to N [12, 48-51]. The latter property simply reflects the experimentally determined fact for the coulombic external potential, of an increase in the successive ionization potentials with decreasing N; this N-dependence can be roughly fitted into the following interpolation formula [12]: $E \approx E(N^0)\exp[-D(N-N^0)]$ with $D \approx 2.2 \pm 0.6$. The concavity with respect to the nuclear charge is also explicitly demonstrated by the approximate curve $E \approx A(Z-B)^C$ [52], where $A < 0$ and $C \approx 2$.

Clearly, fixing N and Z, say for an integer $N = N^0$, uniquely defines the system hamiltonian and thus its ground-state energy. For a non-integer N the ground-state must be understood in terms of density matrices, as the mixture of the pure ground-states with integer (eventually neighbouring) N^0-values [12, 48, 51], which gives the lowest energy. As demonstrated previously [44, 48, 50, 51], $E(N, Z)$ is a convex piecewise linear function of N for every given Z, with the N-slope discontinuity for integer number of electrons at T = 0^0 K.

When N and Z are used to identify the ground-state of an atomic system, the remaining state parameters μ and ζ are also uniquely determined. Namely, ζ directly follows from the ground-state density while the atomic chemical potential can be thought of as that of an external electron reservoir, coupled to the system, for which average number of electrons on the atom reaches the prescribed value at equilibrium. By replacing N by μ the associated values of N and ζ result from a similar thought experiment involving an external reservoir. In order to identify the system one could also replace Z by ζ; the system nuclear charge would then be uniquely identified by the Z-value reproducing ζ and N (or μ).

The atomic LTR corresponding to the following alternative sets of the state parameters:

$$(X, Y) = \{(N, Z), (\mu, Z), (N, \zeta), (\mu, \zeta)\}, \tag{32}$$

are summarized in Table 1. The exact differentials of the corresponding "thermodynamic" potentials,

---

[1] This classification differs from the one in the Ref. 45 and it follows that adopted in Ref. 47.

TABLE 1. Atomic potentials, derivatives, Maxwell relations, and reduced expressions

Potential	Exact Differential	Derivative Identities
$E(N, Z)$	$dE = \mu\, dN + \zeta\, dZ$	$\frac{\partial^2 E}{\partial N^2} = \left(\frac{\partial \mu}{\partial N}\right)_Z \equiv \eta = E_{NN}$ $\frac{\partial^2 E}{\partial N \partial Z} = \left(\frac{\partial \mu}{\partial Z}\right)_N = \left(\frac{\partial \zeta}{\partial N}\right)_Z \equiv \alpha = E_{NZ} = E_{ZN}$ $\frac{\partial^2 E}{\partial Z^2} = \left(\frac{\partial \zeta}{\partial Z}\right)_N \equiv \beta = E_{ZZ}$
$Q(\mu, Z)$ $= E - \mu N$	$dQ = -N\, d\mu + \zeta\, dZ$	$\frac{\partial^2 Q}{\partial \mu^2} = \left(\frac{\partial N}{\partial \mu}\right)_Z = \frac{-1}{\eta} = Q_{\mu\mu}$ $\frac{\partial^2 Q}{\partial \mu \partial Z} = -\left(\frac{\partial N}{\partial Z}\right)_\mu = \left(\frac{\partial \zeta}{\partial \mu}\right)_Z = \frac{\alpha}{\eta} = Q_{\mu Z}$ $\frac{\partial^2 Q}{\partial Z^2} = \left(\frac{\partial \zeta}{\partial Z}\right)_\mu = \frac{\eta\beta - \alpha^2}{\eta} = Q_{ZZ}$
$F(N, \zeta)$ $= E - \zeta Z$	$dF = \mu\, dN - Z\, d\zeta$	$\frac{\partial^2 F}{\partial N^2} = \left(\frac{\partial \mu}{\partial N}\right)_\zeta = \frac{\eta\beta - \alpha^2}{\beta} = F_{NN}$ $\frac{\partial^2 F}{\partial N \partial \zeta} = \left(\frac{\partial \mu}{\partial \zeta}\right)_N = -\left(\frac{\partial Z}{\partial N}\right)_\zeta = \frac{\alpha}{\beta} = F_{N\zeta}$ $\frac{\partial^2 F}{\partial \zeta^2} = -\left(\frac{\partial Z}{\partial \zeta}\right)_N = \frac{-1}{\beta} = F_{\zeta\zeta}$
$R(\mu, \zeta)$ $= E - \mu N$ $- \zeta Z$	$dR = -N\, d\mu - Z\, d\zeta$	$\frac{\partial^2 R}{\partial \mu^2} = -\left(\frac{\partial N}{\partial \mu}\right)_\zeta = \frac{\beta}{\alpha^2 - \eta\beta} = R_{\mu\mu}$ $\frac{\partial^2 R}{\partial \mu \partial \zeta} = -\left(\frac{\partial N}{\partial \zeta}\right)_\mu = -\left(\frac{\partial Z}{\partial \mu}\right)_\zeta = \frac{\alpha}{\eta\beta - \alpha^2} = R_{\mu\zeta}$ $\frac{\partial^2 R}{\partial \zeta^2} = -\left(\frac{\partial Z}{\partial \zeta}\right)_\mu = \frac{\eta}{\alpha^2 - \eta\beta} = R_{\zeta\zeta}$

$$\mathcal{L}(X, Y) = \{E(N, Z)\ ,\ Q(\mu, Z)\ ,\ F(N, \zeta)\ ,\ R(\mu, \zeta)\}\ , \tag{33}$$

generate the second order derivatives representing various atomic sensitivities, related by the respective cross-differentiation Maxwell relations; they can be reduced, using familiar Jacobian transformation technique [53], in terms of the canonical derivatives η, α and β, as also explicitly shown in Table 1.

The canonical derivatives can be approximated by the finite-difference expressions in terms of the ground-state energies $\{E(N^0, Z^0\}$ [44]:

$$\mu(N^0, Z^0) \approx [E(N^0 + 1, Z^0) - E(N^0 - 1, Z^0)]/2$$
$$= -[I(N^0, Z^0) + A(N^0, Z^0)]/2\ , \tag{34}$$

$$\zeta(N^0, Z^0) \approx [E(N^0, Z^0 + 1) - E(N^0, Z^0 - 1)]/2\ , \tag{35}$$

$$\eta(N^0, Z^0) \approx E(N^0 + 1, Z^0) + E(N^0 - 1, Z^0) - 2E(N^0, Z^0)$$
$$= I(N^0, Z^0) - A(N^0, Z^0)\ , \tag{36}$$

$$\alpha(N^0, Z^0) \approx E(N^0 + 1, Z^0 + 1) + E(N^0, Z^0)$$
$$- E(N^0 + 1, Z^0) - E(N^0, Z^0 + 1)$$
$$= A(N^0, Z^0) - I(N^0 + 1, Z^0 + 1)\ , \tag{37}$$

$$\beta(N^0, Z^0) \approx E(N^0, Z^0 + 1) + E(N^0, Z^0 - 1) - 2E(N^0, Z^0)\ , \tag{38}$$

where $I(N^0, Z^0) = E(N^0 - 1, Z^0) - E(N^0, Z^0)$ and $A(N^0, Z^0) = E(N^0, Z^0) - E(N^0 + 1, Z^0)$, respectively, denote the ionization potential and electron affinity of the system identified by $H(N^0, Z^0)$. In Eq. 34 the Mulliken-type [54] finite difference expression for the atomic negative chemical potential (electronegativity) is the arithmetic average of the left and right negative slopes of the ground-state energy function. Similarly, the hardness expression of Eq. 36 represents the N-curvature of the corresponding three-point parabolic fit [13].

Following the above thermodynamic analogy [53] we use the principal potential $F(\boldsymbol{X})$, in which all state parameters have exclusively extensive character, to define the vector of restoring forces,

$$\boldsymbol{f} = -\,\partial F/\partial \boldsymbol{X} = -\boldsymbol{P} = (\chi, Z)\ , \tag{39}$$

grouping the atomic electronegativity [11, 12]

$$\chi = -\mu = -\,\partial F/\partial N = \partial E/\partial q\ , \tag{40}$$

and $Z = -\partial F/\partial\zeta$. Since the second partials of $F(\boldsymbol{X})$ (see Table 1),

$$\mathbf{F} = -\,\partial \boldsymbol{f}/\partial \boldsymbol{X} = \partial \boldsymbol{P}/\partial \boldsymbol{X} = \{\mathrm{F}_{i,j}\}\,, \tag{41}$$

relate forces to displacements of extensive state parameters, the matrix $\mathbf{F}$ can be interpreted as representing the *atomic stiffness moduli*. Its inverse (see Table 1), defined in the $Q(\mu, Z) = Q(\boldsymbol{P})$ representation using only the intensive state parameters,

$$\mathbf{Q} = -\,\partial \boldsymbol{X}/\partial \boldsymbol{f} = \partial \boldsymbol{X}/\partial \boldsymbol{P} = \{\mathrm{Q}_{i,j}\}\,, \tag{42}$$

can therefore be considered, by analogy to the theory of elasticity [53], as representing the *atomic compliance matrix.*

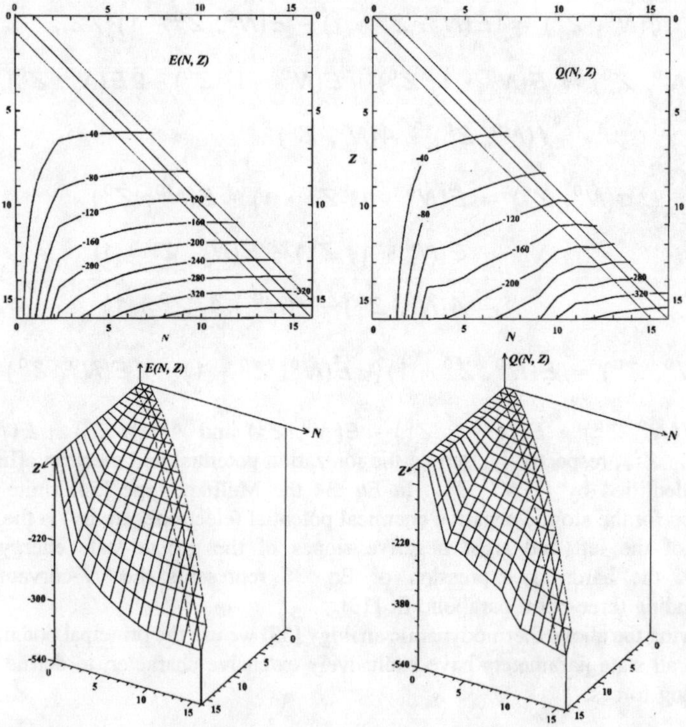

Figure 3. Perspective views and contour maps of the $(N \leq Z+1, Z \leq 16)$-fragments of alternative equilibrium ground-state potentials of atomic systems: $E(N, Z)$, $Q(\mu, Z) \equiv Q[N(\mu, Z), Z]$, $F(N, \zeta) \equiv F[N, Z(N, \zeta)]$, $R(\mu, \zeta) \equiv R[N(\mu, \zeta), Z(\mu, \zeta)]$.

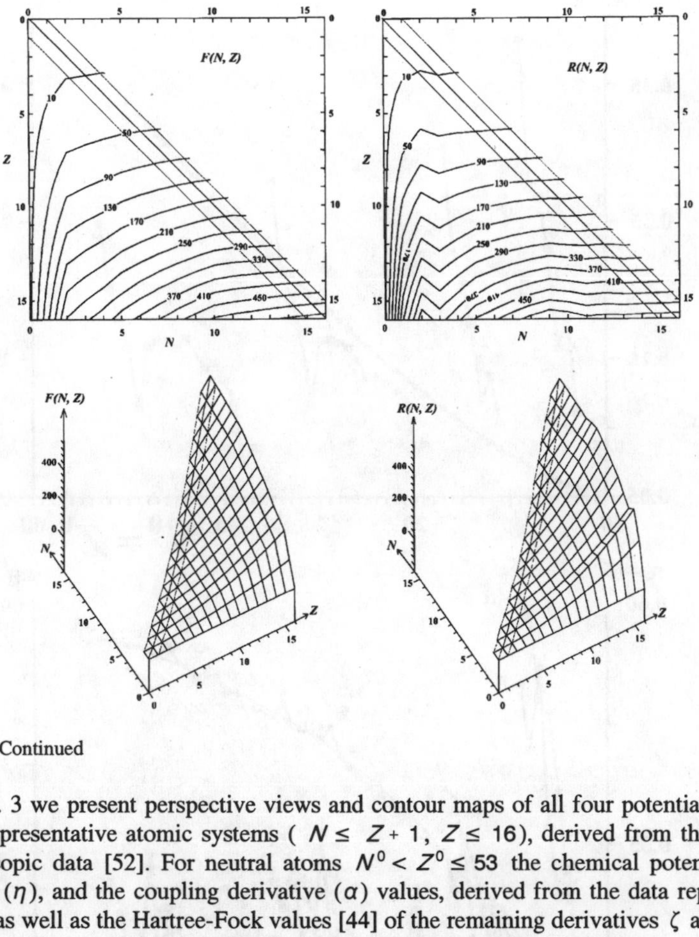

Figure 3. Continued

In Fig. 3 we present perspective views and contour maps of all four potentials of Eq. 33 for representative atomic systems ($N \leq Z + 1$, $Z \leq 16$), derived from the atomic spectroscopic data [52]. For neutral atoms $N^0 < Z^0 \leq 53$ the chemical potential (μ), hardness (η), and the coupling derivative (α) values, derived from the data reported in Ref. 12, as well as the Hartree-Fock values [44] of the remaining derivatives ζ and β are reported in Fig. 4. As seen in the figure the N-type derivative quantities, μ, η and α, exhibit a periodic behaviour, while smooth trends are observed for the Z-type derivatives ζ and β. The periodic structure of the chemical potential is also visible on the $Q(\mu, Z)$ and $R(\mu, \zeta)$ surfaces of Fig. 3.

Consider a given atomic system defined by the relevant two g.s. parameters of the set: N_0, μ_0, Z_0 and ζ_0. The Legendre transforms of the system energy can also be interpreted as the stationary points of the following *non-equilibrium* functions of the parameters which remain unconstrained in a given specification of atomic state:

$$Q(N) = E(N, Z_0) - \mu_0 N \equiv Q(N; \mu_0, Z_0) , \tag{43}$$

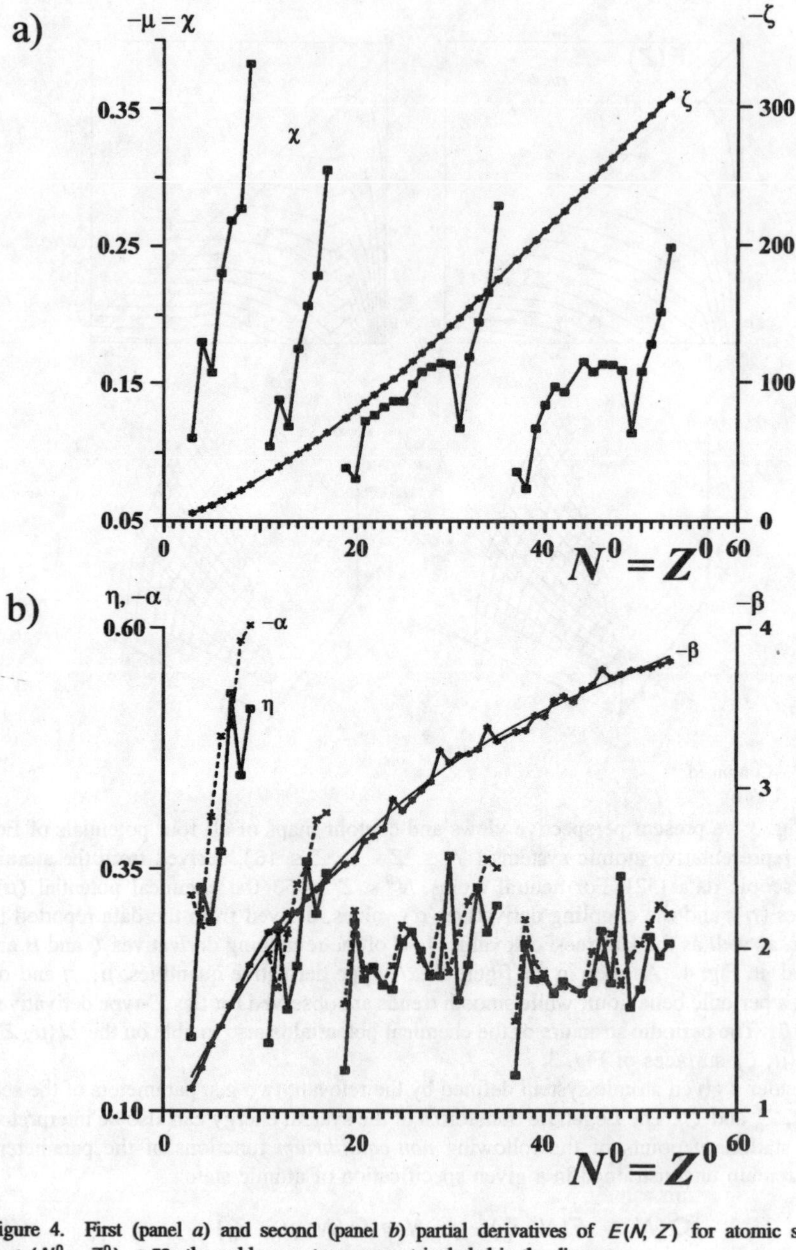

Figure 4. First (panel *a*) and second (panel *b*) partial derivatives of $E(N, Z)$ for atomic systems: $3 \leq (N^0 = Z^0) \leq 53$; the noble gas atoms are not included in the diagrams.

$$F(Z) = E(N_0, Z) - \zeta_0 Z \equiv F(Z; N_0, \zeta_0)\ , \tag{44}$$

$$R(N, Z) = E(N, Z) - \zeta_0 Z - \mu_0 N \equiv R(N, Z; \mu_0, \zeta_0)\ . \tag{45}$$

Obviously, such functions have vanishing derivatives at $N = N_0$ (Eq. 43), $Z = Z_0$ (Eq. 44) and ($N = N_0$, $Z = Z_0$) (Eq. 45), so that the associated equilibrium values of the unconstrained variables are uniquely determined by the extrema of the corresponding cuts of functions $Q(N)$, $F(Z)$, and $R(N, Z)$. This property of the above Legendre transforms is illustrated in Fig. 5 for O^{+2} ion. A reference to this figure shows, that

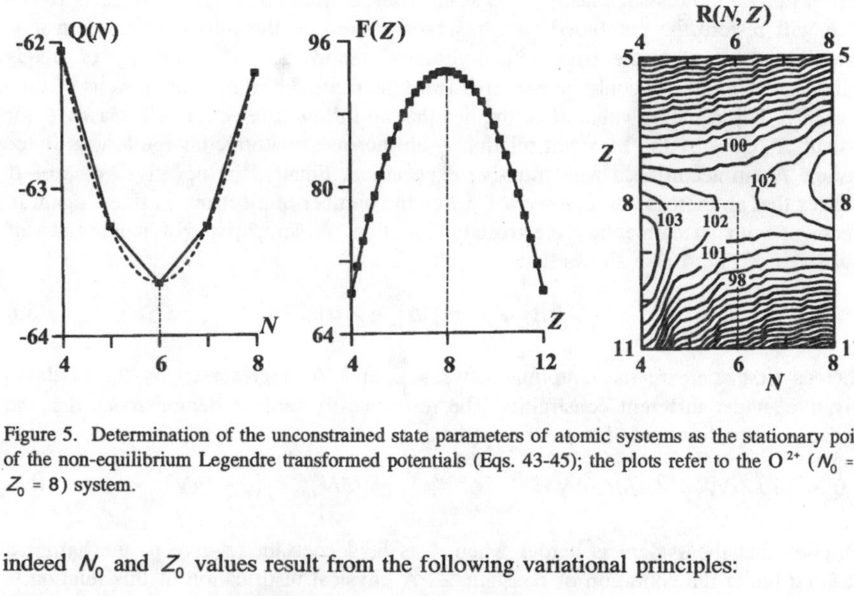

Figure 5. Determination of the unconstrained state parameters of atomic systems as the stationary points of the non-equilibrium Legendre transformed potentials (Eqs. 43-45); the plots refer to the O^{2+} (N_0 = 6, Z_0 = 8) system.

indeed N_0 and Z_0 values result from the following variational principles:

$$Q(N_0) = \min_N Q(N) = Q(\mu_0, Z_0)\ , \tag{46}$$

$$F(Z_0) = \max_Z F(Z) = F(N_0, \zeta_0)\ , \tag{47}$$

$$R(N_0, Z_0) = \min_N \max_Z R(N, Z) = R(\mu_0, \zeta_0)\ , \tag{48}$$

which identify the corresponding *equilibrium* Legendre transforms of the system energy. One could summarize these stationary principles in the following general statement [21, 53]: the equilibrium value of any unconstrained parameter $\mathbf{x}_0$ of the ground-state atomic system, defined by the electron population (N or μ) and external potential (Z or ζ) variables $\mathbf{y}_0$, is determined by the extremum principle:

$$\delta_{\mathbf{x}}\, \mathscr{L}(\mathbf{x};\, \mathbf{y}_0)|_{\mathbf{x}_0} = 0\,, \tag{49}$$

where $\mathscr{L}(\mathbf{x}_0;\, \mathbf{y}_0) = \mathscr{L}(\mathbf{y}_0)$ is the equilibrium Legendre transformed potential for $\mathbf{y}_0$.

2. Qualitative Discussion of Atomic Sensitivities

It follows from Fig. 4 that: $\mu < 0$, $\zeta < 0$, $\eta \approx -\alpha > 0$, and $\beta < 0$. The positive character of the atomic hardness reflects the stability criterion (Le Châtelier rule) [47, 53, 55-57] of the equilibrium state of an atom coupled to the electron reservoir. Namely, it implies that an inflow of electrons to the system ($dN > 0$) increases μ, and an outflow of electrons ($dN < 0$) decreases μ, when Z is held constant. This ensures that a forced electron population displacement of the atom from its initial equilibrium triggers forces which will restore the combined atom-reservoir system to the initial state, when it is allowed to relax after the hypothetical primary perturbation. A similarity of trends exhibited by η and $-\alpha$ could be expected from the finite difference expressions of Eqs. 36 and 37. The negative value of α implies that an inflow of electrons decreases ζ for constant Z, or — by the Maxwell relation — an increase in atomic charge lowers μ for constant N, in accordance with intuitive expectation. Finally, the negative value of β indicates that an increase in Z lowers ζ when the number of electrons is fixed, again in agreement with an elementary electrostatic intuition. A comparison of magnitudes of derivatives of Fig. 4 also shows that:

$$|\zeta| \gg |\beta| \gg \eta \approx |\alpha| \gtrsim |\mu|\,. \tag{50}$$

Let us first compare the coupling between μ and N, represented by the hardness derivative, under different constraints. The reference to Table 1 demonstrates that the following inequality holds for atomic systems:

$$0 < (\partial\mu/\partial N)_Z < (\partial\mu/\partial N)_\zeta \quad \text{or} \quad (\partial\chi/\partial N)_\zeta < (\partial\chi/\partial N)_Z < 0\,. \tag{51}$$

It implies that the system is harder when ζ is held constant relative to the hardness measured under the condition of constant Z. A physical justification of this relation is as follows. An increase in the number of electrons rises μ and lowers ζ. Therefore, the constraint of constant ζ calls for a change in the external potential to offset the change of ζ induced by dN. For example, one can lower Z to bring ζ to its initial, constrained value; this causes an extra increase in μ, relative to the fixed Z case, in accordance with Eq. 51.

It also follows from Table 1 that ζ is more sensitive to changes in Z when N is allowed to relax:

$$(\partial\zeta/\partial Z)_\mu < (\partial\zeta/\partial Z)_N < 0\,. \tag{52}$$

This inequality has a physical rationalization similar to that of Eq. 51. Namely, The constraint of constant μ requires an extra inflow of electrons from the reservoir (an increase in μ and a decrease in ζ) to cancel the decrease in μ induced by the $dZ > 0$. Consequently, the resultant magnitude of the ζ-response must be larger when μ is held constant, in comparison with that for the N = const. constraint.

Relations 51 and 52 compare "diagonal" couplings, between conjugate variables (Eqs. 30 , 31 and 39). Of similar character are the remaining two diagonal inequalities implied by the reduced expressions of Table 1:

$$0 < (\partial N/\partial \mu)_\zeta < (\partial N/\partial \mu)_Z \quad \text{or} \quad (\partial N/\partial \chi)_Z < (\partial N/\partial \chi)_\zeta < 0\ , \tag{53}$$

$$(\partial Z/\partial \zeta)_N < (\partial Z/\partial \zeta)_\mu < 0\ . \tag{54}$$

These inequalities directly follow from Eqs. 51 and 52. Equation 53 indicates that electrons are softer when Z = const. in comparison with the other constraint, while Eq. 54 demonstrates that the effective nuclear charge is more sensitive to changes in ζ when N is fixed.

Using the reduced expressions one similarly arrives at the relations between the corresponding "non-diagonal" derivatives:

$$(\partial \zeta/\partial N)_\mu < (\partial \zeta/\partial N)_Z < 0\ , \tag{55}$$

$$(\partial N/\partial \zeta)_Z < (\partial N/\partial \zeta)_\mu < 0\ , \tag{56}$$

which can be rewritten using the relevant Maxwell relations as :

$$(\partial \mu/\partial Z)_\zeta < (\partial \mu/\partial Z)_N < 0 \quad \text{or} \quad 0 < (\partial \chi/\partial Z)_N < (\partial \chi/\partial Z)_\zeta\ , \tag{57}$$

$$(\partial Z/\partial \mu)_N < (\partial Z/\partial \mu)_\zeta < 0 \quad \text{or} \quad 0 < (\partial Z/\partial \chi)_\zeta < (\partial Z/\partial \chi)_N\ . \tag{58}$$

In Eqs. 55 and 56 couplings between the extensive parameters are compared for different constraints of constant intensive parameters (forces), while in Eqs. 57 and 58 one relates couplings between forces measured under constraints of constant extensive variables. It follows from these inequalities that, e.g., the nuclear charge is more sensitive to change in the system chemical potential when the number of electrons is fixed, in comparison to the fixed ζ case (Eq. 58). This can be physically justified: a given displacement $d\mu > 0$ implies $d\zeta(d\mu) < 0$ and $dN(d\mu) > 0$. In order to generate the compensating shift $-d\zeta(d\mu)$ the atom has to change its number of electrons, $dN_r(-d\zeta) < 0$; this moderating charge transfer explains the inequality of Eq. 58. When examining a physical content of Eq. 56 one observes that a given $d\zeta > 0$ (a weaker attraction of electrons) is associated with $d\mu_r > 0$ and $dN < 0$, as demonstrated by the signs of corresponding derivatives in Table 1. Thus, to keep μ = const., in the second derivative, $dZ(-d\mu_r) > 0$ is required; this change in the external potential creates an induced flow of electrons $dN_r > 0$ which moderates the populational displacement due to the primary shift in $d\zeta$.

Inequalities 55 and 57, involving derivatives inverse to those in Eqs. 56 and 58, respectively, can be given a similar interpretation. The ordering of derivatives in Eq. 55 is because the requirement of constant μ implies that a given displacement dN must be accompanied by a simultaneous change in Z. For example, when $dN > 0$ one also has $d\mu_r(dN) > 0$. In order to diminish μ one requires $dZ(-d\mu_r) > 0$, and this gives rise to an extra lowering of ζ, which explains the inequality. Interpretation of Eq. 57 is analogous. In order to keep ζ fixed when, say, $dZ > 0$ [i.e., $d\zeta(dZ) < 0$] one has to transfer electrons to the reservoir, $dN_r > 0$, to generate the counter-change $-d\zeta > 0$; this moderating flow of electrons generates an extra lowering of chemical potential which justifies the inequality.

Yet another group of inequalities corresponds to the charge sensitivities measuring responses in extensive parameters per unit displacement in the non-conjugate intensive parameter, under the constraint of the other intensive parameter,

$$(\partial N/\partial Z)_{\mu} > 0\ , \tag{59}$$

$$(\partial \zeta/\partial \mu)_Z < 0\ . \tag{60}$$

Indeed, in order to keep μ constant when $dZ > 0$ [i.e., $d\mu(dZ) < 0$] one must have $dN > 0$; similarly, $d\mu > 0$ for constant Z must be a result of $dN > 0$ and this displacement generates $d\zeta(dN) < 0$.

The final two inequalities involve similar derivatives under the constraint of a constant extensive parameter, conjugate of the primarily displaced intensive parameter:

$$(\partial N/\partial Z)_{\zeta} < 0\ , \tag{61}$$

$$(\partial \zeta/\partial \mu)_N > 0\ . \tag{62}$$

Namely, the constraint ζ = const. in Eq. 61 calls for $dN < 0$, which creates $d\zeta(dN) > 0$ balancing $d\zeta(dZ) < 0$ (from the primary perturbation $dZ > 0$). Finally, in Eq. 62 $d\mu > 0$ for constant N is caused by $dZ < 0$ or $d\zeta(dZ) > 0$.

Equations 51 and 54 are examples of the generalized Le Châtelier principle [47, 53, 56, 58]: the response in the intensive parameter P_i, conjugated to the displaced extensive variable X_i, is the largest if all the remaining extensive variables are fixed, and it decreases upon a relaxation of each constraint that frees a variable X_k by coupling the system to a reservoir characterized by the intensive parameter P_k:

$$(\partial P_i/\partial X_i)_{X_{\neq i}} \geq (\partial P_i/\partial X_i)_{P_1,\, X_{\neq(1,i)}} \geq \ldots$$
$$\geq (\partial P_i/\partial X_i)_{P_1,\ldots,P_{i-1},\, X_{\neq(1,\ldots,i)}} > 0\ , \quad i = 1, 2, \ldots, m\ , \tag{63}$$

where m denotes the number of degrees of freedom. Namely, in accordance with our previous identification of extensive and intensive parameters of atomic systems, one directly recognizes the left inequality of Eq. 51 as the generalized Le Châtelier rule; since

ζ is the conjugate of $(-Z)$ one has to multiply Eq. 54 by (-1) to bring this inequality into the form of Eq. 63.

Consider now the AIM in the fixed external potential. Let $\boldsymbol{N} = (N_1, ..., N_m)$ and $\boldsymbol{\mu} = \partial E/\partial \boldsymbol{N} = (\mu_1, ..., \mu_m)$ group the AIM electron populations and chemical potentials, respectively. At global equilibrium state, when there are no restrictions on a flow of electrons between constituent atoms, the atomic chemical potentials are equalized at the global chemical potential level: $\mu_i = \mu = \partial E/\partial N$, $i = 1, 2, ..., m$; here $N = \Sigma_i N_i$ is the overall number of electrons in a molecule M. Besides the global equilibrium state one can also envisage various constrained equilibrium states, when same restrictions on intramolecular flows of electrons are introduced through a given partitioning of M into mutually closed subsystems. One can assume a partitioning of M into two complementary subsystems, $M = (A|B)$, where in particular A can be limited to a single atom X with B(X) representing the molecular remainder. Let d denotes the number of hypothetical "walls" within B, dividing this fragment into smaller mutually closed subunits. Then, the generalized Le Châtelier principle of Eq. 63, comparing the rigid and relaxed hardnesses $\eta_X(d) = (\partial\mu_X/\partial N_X)_d$ of X in M, reads:

$$\eta_X(d_{max}) \geq \eta_X(d_{max}-1) \geq ... \geq \eta_X(1) \geq \eta_X(0) \equiv \partial\mu/\partial N \,, \tag{64}$$

where d_{max} corresponds to all AIM in B being mutually closed, and lower values of d denoting the partitionings with collections of mutually opened AIM in B. The physical content of these inequalities is directly related to the Le Châtelier-Braun principle [20, 34, 41, 53, 56, 59]. Namely, the more relaxed state of B(X) implies the stronger moderating (softening) influence on hardness of X, through the indirectly induced flows of electrons in B(X), due to the primary populational displacement in X. These relaxational flows in B(X) are triggered by the differences in the AIM chemical potentials generated by dN_X [30, 34, 59-61] . The maximum partitioning of a given system in the AIM resolution defines the hardness matrix $\boldsymbol{\eta} = \partial\boldsymbol{\mu}/\partial\boldsymbol{N} = \{\eta_{X,Y}\}$ with the rigid atomic hardness $\eta_X(d_{max}) = \eta_{X,X}$.

Clearly, when the effective atomic charges are used to characterize the effects of changing external potential in a molecule, the previously discussed atomic inequalities remain valid for the corresponding diagonal AIM sensitivities; e.g., for the diagonal hardness of X one has:

$$(\partial\mu_X/\partial N_X)_{\zeta_X, C} \geq (\partial\mu_X/\partial N_X)_{Z_X, C} > 0 \,, \tag{65}$$

for all possible constraints C imposed on the remaining atoms (see Eq. 51).

Equation 64 can be rewritten in a form directly related to the Le Châtelier principle of Eq. 63,

$$\eta_{X,X} \equiv (\partial\mu_X/\partial N_X)_{B(X):\boldsymbol{N}} \geq (\partial\mu_X/\partial N_X)_{B(X):\, \mu_i = \mu_j = \mu_{ij};\, \boldsymbol{N}[\neq (i,j)]} \geq ...$$

$$\geq (\partial\mu_X/\partial N_X)_{B(X):\, \mu_i = \mu_j = ... = \mu_{B(X)}} \equiv \eta^{rel}_{X,X} \,, \quad X = 1, 2, ..., m \,, \tag{66}$$

providing the relaxational hierarchy of the AIM hardness parameters.

As we have argued in Chapter 1 (Eqs. 21-25), the atomic diagonal hardnesses can be related to the one-centre valence-shell electron repulsion integrals; the finite difference expression of Eq. 36 can be recognized as the familiar Pariser formula [36] providing an estimate of the two-electron coulomb integral within the classical semiempirical SCF MO theories,

$$\eta_{X,X} \approx \gamma_{X,X} = \int d\mathbf{r}_1 \int d\mathbf{r}_2 \, |s_X(\mathbf{r}_1)|^2 \frac{1}{r_{12}} |s_X(\mathbf{r}_2)|^2 , \tag{67}$$

where s_X is the valence-shell s orbital of atom X and $r_{12} = |\mathbf{r}_1 - \mathbf{r}_2|$. This association has been extended [33, 34] by similarly approximating the off-diagonal AIM hardnesses, $\eta_{X,Y}$, by the corresponding two-centre electron repulsion integrals (see Eqs. 24, 25),

$$\eta_{X,Y} \approx \gamma_{X,Y} = \int d\mathbf{r}_1 \int d\mathbf{r}_2 \, |s_X(\mathbf{r}_1)|^2 \frac{1}{r_{12}} |s_Y(\mathbf{r}_2)|^2 , \tag{68}$$

using standard interpolation formulas of the semiempirical SCF MO theories. These electron repulsion parameters, both *orbitally rigid* (unrelaxed orbitals) and *orbitally relaxed*, can also be determined from the X_α calculations [62] or estimated from the so-called *Valence State Ionization Energies* (VSIE); the latter have been developed using the average atomic spectroscopic data for the *Self-Consistent Charge and Configuration* (SCCC) implementation of the molecular orbital theory [63].

3. Reduction of Derivatives: Equipotential Sensitivities

The reduction of the second-order derivatives of Table 1 to identities in terms of basic derivatives α, β, and η, is equivalent to the derivative transformation to the independent variables N and Z. The procedure is based upon the mathematical properties of the Jacobians [53]. These reduced expressions can be further applied to express any sensitivity in terms of basic derivatives. Let us consider as an illustrative example the set of six sensitivities corresponding to the constraints that require the system to conserve values of the respective potentials $\mathcal{L}(X, Y)$. We call them *equipotential* or *horizontal* atomic sensitivities. In Fig. 2 a particular example of $(\partial Z/\partial N)_E = -\mu/\zeta$ is shown on a schematic contour of the $E(N, Z)$ surface.

The reduced expressions for all atomic horizontal sensitivities are listed in Table 2. They immediately follow from their definitions and the reduced expressions of Table 1, when a given equipotential sensitivity $(\partial t_i/\partial t_j)_{\mathcal{L}(X,Y)}$ is transformed to the (X, Y) representation; here $\boldsymbol{t} = (\boldsymbol{X}, \boldsymbol{P})$ groups all four state variables. For example,

$$\left(\frac{\partial Z}{\partial N}\right)_{R(\mu,\zeta)} = \frac{\partial(Z, R)}{\partial(N, R)} = \frac{\partial(Z, R)}{\partial(\mu, \zeta)} \Bigg/ \frac{\partial(N, R)}{\partial(\mu, \zeta)} . \tag{69}$$

TABLE 2. Atomic equipotential sensitivities and their reduced expressions[2]

Sensitivity	Potential ($\mathcal{L}$)			
	$E(N,Z)$	$Q(\mu,Z)$	$F(N,\zeta)$	$R(\mu,\zeta)$
$\left(\frac{\partial Z}{\partial N}\right)_{\mathcal{L}}$	$-\frac{\mu}{\zeta}$	$\frac{N\eta}{\zeta - N}$	$\frac{\mu - \alpha Z}{\beta Z}$	$\frac{\eta N - Z\alpha}{\alpha N + Z\beta}$
$\left(\frac{\partial Z}{\partial \mu}\right)_{\mathcal{L}}$	$\frac{\mu}{\alpha\mu - \eta\zeta}$	$\frac{N}{\zeta}$	$\frac{Z\alpha - \mu}{Z(\alpha^2 - \eta\beta) - \mu\alpha}$	$\frac{\eta N + Z\alpha}{Z(\alpha^2 - \eta\beta)}$
$\left(\frac{\partial Z}{\partial \zeta}\right)_{\mathcal{L}}$	$\frac{\mu}{\beta\mu - \alpha\zeta}$	$\frac{-N\eta}{N(\alpha^2 - \eta\beta) - \alpha\zeta}$	$\frac{\mu - \alpha Z}{\mu\beta}$	$-\frac{\eta N + Z\alpha}{N(\alpha^2 - \eta\beta)}$
$\left(\frac{\partial N}{\partial \mu}\right)_{\mathcal{L}}$	$\frac{\zeta}{\eta\zeta - \alpha\mu}$	$\frac{\zeta - \alpha N}{\eta\zeta}$	$\frac{Z\beta}{Z(\eta\beta - \alpha^2) + \alpha\mu}$	$-\frac{\alpha N + Z\beta}{Z(\alpha^2 - \eta\beta)}$
$\left(\frac{\partial N}{\partial \zeta}\right)_{\mathcal{L}}$	$\frac{\zeta}{\alpha\zeta - \beta\mu}$	$\frac{\zeta - \alpha N}{\alpha\zeta + N(\eta\beta - \alpha^2)}$	$\frac{Z}{\mu}$	$\frac{\alpha N + Z\beta}{N(\alpha^2 - \eta\beta)}$
$\left(\frac{\partial \mu}{\partial \zeta}\right)_{\mathcal{L}}$	$\frac{\eta\zeta - \alpha\mu}{\alpha\zeta - \beta\mu}$	$\frac{\zeta\eta}{\alpha\zeta + N(\eta\beta - \alpha^2)}$	$\frac{Z(\eta\beta - \alpha^2) + \alpha\mu}{\mu\beta}$	$\frac{-Z}{N}$

Expanding the Jacobians as determinants and using the definitions and reduced derivatives of Table 1 for the $R(\mu, \zeta)$ representation give the corresponding reduced expression shown in Table 2.

It follows from Eq. 50 and Table 2 that:

$$(\partial Z/\partial N)_{\mathcal{L}} < 0\,, \quad \text{all } \mathcal{L}\,; \tag{70}$$

[2] For the definitions of basic derivatives η, α and β, see Table 1.

$$(\partial Z/\partial \mu)_{\mathcal{L}} \begin{cases} < 0\,, & \mathcal{L} = \{E,\, Q,\, F,\, R(\text{if } \eta < |\alpha|)\} \\ > 0\,, & \mathcal{L} = R(\text{if } \eta > |\alpha|) \end{cases} ; \tag{71}$$

$$(\partial Z/\partial \zeta)_{\mathcal{L}} \begin{cases} > 0\,, & \mathcal{L} = \{E,\, Q,\, F,\, R(\text{if } \eta < |\alpha|)\} \\ < 0\,, & \mathcal{L} = R(\text{if } \eta > |\alpha|) \end{cases} ; \tag{72}$$

$$(\partial N/\partial \mu)_{\mathcal{L}} > 0\,, \quad \text{all } \mathcal{L}\,; \tag{73}$$

$$(\partial N/\partial \zeta)_{\mathcal{L}} < 0\,, \quad \text{all } \mathcal{L}\,; \tag{74}$$

$$(\partial \mu/\partial \zeta)_{\mathcal{L}} < 0\,, \quad \text{all } \mathcal{L}\,. \tag{75}$$

For positive ions, i.e., $Z > N$, one should expect in most cases $(\partial Z/\partial \mu)_R < 0$ and $(\partial Z/\partial \zeta)_R > 0$.

Within the $Q(\mu, Z)$ and $R(\mu, \zeta)$ representations the horizontal sensitivities: $(\partial Z/\partial N)_{Q(R)}$, $(\partial \mu/\partial N)_{Q(R)}$, and $(\partial \zeta/\partial N)_{Q(R)}$, appear when one considers displacements in the number of particles from the equilibrium state of an atomic system coupled to an electron reservoir. This is because such a state corresponds to a minimum of Q or R with respect to the flow of electrons (Eqs. 46 and 48), $dQ/dN = 0$ and $(\partial R/\partial N)_Z = 0$, respectively, so that a displacement dN does not change these potentials in the first-order. Similarly, at equilibrium, the potentials $F(Z)$ and $R(N, Z)$ are stationary with respect to the first-order displacements in Z. Therefore, the contour following aspect of horizontal sensitivities for constant $\mathcal{L}(E, Q, F, R)$ has also the stationary point content when one deals with non-equilibrium functions $\mathcal{L} = (Q, F, R)$.

In Table 3 we have displayed illustrative numerical values of the horizontal derivatives of Table 2, or their inverses, for selected neutral alkali metals and halogens. When interpreting these results one should remember that F is exactly the ground-state sum of the electron kinetic and repulsion energies, with the electron-nucleus attraction energy being removed by the Legendre transformation (Table 1). We also observe at this point that such a transformation removes in Q and R (F and R) energy contributions that are homogeneous of degree 1 in N (Z). This removal of a linear part increases a degree of convexity (concavity) with respect to N (Z) in these three transforms of the system energy, as explicitly shown in Figure 5. The above general observations are important in a physical rationalization of trends exhibited by the numerical results of Table 3.

It follows from the $(\partial Z/\partial N)_{\mathcal{L}}$ data that for the fixed $\mathcal{L}$ and the group of the periodic table, the magnitude of an equilibrium horizontal response in Z matching a unit displacement in N decreases with increasing atomic size (atomic number). One also detects a strong dependence of this sensitivity upon the atom electronegativity (hardness). For example, the responses predicted for halogens are stronger than those for alkali metals with a smaller value of atomic number. All these responses are seen to be negative for neutral atoms, in accordance with the slopes of the contours in Fig. 3. Clearly, in the

TABLE 3. Equipotential (horizontal) sensitivities of Table 2 for selected neutral alkali metals and halogens. The relevant numerical values of μ, ζ, α, β, and η (see also Fig.3), used to calculate the derivatives, are listed at the bottom of the table.

$\mathcal{L}$	Sensitivity	M				X			
		Li	Na	K	Rb	F	Cl	Br	I
E	$\partial Z/\partial N$	-0.019	-0.003	-0.001	0.000	-0.014	-0.005	-0.002	-0.001
	$\partial Z/\partial \mu$	-0.107	-0.017	-0.008	-0.003	-0.028	-0.014	-0.005	-0.003
	$\partial \zeta/\partial Z$	15.47	86.09	173.9	419.7	39.85	83.70	218.2	372.9
	$\partial N/\partial \mu$	5.519	5.889	7.074	7.350	1.909	2.879	3.203	3.674
	$\partial \zeta/\partial N$	-0.300	-0.255	-0.206	-0.190	-0.575	-0.396	-0.346	-0.300
	$\partial \zeta/\partial \mu$	-1.657	-1.500	-1.461	-1.396	-1.098	-1.141	-1.107	-1.101
Q	$\partial Z/\partial N$	-0.060	-0.040	-0.029	-0.022	-0.131	-0.072	-0.052	-0.040
	$\partial Z/\partial \mu$	-0.526	-0.311	-0.254	-0.195	-0.340	-0.264	-0.199	-0.172
	$\partial \zeta/\partial Z$	1.770	2.463	3.012	3.611	0.844	1.567	1.973	2.365
	$\partial N/\partial \mu$	4.747	5.436	6.710	7.080	1.544	2.583	2.985	3.486
	$\partial \zeta/\partial N$	-0.196	-0.141	-0.114	-0.099	-0.186	-0.160	-0.131	-0.116
	$\partial \zeta/\partial \mu$	-0.931	-0.765	-0.764	-0.703	-0.287	-0.414	-0.392	-0.406
F	$\partial Z/\partial N$	-0.251	-0.120	-0.081	-0.056	-0.296	-0.161	-0.104	-0.079
	$\partial Z/\partial \mu$	-0.981	-0.597	-0.512	-0.385	-0.426	-0.391	-0.299	-0.266
	$\partial \zeta/\partial Z$	0.147	0.080	0.058	0.041	0.144	0.112	0.077	0.060
	$\partial N/\partial \mu$	3.909	4.993	6.326	6.813	1.442	2.433	2.872	3.381
	$\partial \zeta/\partial N$	-0.037	-0.010	-0.005	-0.002	-0.043	-0.018	-0.008	-0.005
	$\partial \zeta/\partial \mu$	-0.144	-0.048	-0.030	-0.016	-0.061	-0.044	-0.023	-0.016
R	$\partial Z/\partial N$	-0.341	-0.182	-0.128	-0.092	-0.448	-0.266	-0.182	-0.140
	$\partial Z/\partial \mu$	-0.487	-0.217	-0.171	-0.113	-0.065	-0.062	-0.034	-0.028
	$\partial \zeta/\partial Z$	2.055	4.607	5.863	8.869	15.32	16.07	29.25	36.36
	$\partial N/\partial \mu$	4.820	5.581	6.833	7.196	1.865	2.821	3.170	3.647
	$\partial \zeta/\partial N$	-0.207	-0.179	-0.146	-0.139	-0.536	-0.354	-0.315	-0.274
	$\partial \zeta/\partial \mu$	-1.000	-1.000	-1.000	-1.000	-1.000	-1.000	-1.000	-1.000
	μ	-0.111	-0.105	-0.089	-0.086	-0.383	-0.305	-0.279	-0.248
	ζ	-5.700	-35.40	-74.90	-190.0	-26.50	-64.40	-176.0	-309.0
	α	-0.322	-0.261	-0.209	-0.191	-0.603	-0.408	-0.351	-0.303
	β	-1.137	-2.101	-2.529	-3.350	-1.895	-2.425	-3.295	-3.790
	η	0.175	0.169	0.141	0.136	0.515	0.345	0.312	0.272

$\mathcal{L} = E$ case (see Figs. 2 and 3) $dN > 0$ implies a decrease of E for constant Z, if no compensating lowering of Z takes place, to keep the energy constant. The same property is exhibited by the contour maps of $\mathcal{L} = Q$ (Fig. 3). The opposite trend is seen for the remaining potentials, $\mathcal{L} = (F, R)$, which increase their values with an inflow of electrons to the atom. Again, in order to keep $F =$ constant, one has to expand the electron distribution of an atom by lowering Z, to keep the sum of the electronic kinetic and repulsion energies at the initial level. The enhanced convexity (concavity) with respect to N (Z) of the energy Legendre transforms explains the observed larger magnitudes of $(\partial Z/\partial N)_{\mathcal{L}}$ sensitivities for $\mathcal{L} = (Q, F$, and $R)$. This also explains why the predicted magnitudes of this derivative for constant R are larger than those of the corresponding values for constant Q or F.

A reference to Table 3 also shows that the horizontal responses in Z per unit displacement in the system chemical potential are also negative for all $\mathcal{L}$, with the largest magnitude predicted for constant F. These derivatives represent horizontal equivalents of the derivative α^{-1} (Table 1), also negative. A relatively large response for the fixed F can be justified by realizing that $d\mu > 0$ implies $dN > 0$; as seen in Fig. 3 this displacement is associated with an increase in F, which must therefore be compensated by $dZ < 0$, to keep F at its initial value.

The horizontal analogs of the derivative β, $(\partial\zeta/Z)_{\mathcal{L}}$, are all positive in contrast to $\beta < 0$. The physical reason for this change of sign can be found by considering changes in ζ required to satisfy the constraint. For example, in the $\mathcal{L} = E$ case, $dZ > 0$ decreases energy, so that $d\zeta > 0$ is required to raise the energy back to its initial level. Similarly, in the case of $\mathcal{L} = F$, $dZ > 0$ implies $dF > 0$ (Fig. 3); this shift can be reversed only by an expansion of the density distribution associated with $d\zeta > 0$.

Let us consider now the equipotential softnesses in Table 3, $(\partial N/\partial\mu)_{\mathcal{L}}$, which parallel the η^{-1} derivatives, all of them positive. As a general rule one observes, that for a given atom its horizontal softness changes little with the nature of the equipotential constraint.

The next horizontal derivative reported in Table 3, $(\partial\zeta/\partial N)_{\mathcal{L}}$, is analogous to the α derivative of Table 1; both exhibit the same sign. The physical content of this observation is that, e.g., $dN > 0$ implies $dE < 0$, to be compensated by $dZ < 0$ to increase energy; this in turn implies $d\zeta < 0$ since $\beta < 0$. As a result one finds that $(\partial\zeta/\partial N)_E < 0$.

Finally, the equipotential response in ζ per unit displacement in μ is also seen to be negative, irrespective of the constraint. The sign of this derivative for $\mathcal{L} = E$ can be justified as follows: a given displacement $d\mu > 0$ can be obtained by $dZ < 0$ (due to $\alpha < 0$) and this implies $dE > 0$; thus, to lower the energy one has to increase the number of electrons, $dN > 0$ (since $\mu < 0$), which changes ζ, $d\zeta < 0$ (due to $\alpha < 0$).

4. Diatomic Donor-Acceptor Systems

4.1. Harpoon Mechanism

The non-adiabatic, alkali metal - halogen reactions producing alkali halide have no observable energetical thresholds and exhibit large reaction cross sections [64]. The microscopic description of these reactions is via the so-called *harpoon mechanism* of

Polanyi [65]. As the neutral reactants (metal atom, M, and halogen molecule, X_2) approach along the nearly flat (week long-range dispersion interaction) covalent potential energy surface, at the critical, relatively large separation R_c a transfer of a single electron from M to X_2 takes place (the alkali metal atom throws the electron "harpoon" into the halogen) to form an ion pair. This CT practically assures the reactive collision, since the strong coulombic attractive force (on a lower, ionic potential energy surface) accelerates further approach of reactants (pulling the harpoon). This harpoon-like use of a metal valence electron allows the formation of a neutral alkali halide MX with a simultaneous release of X. Large values of R_c imply relatively large reaction cross sections, $\sigma_{react} \approx \pi R_c^2$, ranging from 100 Å^2 (Li + I_2) to 200 Å^2 (Cs + Br_2) [64].

In this section we shall interpret the critical distance and the underlying energy balance equations in terms of the atomic chemical potentials and charge sensitivities. To simplify the problem we consider a hypothetical reactive approach of atomic neutral reactants M and X, for which the basic mechanism remains unchanged:

$$M + X \rightarrow [M^+ \text{---} X^-] \rightarrow MX . \qquad (76)$$

If one neglects weak van der Waals interactions between neutral reactants, the energy balance equation (a.u.) marking the intersection of *ionic* and *covalent* potential energy curves at R_c is:

$$E_{CT}(N_{CT} = 1;\ R = R_c) \cong I_M - A_X - 1/R_c = 0\ ; \qquad (77)$$

here E_{CT} is the CT energy and N_{CT} denotes the amount of CT. This equation simply reads that the energy required to remove one electron from M must be balanced at R_c by the energy release due to the electron capture by X and the extra stabilization due to coulombic interaction between the newly formed ionic pair.

A closer examination of an interaction between atomic donor and acceptor reactants requires the relevant energy expression. In the next subsection we examine the interaction energy in such systems using concepts and ideas of Section 1 of this chapter [12, 25, 44, 56, 57, 59, 61].

The CSA interpretation of Eq. 77 can be carried out using either the *biased* (b) or *unbiased* description of reactants. The former recognizes the *basic* (B) character of M and the *acidic* (A) role of X in the harpoon reaction mechanism, while the latter does not *a priori* predetermine their A or B characters. Clearly, the finite difference formulas for atomic chemical potential and hardness of Section 1 fall into the unbiased category, since these derivatives have been estimated by the corresponding ratios involving energy differences related to an adding and removing of an electron to/from the atom or ion under consideration, with both these hypothetical "experiments" being given the same weighting factors in the averages. This unbiased approach is adequate when both A and B characters are in principle admissible in a molecular environment, so that there is no *a priori* reason to bias one of these alternatives relative to the other. In this respect the harpoon reaction is somewhat atypical, since the mechanism clearly predetermines the donor/acceptor character of reactants, thus indicating the biased description as the one which conforms to the full *a priori* information about the system.

The biased chemical potential of M and X in the harpoon reaction are [12, 48]:

$$\mu_M^b = \mu_B^{AB} = -I_M \quad \text{and} \quad \mu_X^b = \mu_A^{AB} = -A_X . \tag{78}$$

These constant slopes of atomic energies further imply that the reactant diagonal hardnesses have to vanish identically in the biased description:

$$\eta_M^b = \partial\mu_M^b/\partial N_M = 0 \quad \text{and} \quad \eta_X^b = \partial\mu_X^b/\partial N_X = 0 . \tag{79}$$

We also observe at this point that at distances $R > R_c$, i.e., much larger that equilibrium bond length R_0 in MX, the off-diagonal hardness of Eq. 68 already approaches the asymptotic behaviour,

$$\eta_{M,X} \overset{R\to\infty}{\approx} 1/R_{M,X} . \tag{80}$$

This immediately follows from Eq. 68 and it is explicitly included in the relevant interpolation formulas (see Eqs. 24, 25).

4.2. Interaction Energy

Consider now the CT energy expression in terms of the atomic data. The M → X CT in the B(M) --- A(X) diatomic system preserves the global number of electrons,

$$N_{CT} = dN_A = -dN_B > 0 . \tag{81}$$

This obvious isoelectronic constraint on atomic populational displacements in the A---B system implies a simple dependence of N_A and N_B on N_{CT}, given by the trivial *in situ* CT FF indices of reactants: $f_A = \partial N_A/\partial N_{CT} = 1$ and $f_B = \partial N_B/\partial N_{CT} = -1$, grouped in the row vector $\boldsymbol{f}$. The electronic energy change, relative to the *separated atoms/ions limit* (SAL) level, is a sum of the familiar *CT energy*, E_{CT} (due to the charge separation), *electrostatic energy*, E_{ES} (due to the interaction of the "frozen" reactant charge distributions), and *polarization energy*, E_P (due to the charge relaxation of reactants, before CT, in the field of the other reactant).

In the diatomic reactive system A---B the atomic second-order charge sensitivities η, α, and β of Section 1 are replaced by the corresponding matrices:

$$\boldsymbol{\eta} = \frac{\partial^2 E}{\partial \boldsymbol{N}\,\partial \boldsymbol{N}} = \frac{\partial \boldsymbol{\mu}}{\partial \boldsymbol{N}} = \left\{ \eta_{i,j} = \frac{\partial \mu_j}{\partial N_i} = \frac{\partial \mu_i}{\partial N_j} = \eta_{j,i} \right\} , \tag{82}$$

$$\boldsymbol{\alpha} = \frac{\partial^2 E}{\partial \boldsymbol{N}\,\partial \boldsymbol{Z}} = \frac{\partial \boldsymbol{\zeta}}{\partial \boldsymbol{N}} = \left(\frac{\partial \boldsymbol{\mu}}{\partial \boldsymbol{Z}}\right)^{\dagger} = \left\{ \alpha_{i,j} = \frac{\partial \zeta_j}{\partial N_i} = \frac{\partial \mu_i}{\partial Z_j} = \alpha_{j,i} \right\} , \tag{83}$$

$$\boldsymbol{\beta} = \frac{\partial^2 E}{\partial \boldsymbol{Z}\, \partial \boldsymbol{Z}} = \frac{\partial \boldsymbol{\zeta}}{\partial \boldsymbol{Z}} = \left\{ \beta_{i,j} = \frac{\partial \zeta_j}{\partial Z_i} = \frac{\partial \zeta_i}{\partial Z_j} = \beta_{j,i} \right\}, \tag{84}$$

where (i, j) = (A, B), $\boldsymbol{N} = (N_A, N_B)$, $\boldsymbol{Z} = (Z_A, Z_B)$, $\boldsymbol{\mu} = (\mu_A, \mu_B)$, and $\boldsymbol{\zeta} = (\zeta_A, \zeta_B)$. Their diagonal (i = j) elements represent charge couplings in the spherical AIM, while the off-diagonal (i ≠ j) derivatives are responsible for the AIM polarization as representing various components of the AIM interaction. For example, the off-diagonal hardness (N-N coupling) is mainly the interaction between the valence electrons on both atoms (Eq. 68). Similarly, the off-diagonal N-Z and Z-Z coupling derivatives (see Eq. 31),

$$\alpha_{A,B} = \frac{\partial \zeta_B}{\partial N_A} = \int \frac{\partial \rho_B(\mathbf{r})}{\partial N_A} \frac{1}{|\mathbf{r} - \boldsymbol{R}_B|} d\mathbf{r}, \tag{85}$$

$$\beta_{A,B} = \frac{\partial \zeta_B}{\partial Z_A} = \int \frac{\partial \rho_B(\mathbf{r})}{\partial Z_A} \frac{1}{|\mathbf{r} - \boldsymbol{R}_B|} d\mathbf{r}, \tag{86}$$

where $\boldsymbol{R}_B$ stands for the position of nucleus B, probe the corresponding responses of the electron density on atom B, $\rho_B(\mathbf{r})$, to the N and Z displacements on the other atom. Clearly, both such perturbations must polarize the spherical densities of the isolated atoms.

The second-order Taylor expansion of the system energy in terms of N_{CT}, ΔZ_A, and ΔZ_B, is given by the following expression:

$$E_{AB}(N_{CT}, \Delta \boldsymbol{Z}) = E(\Delta \boldsymbol{N}, \Delta \boldsymbol{Z}) \cong \frac{\partial E}{\partial \boldsymbol{N}} \boldsymbol{f}^{\dagger} N_{CT} + \frac{1}{2} \boldsymbol{f} \frac{\partial^2 E}{\partial \boldsymbol{N}\, \partial \boldsymbol{N}} \boldsymbol{f}^{\dagger} N_{CT}^2$$

$$+ \frac{1}{2}\left[N_{CT} \boldsymbol{f} \frac{\partial^2 E}{\partial \boldsymbol{N}\, \partial \boldsymbol{Z}} \Delta \boldsymbol{Z}^{\dagger} + \Delta \boldsymbol{Z} \frac{\partial^2 E}{\partial \boldsymbol{Z}\, \partial \boldsymbol{N}} \boldsymbol{f}^{\dagger} N_{CT} \right] + \frac{\partial E}{\partial \boldsymbol{Z}} \Delta \boldsymbol{Z}^{\dagger} + \frac{1}{2} \Delta \boldsymbol{Z} \frac{\partial^2 E}{\partial \boldsymbol{Z} \partial \boldsymbol{Z}} \Delta \boldsymbol{Z}^{\dagger}$$

$$\equiv \left(\boldsymbol{\mu}^0 + \frac{1}{2} \Delta \boldsymbol{Z}\, \boldsymbol{\alpha}^{\dagger} \right) \boldsymbol{f}^{\dagger} N_{CT} + \frac{1}{2} \boldsymbol{f}\, \boldsymbol{\eta}\, \boldsymbol{f}^{\dagger} N_{CT}^2$$

$$+ \left(\boldsymbol{\zeta}^0 + \frac{1}{2} N_{CT} \boldsymbol{f}\, \boldsymbol{\alpha} \right) \Delta \boldsymbol{Z}^{\dagger} + \frac{1}{2} \Delta \boldsymbol{Z}\, \boldsymbol{\beta}\, \Delta \boldsymbol{Z}^{\dagger}$$

$$\equiv \mu_{CT}^{(+)} N_{CT} + \frac{1}{2} \eta_{CT} N_{CT}^2 + \boldsymbol{\zeta}^{(+)} \Delta \boldsymbol{Z}^{\dagger} + \frac{1}{2} \Delta \boldsymbol{Z}\, \boldsymbol{\beta}\, \Delta \boldsymbol{Z}^{\dagger}, \tag{87}$$

where

$$\mu_{CT}^{(+)} = \mu_A^{(+)} - \mu_B^{(+)} ,$$

$$\mu_A^{(+)} = \mu_A^0 + \frac{1}{2}(\Delta Z_A\, \alpha_{A,A} - \Delta Z_B\, \alpha_{B,A}) ,$$

$$\mu_B^{(+)} = \mu_B^0 + \frac{1}{2}(\Delta Z_A\, \alpha_{A,B} - \Delta Z_B\, \alpha_{B,B}) , \qquad (88)$$

$$\eta_{CT} = (\eta_{A,A} - \eta_{B,A}) + (\eta_{B,B} - \eta_{A,B}) \equiv \eta_A^{AB} + \eta_B^{AB} , \qquad (89)$$

$$\zeta_A^{(+)} = \zeta_A^0 + \frac{1}{2}(\alpha_{A,A} - \alpha_{B,A})\, N_{CT} \equiv \zeta_A^0 + \frac{1}{2}\alpha_A^{AB}\, N_{CT} ,$$

$$\zeta_B^{(+)} = \zeta_B^0 + \frac{1}{2}(\alpha_{A,B} - \alpha_{B,B})\, N_{CT} \equiv \zeta_B^0 + \frac{1}{2}\alpha_B^{AB}\, N_{CT} . \qquad (90)$$

The ES, P and CT contributions of the quadratic interaction energy of Eq. 87 are given by the following expressions:

$$E_{ES} = \zeta^0\, \Delta Z^\dagger , \qquad (91)$$

$$E_P = \frac{1}{2}\Delta Z\, \boldsymbol{\beta}\, \Delta Z^\dagger , \qquad (92)$$

$$E_{CT} = \left(\mu_{CT}^{(+)} + \frac{1}{2}\boldsymbol{f}\,\boldsymbol{\alpha}\,\Delta Z^\dagger\right) N_{CT} + \frac{1}{2}\eta_{CT}\, N_{CT}^2$$
$$\equiv \overset{+}{\mu}_{CT}\, N_{CT} + \frac{1}{2}\eta_{CT}\, N_{CT}^2 , \qquad (93)$$

where

$$\overset{+}{\mu}_{CT} \equiv \overset{+}{\mu}_A - \overset{+}{\mu}_B ,$$

$$\overset{+}{\mu}_A = \mu_A^0 + [\alpha_{A,A} - \frac{1}{2}(\alpha_{A,B} + \alpha_{B,A})]\, \Delta Z_A \equiv \mu_A^0 + \overset{+}{\alpha}_A\, \Delta Z_A ,$$

$$\overset{+}{\mu}_B = \mu_B^0 + [\alpha_{B,B} - \frac{1}{2}(\alpha_{A,B} + \alpha_{B,A})]\, \Delta Z_B \equiv \mu_B^0 + \overset{+}{\alpha}_B\, \Delta Z_B . \qquad (94)$$

Modelling of ΔZ_i as function of internuclear distance is not unique. Some ideas in this direction have been reported in Refs. 25 and 44.

The *in situ* chemical potential for the isoelectronic B → A CT, $\overset{+}{\mu}_{CT}$, corresponds to the *polarized* atoms. Their chemical potentials, $\overset{+}{\mu}_X$ and $\overset{+}{\mu}_M$ differ from the respective isolated atom potentials, μ_X^0 and μ_M^0 due to contributions resulting from changes in the external potential on both reactants; in our phenomenological description of Section 1 these external potential effects are modelled by displacements in the atomic nuclear charges: ΔZ_X (due to M) and ΔZ_M (due to X), both depending on the internuclear distance R:

$$\mu_X^+(R) = \mu_X^0 + \alpha_X^+ \Delta Z_X(R) \quad \text{and} \quad \mu_M^+(R) = \mu_M^0 + \alpha_M^+ \Delta Z_M(R) \, . \tag{95}$$

Similarly, the derivatives of the (ES + P)-energy with respect to the effective nuclear charge of atomic reactants in A---B, representing the ζ-potentials of polarized reactants before CT are modified with respect to the separated atom values:

$$\zeta_X^+(R) = \zeta_X^0 + \beta_{X,X} \Delta Z_X(R) + \frac{1}{2}(\beta_{M,X} + \beta_{X,M}) \Delta Z_M(R) \, ,$$

$$\zeta_M^+(R) = \zeta_M^0 + \beta_{M,M} \Delta Z_M(R) + \frac{1}{2}(\beta_{X,M} + \beta_{M,X}) \Delta Z_X(R) \, . \tag{96}$$

We would like to emphasize the differences in definitions of the atomic potentials in Eqs. 88, 90, introduced in Eq. 87 and identified by the superscript $^{(+)}$, and of the quantities in Eqs. 94-96 identified by the superscript $^+$, introduced through Eqs. 91-93 defining the conventional ES, P, and CT contributions to the energy.

The changes in nuclear charges, due to the unscreened nucleus of the other neutral reactant, are expected to decay fast with increasing R and vanish at $R \to \infty$:

$$\Delta Z_X(\infty) = \Delta Z_M(\infty) = 0 \, . \tag{97}$$

Therefore, one may assume with a degree of confidence that these displacements can be neglected at the large critical distances of the harpoon mechanism:

$$\mu_{CT}^+(R_c) \simeq \mu_{CT}^0 \equiv \mu_X^0 - \mu_M^0 \, , \qquad \zeta^+(R_c) = \zeta^0 \, . \tag{98}$$

The optimum (fractional) value of N_{CT}, $N_{CT}(R)$, can be determined from the minimization of $E_{CT}(N_{CT}, R)$ for the fixed R (Eq. 93):

$$\partial E_{CT}/\partial N_{CT}|_{N_{CT}(R)} \equiv \mu_{CT}|_{N_{CT}(R)} = 0 \, ; \tag{99}$$

hence

$$N_{CT}(R) = -\mu_{CT}^+(R)/\eta_{CT}(R) \, , \tag{100}$$

$$E_{CT}[N_{CT}(R), R] \equiv E_{CT}(R) = \frac{1}{2}\mu_{CT}^+(R)\, N_{CT}(R) = -\frac{[\mu_{CT}^+(R)]^2}{2\,\eta_{CT}(R)} \, . \tag{101}$$

This amount of CT equalizes the equilibrium AIM chemical potentials in the A---B system at the global chemical potential level, $\mu_{AB} = \partial \mu_{AB}/\partial N_{AB}$:

$$\mu_{AB} = \mu_A = \overset{+}{\mu}_A + (f_A \partial\mu_A/\partial N_A + f_B \partial\mu_A/\partial N_B) N_{CT}$$
$$= \overset{+}{\mu}_A + (\eta_{A,A} - \eta_{B,A}) N_{CT} \equiv \overset{+}{\mu}_A + \eta_A^{AB} N_{CT}$$
$$= \mu_B = \overset{+}{\mu}_B + (f_B \partial\mu_B/\partial N_B + f_A \partial\mu_B/\partial N_A) N_{CT}$$
$$= \overset{+}{\mu}_B - (\eta_{B,B} - \eta_{A,B}) N_{CT} \equiv \overset{+}{\mu}_B - \eta_B^{AB} N_{CT} \,. \tag{102}$$

Clearly, the harpoon reaction present somewhat inappropriate case for testing the above general expressions for $E_{CT}(R)$ and $N_{CT}(R)$, since the real CT at $R = R_c$ involves a single electron, while Eq. 100 must generate (within the unbiased approach) fractional values of CT. Therefore, in Fig. 6 we have compared instead the predicted $E_{CT}(R_e)$ values against the experimental bond dissociation energies D_e for alkali metal halides at equilibrium bond length $R = R_e$. Also shown in the figure are the fractional $N_{CT}(R_e)$ values. These predictions from Ref. [66] are based upon the semiempirical hardness matrix from the Pariser/Ohno formulas, and the experimental equilibrium internuclear distances [67]. Reference to Fig. 6 shows a semiquantitative reproduction of the experimental trends exhibited by dissociation energies, by the predicted CT stabilization energies. It also follows from the figure that the optimum amounts of CT are indeed very close to a single electron transfer.

Figure 6 demonstrates that the atomic charge sensitivities can be successfully applied to rationalize trends in bond energies. Their use in generating the diatomic potential energy curves and relating force constants can also be envisaged [12, 44], although a more reliable R-dependence of the model ΔZ_i parameters is required for this purpose.

4.3. Chemical Potential Equalization Perspective on Harpoon Mechanism

Let us return now to the harpoon reaction of Section 4.1. One can quickly verify using Eqs. 34, 36, 80 and 93 that for $N_{CT} = 1$ one obtains the energy balance Eq. 77; thus, the unbiased approach correctly identifies the R_c value for the assumed single electron transfer. The biased description gives the same result, when one again fixes $N_{CT} = 1$ (Eqs. 78-80 and 93). However, the biased treatment additionally provides an interesting chemical potential/electronegativity equalization interpretation of the critical distance at which the M → X CT takes place [32, 61]. We call this the CT *forward* (f) direction. In what follows we shall also consider the *reverse* (r) CT, $M^+ \leftarrow X^-$, between ionic pair resulting from the forward CT.

One could formally consider $N_{CT} > 0$ as a continuous variable. Then the corresponding expression for the CT energy at large distances (see Eqs. 78-81 and 93) in the biased description becomes:

$$E_{CT}^{b}(N_{CT}, R) = (I_M - A_X)\, N_{CT} - N_{CT}^2/R$$
$$= -I_M\, dN_M - A_X\, dN_X + dN_M\, dN_X/R \,. \tag{103}$$

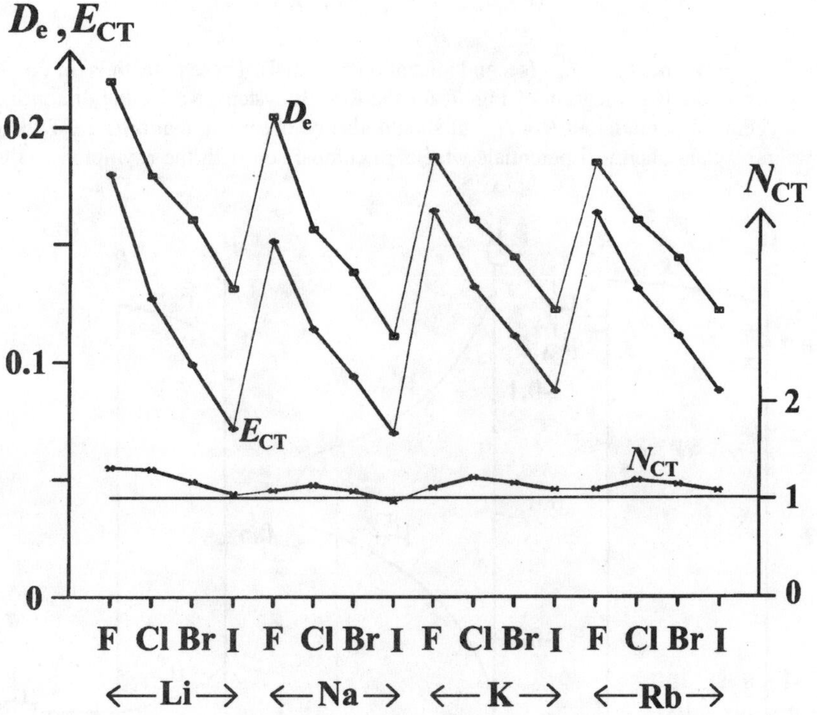

Figure 6. A comparison between bond dissociation energies D_e and $E_{CT}(R_e)$ values for alkali metal halides. The corresponding $N_{CT}(R_e)$ quantities, from the electronegativity equalization principle in the quadratic approximation, are also shown.

The corresponding chemical potentials of reactants are obtained by differentiating Eq. 103 with respect to N_A and N_B (see Eq. 81):

$$\mu_X^f(dN_M, dN_X, R) = -A_X + dN_M/R, \quad \mu_M^f(dN_M, dN_X, R) = -I_M + dN_X/R, \tag{104}$$

include obvious electrostatic contributions due to the other reactant. At $R \to \infty$: $\mu_X^f = -A_X > \mu_M^f = -I_M$, which prevents a spontaneous CT to the halogen at SAL. It follows from Eq. 104 that a mutual approach of both reactants increases (decreases) the chemical potential of M (X), when both reactants exhibit net fractional charges. Such a smooth change is not allowed in real systems, for which only $N_{CT} = 1$ is energetically allowed at R_c. Therefore, at $R > R_c$ μ_X^f and μ_M^f are equal to their asymptotic values. After the f-CT ($N_{CT} = 1$) the chemical potential of reactants for the same direction of CT are (see Eq. 104):

$$\mu^{f}_{X^-}(1, R) = -A_{X^-} - 1/R, \qquad \mu^{f}_{M^+}(1, R) = -I_{M^+} + 1/R, \tag{105}$$

where $A_{X^-} \approx 0$ and $I_{M^+} = I_M^{(2)}$ (second ionization potential). These potentials for the f-CT are shown in the first diagram of Fig. 7, for the K + Br system. Notice the discontinuity of the chemical potential at $R = R_c$. It should also be observed, that after f-CT the gap between reactant chemical potentials widens in comparison with the asymptotic value.

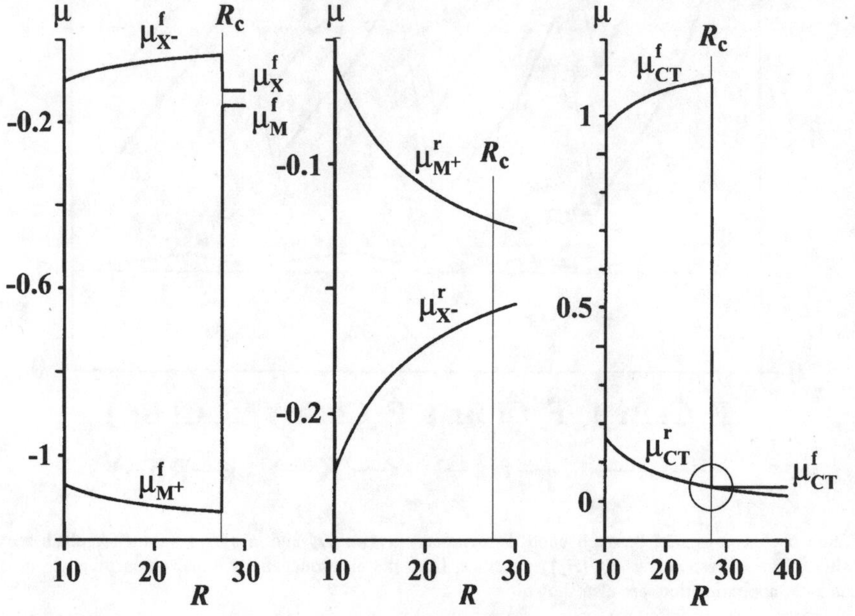

Figure 7. The biased chemical potentials of reactants in the K + Br harpoon reaction.

Consider now the r-CT between ions resulting from the f-CT, in which M^+ and X^- act as the electron acceptor and donor, respectively. Their respective biased chemical potential functions for $N_{CT} = 1$ are:

$$\mu_{M^+} = -A_{M^+} + 1/R = -I_M + 1/R, \quad \mu_{X^-} = -I_{X^-} - 1/R = -A_X - 1/R. \tag{106}$$

They are shown in the second diagram of Fig. 7. The third diagram of the figure shows plots of the corresponding *in situ* chemical potentials for the f- and r-CT. Notice, that

$$\mu^{f}_{CT}(0, R_c) = \mu^{r}_{CT}(1, R_c). \tag{107}$$

Thus, the critical distance marks the equalization of the *in situ* chemical potentials for the CT in the forward and reverse directions. It should be observed, that there is no equalization of the biased chemical potentials of reactants in the harpoon mechanism.

5. Combination Rules

In this section we examine simple formulas for combining the AIM (fragment) parameters into those characterizing larger molecular fragments and, eventually, the whole molecule [33, 41, 44, 59]. These combination rules are useful in a qualitative thinking about molecular charge couplings and the donor–acceptor bond.

The simplest case is the partitioning of a molecular system M into two complementary subsystems, e.g., the acidic and basic reactants in the A---B system: M = (A|B). Following this example we assume, for reasons of definiteness, that $\mu_B^+ > \mu_A^+$ (or $\chi_B^+ < \chi_A^+$) and $\eta_{B,B}^M < \eta_{A,A}^M$ (Fig. 8), were the superscript M specifies the system for which the hardnesses are defined. Let us consider the equilibrium chemical potential of M:

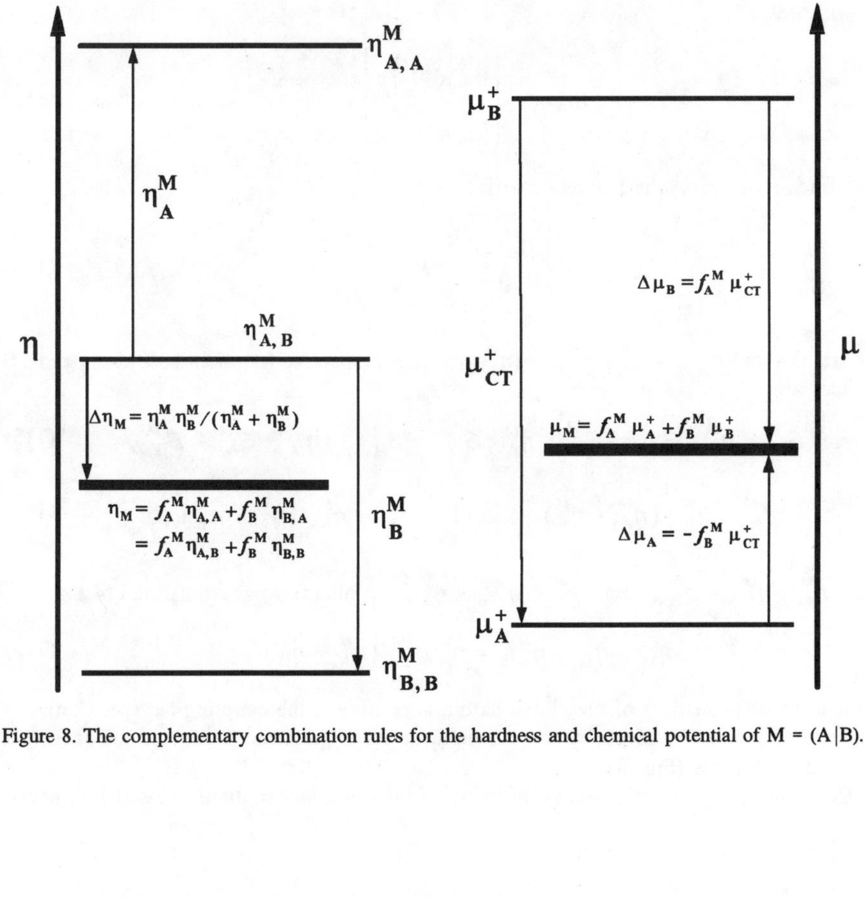

Figure 8. The complementary combination rules for the hardness and chemical potential of M = (A|B).

$$\mu_M = \frac{\partial E_M}{\partial N_M} = \frac{\partial E_M}{\partial N_A}\frac{\partial N_A}{\partial N_M} + \frac{\partial E_M}{\partial N_B}\frac{\partial N_B}{\partial N_M} \equiv \overset{+}{\mu}_A f_A^M + \overset{+}{\mu}_B f_B^M . \tag{108}$$

The displacements of this equalized level from the initial chemical potentials of A and B can be expressed in terms of $\overset{+}{\mu}_{CT} = \overset{+}{\mu}_A - \overset{+}{\mu}_B < 0$ and the fragment FF indices f_A^M and f_B^M (see Figure 8):

$$\Delta\mu_B = \mu_M - \overset{+}{\mu}_B = f_A^M \overset{+}{\mu}_{CT} \quad \text{and} \quad \Delta\mu_A = \mu_M - \overset{+}{\mu}_A = -f_B^M \overset{+}{\mu}_{CT} . \tag{109}$$

The FF indices are ratios of the fragment and global softnesses:

$$f_X^M = \frac{\partial N_X^M}{\partial N_M} = \frac{\partial N_X^M}{\partial \mu_M} \Bigg/ \frac{\partial N_M}{\partial \mu_M} \equiv \frac{S_X^M}{S_M} . \tag{110}$$

In order to obtain the softness (FF) parameters from the fragment hardness matrix $\{\eta_{X,Y}^M = \partial\mu_Y/\partial N_X\}$,

$$\boldsymbol{\eta}_{(A|B)} = \begin{bmatrix} \eta_{A,A}^M & \eta_{A,B}^M \\ \eta_{B,A}^M & \eta_{B,B}^M \end{bmatrix} , \tag{111}$$

one finds its inverse, the softness matrix:

$$\boldsymbol{\sigma}_{(A|B)} = \begin{bmatrix} \eta_{B,B}^M/G & -\eta_{A,B}^M/G \\ -\eta_{A,B}^M/G & \eta_{A,A}^M/G \end{bmatrix} , \tag{112}$$

where $G = \det \boldsymbol{\eta}_{(A|B)} = \eta_{A,A}^M \eta_{B,B}^M - (\eta_{A,B}^M)^2$. The respective fragment softnesses and FF indices are:

$$S_A^M = \eta_B^M/G , \qquad S_B^M = \eta_A^M/G , \qquad S_M = 1/\eta_M = S_A^M + S_B^M , \tag{113}$$

$$f_A^M = \eta_B^M/(\eta_A^M + \eta_B^M) , \quad \text{and} \quad f_B^M = \eta_A^M/(\eta_A^M + \eta_B^M) ; \tag{114}$$

here $\eta_A^M = \eta_{A,A}^M - \eta_{B,A}^M$ and $\eta_B^M = \eta_{B,B}^M - \eta_{A,B}^M$. It also follows from Eq. 114 that

$$\Delta\eta_M = \eta_M - \eta_{A,B}^M = \eta_A^M \eta_B^M/(\eta_A^M + \eta_B^M) , \tag{115}$$

so that the displacement of the global hardness relative to the coupling hardness between the two fragments is given by the harmonic average (reduced value) of the fragment relative hardnesses (Fig. 8).

Consider now a larger system consisting of three initially mutually closed fragments:

$M' = (M|X) = (A|B|X)$. By straightforward chain rule transformations one finds:

$$\eta^{M'}_{X,M} = \eta^{M'}_{X,A} f^M_A + \eta^{M'}_{X,B} f^M_B = \left(\eta^{M'}_{X,A}\eta^M_B + \eta^{M'}_{X,B}\eta^M_A\right)/\left(\eta^M_A + \eta^M_B\right), \qquad (116)$$

Of great interest is also the relaxational correction to $\eta^{M}_{X,X}$ due to the spontaneous charge response in M, $\delta N_M = dN_A(dN_X) = -dN_B(dN_X)$, to the primary populational displacement on X [34]:

$$\delta\eta^{M'}_{X,X}(\delta N_M) = \left(\frac{\partial\mu_X}{\partial N_A}\right)\left(\frac{\partial N_A}{\partial N_X}\right)_{N_M} + \left(\frac{\partial\mu_X}{\partial N_B}\right)\left(\frac{\partial N_B}{\partial N_X}\right)_{N_M} = -\frac{\left(\eta^{M'}_{X,A} - \eta^{M'}_{X,B}\right)^2}{\eta^M_A + \eta^M_B}. \qquad (117)$$

One arrives at this expression by considering the chemical potential differentiation between the two fragments in M generated by dN_X,

$$\mu^M_{CT}(dN_X) \equiv d\mu_A(dN_X) - d\mu_B(dN_X) = \left(\eta^{M'}_{X,A} - \eta^{M'}_{X,B}\right)dN_X, \qquad (118)$$

and determining the internal CT in M, triggered by this force, which equalizes the chemical potential in M. Namely, it follows from Eq. 100 that

$$\left(\partial N_A/\partial N_X\right)_{N_M} = -\mu^M_{CT}/\eta^M_{CT} = -\left(\partial N_B/\partial N_X\right)_{N_M}. \qquad (119)$$

Using Eqs. 118 and 119 in the chain rule of Eq. 117 gives the final expression for the relaxation correction to $\eta_{X,X}$.

The corresponding expression for the relaxational correction to the coupling hardness $\eta^{M''}_{X,Y}$ in $M'' = (X|Y|M) = (X|M')$,

$$\delta\eta^{M''}_{X,Y}(\delta N_M) = \left(\frac{\partial\mu_Y}{\partial N_A}\right)\left(\frac{\partial N_A}{\partial N_X}\right)_{N_M} + \left(\frac{\partial\mu_Y}{\partial N_B}\right)\left(\frac{\partial N_B}{\partial N_X}\right)_{N_M}$$

$$\equiv -\frac{\left(\eta^{M''}_{X,A} - \eta^{M''}_{X,B}\right)\left(\eta^{M''}_{Y,A} - \eta^{M''}_{Y,B}\right)}{\eta^M_A + \eta^M_B}, \qquad (120)$$

also follows from a similar chemical potential equalization argument. Namely, using the previously determined (Eqs. 118 and 119) relaxational flow in M due to the dN_X, and taking into account its effect on μ_Y, as reflected by the coupling hardnesses $\eta^{M''}_{Y,A}$ and $\eta^{M''}_{Y,B}$ (first derivatives in the chain rules of Eq. 120), gives Eq. 120.

It should be observed that for stable electron distributions in M ($\eta^M_{CT} > 0$) the relaxational correction of Eq. 117 is always negative (softening influence). Thus, a spontaneous electron distribution adjustment diminishes the effect of the primary displacement ($dN_X > 0$). This is in accordance with the Le Châtelier-Braun principle, that the indirectly induced flows must always moderate the force created by the primary displacement.

6. Illustrative Qualitative Applications

6.1. Electronegativity Equalization Rules

One of the classical problems in chemistry is to predict the equilibrium chemical potential (electronegativity) of the combined system from those of separated fragments. In case of the two complementary subsystems, M = (A|B), the combination formula 108 applies (see also Fig. 8). It expresses μ_M as the FF "weighted" average of μ_A^+ and μ_B^+; it should be emphasized, however, that these response functions can be negative. Chemists have developed approximate rules in which one averages only electronegativities, i.e., without any reference to the response information required by EE principle.

An example of such a classical rule is given by the *geometric mean* principle of Sanderson [18], expressing $\chi_{AB} = -\mu_{AB}$ in terms of electronegativities $\chi_A^0 = -\mu_A^0$ and $\chi_B^0 = -\mu_B^0$ of the separated fragments:

$$\chi_{AB} \approx \left(\chi_A^0 \chi_B^0\right)^{1/2} . \tag{121}$$

Parr and Bartolotti [68] analyzed the assumptions which are sufficient to obtain this principle; they have demonstrated that it requires the exponential decay of atomic (valence-state) energies with the number of electrons, with a universal falloff parameter for all the atoms. Later, the effects of the external potential have been examined [25, 44]. Using Eqs. 109 (see Fig. 8) one obtains the expression

$$\mu_{AB}^2 = \mu_A^+ \mu_B^+ + \left(\mu_{CT}^+\right)^2 + \Delta\mu_A \, \Delta\mu_B , \tag{122}$$

which includes small correction terms to the geometric mean rule, involving chemical potential differences; the dominant correction is the square of the initial chemical potential difference between both fragments. This equation suggests that one should expect increasing deviations between the predicted equilibrium chemical potential levels (from the geometric average of the μ_A^+ and μ_B^+) and the exact values of μ_{AB}, when $|\mu_{CT}^+|$ is relatively large, e.g., for a hard acid (A) and a soft base (B). These expectations are indeed reflected by numerical results [25].

It has been found [25] that a more realistic estimate of μ_{AB} for diatomic systems follows from the *harmonic mean* law,

$$\mu_{AB} \approx 2\mu_A^+ \mu_B^+ / (\mu_A^+ + \mu_B^+) , \tag{123}$$

which results from Eqs. 108 and 114, when one takes into account the empirically observed proportionality between the atomic chemical potential and hardness: $\mu_X^0 / \eta_X^0 \approx$ const.

6.2. Hard-Soft-Acids-and-Bases Principles

One of the first general rules of chemistry established theoretically within a qualitative CSA has been the *Hard-Soft-Acids-and-Bases* (HSAB) principle [69-71]. It reads [69, 70] that hard (H) Lewis acids (A) prefer to coordinate to hard Lewis bases (B), and soft (S) acids to soft bases, with the soft-soft interaction being largely covalent and the hard-hard interaction - predominantly ionic. The mixed H-S and S-H coordinations have been found to be relatively unstable. Another version, systematizing observations concerning coordination preferences by transition metal ions [71], the so called *symbiosis rule* with respect to the HSAB behaviour, states that hard (soft) ligands enhance a tendency of the central metal ion to coordinate more hard (soft) ligands. The first CSA-type rationalization of the HSAB rule by Parr and Pearson [13] was based solely upon the E_{CT} part of the interaction energy (Eq. 101). Later this explanation has been extended by one of authors [43] by introducing the electrostatic interaction, E_{ES} . The role of electrostatic contributions was stressed earlier by Klopman [72] (for the recent refinements see [12, 29, 56]). The symbiosis rule has been qualitatively justified within CSA [34, 56] and supported by quantitative Fukui function analysis [73]. The HSAB principles are still a subject of investigations using the charge sensitivity and density-functional concepts [29], e.g., in the context of the *maximum hardness principle* [28, 74].

Following Parr and Pearson [13] one deduces the soft-soft preference in the acid-base interactions by examining the E_{CT} expression (Eq. 101). Namely, the *in situ* hardness, η_{CT} , is small only for soft reactants thus giving rise to large CT stabilization, which explains a relative stability of the S-S interactions. This denominator in Eq. 101 is larger for the mixed hardness interactions and reaches its maximum for the H-H combination.

Therefore, a relative stability of complexes between hard reactants must have different origin. A stability of such compounds is mainly due to electrostatic interactions in Eq. 91. Clearly, the magnitude of ΔZ_X for an AIM depends on the nature of its bonding partner, Y. The donor or acceptor atoms of soft species (small η) are of large size and high polarizability (well shielded nucleus). This suggests that for soft X, $\Delta Z_Y(X)$ which it induces on its partner Y, is relatively small. In consequence, the electrostatic energy should indeed contribute the most in the H-H interactions.

Consider now the numerator in the E_{CT} expression (Eq. 101). The hard species, and acids in particular, H(A), should exhibit low levels of chemical potential, having strongly bound valence electrons; by the same chemical intuition one generally associates high levels of chemical potential with soft species (loosely bound valence electrons), and particularly with soft bases, S(B). This general rule is indeed reflected by the neutral atom data of Fig. 4 and Table 3. Therefore, one predicts for the complexes H(A)–S(B) and S(A)–H(B) the largest and smallest magnitudes of $\overset{0}{\mu}_{CT}$ (Eq. 98), respectively.

Let us now examine the influence of the external potential due to the other reactant (atom), which shifts $\overset{0}{\mu}_{CT}$ to $\overset{+}{\mu}_{CT}$ (Eq. 94),

$$\begin{aligned}\overset{+}{\mu}_{CT} &= \overset{0}{\mu}_{CT} + \left[\overset{+}{\alpha}_A \Delta Z_A(B) - \overset{+}{\alpha}_B \Delta Z_B(A)\right] \\ &\approx \overset{0}{\mu}_{CT} + \left[\overset{0}{\alpha}_A \Delta Z_A(B) - \overset{0}{\alpha}_B \Delta Z_B(A)\right] . \end{aligned} \qquad (124)$$

Since the negative coupling atomic sensitivities $(-\overset{0}{\alpha}_X)$ closely follow trends in the AIM hardness values, one predicts a substantial cancellation in the second term since the acid, generally harder than the base, will induce larger ΔZ_B (A) than the corresponding shift in ΔZ_A (B) generated by the base. This cancellation should be particularly significant in the mixed hardness complexes, S(A)−H(B), where both α_X^0 and ΔZ_X should have comparable magnitudes for X = A, B, e.g., I and Li in the MX systems.

It has been predicted qualitatively [56] that at large distances $|\mu_{CT}^+| < |\mu_{CT}^0|$, so that the driving force for the electron transfer becomes diminished in the polarized reactants, relative to the corresponding SAL value. This also follows from the Le Châtelier rule, which justifies a general expectation that spontaneous system responses to a perturbation created by the reactant approach should diminish the force for the CT. This effect of bringing the chemical potential levels of reactants closer together, as a result of the density polarization, should be particularly strong for the hard reactants which induce larger shifts $\Delta\mu_X^+ \equiv \mu_X^+ - \mu_X^0 \approx \alpha_X^0\, \Delta Z_X^0$, due to relatively large values of both factors in the product. It should be stressed that at larger distances $\Delta Z_A(\mathrm{B})$, due to a negatively charged base, say X^-, is also negative, thus giving $\Delta\mu_A^+ > 0$ and $\Delta\mu_B^+ < 0$. As a result, the CT stabilization in the H−H case should be very small due to both a relatively large denominator and a small numerator in Eq. 101.

In a similar qualitative way one can justify the symbiosis principle that the H (S) ligand X enhances a tendency of the central metal ion M to coordinate more H (S) ligands Y. The H−H coordination is realized mostly through the ionic bond (small L → M net donation) while the S−S coordination is mainly through the covalent bond (substantial L → M CT).

Obviously, the symbiosis principle of preferences exhibited by the *next* M---Y coordination in X —— M --- Y as "functions" of the *previous* M — X coordination, is the resultant effect of both the *direct* influence of X on M, and the *indirect* (inductive) effect of X on the approaching Y. Let us examine how the formation of the first bond modifies the central metal atom M and the second ligand Y.

The softness matrix, $\boldsymbol{\sigma} = \boldsymbol{\eta}^{-1} = \partial \boldsymbol{N}/\partial \boldsymbol{\mu}$, of the MX system, can be obtained by inverting the (2 × 2) AIM hardness matrix, $\boldsymbol{\eta}$ (see Section 5 of this Chapter). The inverse of the local softness of M, $S_M^{MX} = \partial N_M/\partial\mu_{MX} = \sigma_{M,M} + \sigma_{M,X}$, is (Eq. 113):

$$(S_M^{MX})^{-1} = \eta_{M,M} + \eta_{M,X}\left(\eta_M^{MX}/\eta_X^{MX}\right) \approx \eta_{M,M} + \eta_{M,X}\,, \tag{125}$$

where the fragment hardnesses, $\eta_M^{MX} = \eta_{M,M} - \eta_{M,X}$ and $\eta_X^{MX} = \eta_{X,X} - \eta_{M,X}$, represent the *in situ* hardnesses in MX. The approximate equality in Eq. 125 follows from the previous HSAB principle which predicts that stable coordinations take place for comparable hardnesses (H−H or S−S) of M and X, so that the ratio in parentheses is close to 1. Therefore, if one considers the inverse of the local softness of M in MX as a measure of the local hardness, one concludes that the increase in hardness of M in MX relative to the isolated M is of the order of the coupling hardness (Eq. 68), which should be large for the H−H pair (contracted electron distributions) and small for the S−S complex (diffused electron distributions). Therefore, when a hard X coordinates to a hard M, the hardness of M is substantially increased, thus facilitating the next H−H coordination of another hard Y to (M−X). Similarly, when a soft X coordinates to a soft M, the central atom becomes only slightly less soft, thus still preferring further coordination of another soft Y.

Consider now the inductive effect of a charge relaxation in MX on the approaching Y, represented by the softening correction (Eq. 117),

$$\delta\eta_{Y,Y}^{XMY} = -(\eta_{Y,M} - \eta_{Y,X})^2/(\eta_M^{MX} + \eta_X^{MX}) \; . \qquad (126)$$

This correction is relatively small for the hard fragments in XM (large denominator), so that its effect should be felt particularly strongly by the soft XM. The numerator of Eq. 126 effectively measures the difference between electron repulsion couplings: Y–M (stronger) and Y–X (weaker), in accordance with the diminishing internuclear distances. Therefore, the relaxation within MX should effectively reduce the hardness of Y; this should facilitate the Y → M–X donation responsible for the new covalent bond between the soft Y and M. In other words, the inductive effect creates conditions which select the soft ligand Y in accordance with the symbiosis HSAB principle.

We conclude that the direct influence, strong for the first H–H coordination, increases the H–H selectivity of the next coordination, while the inductive, indirect influence, strong for the first S–S coordination, increases the S–S preference of the next coordination.

6.3. Directing Influence of Ligands

In complexes ML_n or (AL_n) of both transition metal (M) and the main group (A) elements the mutual influence of ligands is displayed in changes of properties of M–L bonds accompanying a substitution of one ligand by another, X, yielding MXL_{n-1} . In general, the differentiation between the *cis*- and *trans*-bonds in such lower symmetry complexes depend very strongly on the nature of X. The predominant effect is the *trans*-influence, i.e., the $M–L_{trans}$ bond is more sensitive to changes in the character of X, relative to the $M–L_{cis}$ bond; an increase in the donor ability of X, all other conditions being equal, usually strengthens the M–X bond and weakens the $M–L_{trans}$ bond; the most influencing ligands are the strongest donors. However, in some cases, and especially for the main group element complexes, the observed trends are less regular [75].

Consider a quasi-square complex MXL_3 (generalization to quasi-octahedral complexes MXL_5 is straightforward):

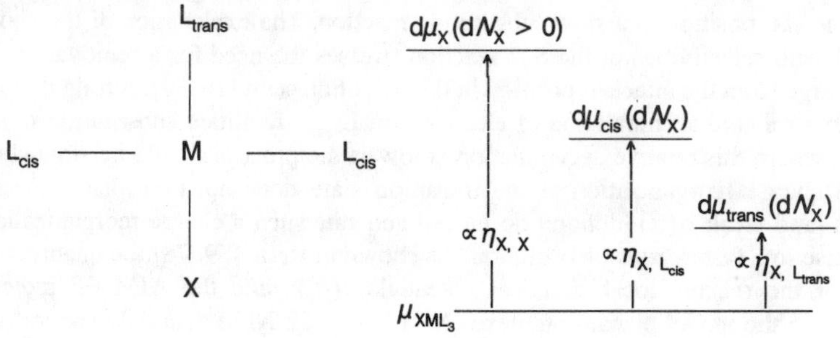

(127)

Let us further assume the atomic character of L and X, for reasons of simplicity. One observes that the hardness coupling between X and L_{cis}, $\eta_{X,cis}$, is stronger then $\eta_{X,trans}$, in accordance with the proportions of relevant internuclear distances. This differentiation is additionally strengthened by relaxational contributions to the coupling hardnesses (Eq. 120):

$$\delta\eta_{X,Y}^{MXL_3} = -(\eta_{X,1} - \eta_{X,2})(\eta_{Y,1} - \eta_{Y,2})/(\eta_1^R + \eta_2^R), \tag{128}$$

where Y = (L_{cis} , L_{trans}), and the relaxing fragments include two remaining ligands (besides X and Y): R = (1|2). Clearly, due to the symmetry, the relaxational contribution of Eq. 128 vanishes identically for Y = L_{trans} , and it is positive (increasing the charge coupling) for Y = L_{cis} [34].

An increase in the donor ability of X (higher chemical potential μ_X) can be attributed to $dN_X > 0$ (Le Châtelier principle). The corresponding shifts in the chemical potentials of each remaining constituent atom will be proportional to the corresponding coupling hardness, as schematically shown in the diagram 127. The differences in couplings between X and ligands in the *cis*- and *trans*-positions, respectively, immediately reflect differences in the ligand chemical potentials relative to the initial equalized value, $d\mu_{cis}(dN_X) > d\mu_{trans}(dN_X)$, as also shown in the diagram. Therefore, after allowing the chemical potential equalization in the "displaced" complex, to a common value $\mu(dN_X)$, one should expect a larger CT M→L_{trans} than M→L_{cis} , since magnitudes of such directly induced flows are proportional to the initial chemical potential differences (Eq. 100). This implies a smaller L_{trans}→M donation (weaker, longer bond) relative to L_{cis}→M. Such an effect is indeed observed experimentally [75]. A similar discussion for X exhibiting a weaker donor ability ($dN_X < 0$) shows that such a "displacement" in the nature of X should lead to a relative strengthening of M–L_{trans} bond in comparison to the M–L_{cis} bond, since the net donation L_{trans}→M is then increased relative to L_{cis}→M one. However, when the M–L bond includes a strong π-acceptor component these predictions may be altered, due to a competition between (σ, π)-donor and π-acceptor bond contributions.

These qualitative predictions of the relaxational flows in the coordination sphere, as functions of the nature of X, have also correct implications for the kinetic effect. It states that a typical π-acceptor X (a weak net donor) has a strong *trans*-directing influence in a fast nucleophilic substitution, while a typical strong donor X directs the approaching ligand into the *cis*-position in a slow substitution reaction. The explanation of the above kinetic trends and selectivities of the S_N2 reaction stresses the need for a removal of the electronic charge from the attacked position in the transition state [76, 77]. Strong donors, by leading to a relative accumulation of electrons on L_{trans} , facilitate substitution in the *cis*-position, where this relative accumulation is lower; the process should be slow since the required charge reorganization at the transition state does not take place. Strong π-acceptors [and weak σ(π)-donors] do indeed generate such a charge reorganization, thus giving rise to a faster *trans*-substitution. As shown in Refs. [59, 78] the quantitative predictions of the relaxed local chemical potentials, $\{\mu_i\}$, and the AIM FF indices, $f_i = \partial N_i/\partial N$, in the model planar complexes $XPtL_3$ (L = Cl, NH_3) generally support the qualitative predictions summarized above. Namely, for a weak donor, X = F, $\mu_{trans} < \mu_{cis}$

($f_{trans}^{MXL_3} < f_{cis}^{MXL_3}$) and for a stronger donor e.g., X = I, $\mu_{trans} > \mu_{cis}$ ($f_{trans}^{MXL_3} > f_{cis}^{MXL_3}$). In other words, strong donors create a relatively soft electron distribution on L_{trans}, while the opposite is the case for weak donors.

7. General Legendre Transformed Representations

In a general treatment of alternative descriptions of ground-state displacements of open molecular systems one has to adopt the *local approach*, which explicitly considers changes in global number of electrons N and the external potential $v(\mathbf{r})$, which together define the system hamiltonian $H(N^0, v)$ and the associated ground-state energy, $\mathscr{E}[N^0, v]$. The relevant Hellmann-Feynman relation,

$$d\mathscr{E}[N, v] = (\partial\mathscr{E}/\partial N)_v\, dN + \int[\delta\mathscr{E}/\delta v(\mathbf{r})]_N\, dv(\mathbf{r})\, d\mathbf{r}$$
$$= \mu\, dN + \int\rho(\mathbf{r})\, dv(\mathbf{r})\, d\mathbf{r}\,, \qquad (129)$$

identifies the set of the energy conjugated global (N, μ) and local [$v(\mathbf{r})$, $\rho(\mathbf{r})$] ground-state variables and functions; here $\rho(\mathbf{r})$ is the g.s. electron density. Again, one recognizes the "extensive" variables, $\mathscr{X} = (N, \rho)$, and their "intensive" conjugates $\mathscr{P} = (\mu, -v)$, defined by the respective derivatives of the principal ground-state potential consisting of the electronic kinetic and electron-repulsion energies:

$$\mathscr{F}[N, \rho] = \mathscr{E} - \int\rho(\mathbf{r})\, v(\mathbf{r})\, d\mathbf{r} = T_e + V_{ee}\,, \qquad (130)$$

$$(\partial\mathscr{F}/\partial N)_\rho = \mu \quad \text{and} \quad [\partial\mathscr{F}/\partial\rho(\mathbf{r})]_N = -v(\mathbf{r})\,. \qquad (131)$$

Since $N = \int\rho\, d\mathbf{r}$ one should read these conjugation relations as partials probing the explicit N- and ρ- dependencies of the universal (v-independent) functional $\mathscr{F}$.

The alternative LTR of this local description and the associated derivative identities have been discussed by Nalewajski et al. [21, 25, 46, 47, 55, 79]. Recently, a new representation for the ground-state and its Legendre transforms have been proposed by Cedillo [80] using the energy per particle as the basic potential, a strict concave functional with respect to both N and ρ, which are not independent.

The local LTR have profound implications for general concepts of DFT [12, 14, 15, 21, 49, 51, 81]. Of particular importance for a phenomenological, thermodynamic-like description of open molecular systems is the dependence of the ground-state energy $\mathscr{E}[N, v]$ and $\mathscr{F}[N, \rho]$ on the particle number N, treated as a continuous, non-negative variable. For *non-interacting* particles moving in an effective external potential v, i.e., within the familiar Kohn-Sham picture [15], $\mathscr{E}[N, v]$ is a convex, piecewise linear function of N for every given v [48] (Fig. 9). It is still an open theoretical problem [49] whether $\mathscr{E}[N, v]$ is convex in N for the coulomb interaction.

The local equivalents of the E, Q, F and R potentials of Table 1 (see also Eqs. 46-48) are [21, 49, 51]:

$$\mathscr{E}[N^0, v] = \inf_{\psi_{N^0}} \langle\psi_{N^0}|H(N^0, v)|\psi_{N^0}\rangle = \langle\Psi_{N^0}|H|\Psi_{N^0}\rangle\,, \qquad (132)$$

$$\mathcal{Q}[\mu,\ \nu] = \inf_N\{\mathscr{E}[N,\nu] - \mu N\}\ , \tag{133}$$

$$\mathscr{F}[N,\rho] = \sup_\nu\{\mathscr{E}[N,\nu] - \int \rho\nu d\mathbf{r}\}\ , \tag{134}$$

$$\begin{aligned}\mathscr{R}[\mu,\rho] &= \inf_N\{\mathscr{F}[N,\rho] - \mu N\}\\ &= \sup_\nu\{\mathcal{Q}[\mu,\nu] - \int \rho\nu\, d\mathbf{r}\}\ .\end{aligned} \tag{135}$$

For non-integer N, the ground-state must be understood in terms of density matrices, i.e., the mixing of pure ground-states with integer N-values (neighbouring), which gives the lowest energy.

For $\mathscr{E}$ convex in N one obtains the convex envelope to $\mathscr{E}[N,\ \nu]$ as a function of N, for a given ν, defined by the following variational principles [51]:

$$\begin{aligned}\mathscr{E}[N,\nu] &= \sup_\mu\{\mathcal{Q}[\mu,\nu] + \mu N\}\\ &= \inf_\rho\{\mathscr{F}[N,\rho] + \int \rho\,\nu\, d\mathbf{r}\}\\ &= \sup_\mu \inf_\rho\{\mathscr{R}[\mu,\rho] + \int \rho\,\nu\, d\mathbf{r} + \mu N\}.\end{aligned} \tag{136}$$

Clearly, one can also express $\mathcal{Q}[\mu,\nu]$ of Eq. 133 in terms of $\mathscr{F}[N,\rho] = \mathrm{F}[\rho]$ (see Eq. 136):

$$\mathcal{Q}[\mu,\nu] = \inf_\rho\{\mathrm{F}[\rho] + \int \rho(\nu-\mu)\, d\mathbf{r}\} \equiv \mathcal{Q}[\nu-\mu]\ , \tag{137}$$

where $\mathrm{F}[\rho]$ is the universal (ν-independent) density functional [14] generating for the g.s. density the sum of electronic kinetic and repulsion energies. This variational principle is reminiscent of the fundamental equation of the HK DFT:

$$\inf_\rho\{E_\nu[\rho] - \mu(\mathrm{N}[\rho] - N)\} = \delta\mathcal{Q}[\nu-\mu] = 0\ , \tag{138}$$

with the density functional for the system electronic energy: $E_\nu[\rho] = \mathrm{F}[\rho] + \int \rho\,\nu\, d\mathbf{r}$.

Finally, it is of interest to examine general properties of $\mathcal{Q}[\nu-\mu]$. For the Kohn-Sham model of non-interacting electrons, which provides the standard reference for constructions of density functionals, the energy $\mathscr{E}[N,\nu]$ is convex in N for a given fixed ν [12, 48-51]; one obtains in this case

$$\mathcal{Q}[\nu-\mu] = \sum_{N=0}^{N_\mu-1}\mu_N - N_\mu\mu\ ,\qquad \mu_{N_\mu-1} \le \mu \le \mu_{N_\mu}\ , \tag{139}$$

where N_μ is the equilibrium (integer) number of electrons for specified μ (and ν), while the slopes of $\mathscr{E}[N,\ \nu]$ are:

$$\mu_N \equiv \mathscr{E}[N+1,\nu] - \mathscr{E}[N,\nu] \equiv I_{N+1} \equiv A_N\ ; \tag{140}$$

here I_N and A_N are the ionization potential and electron affinity for the system consisting of N electrons moving in the external potential ν. This general situation is sketched in Fig. 9.

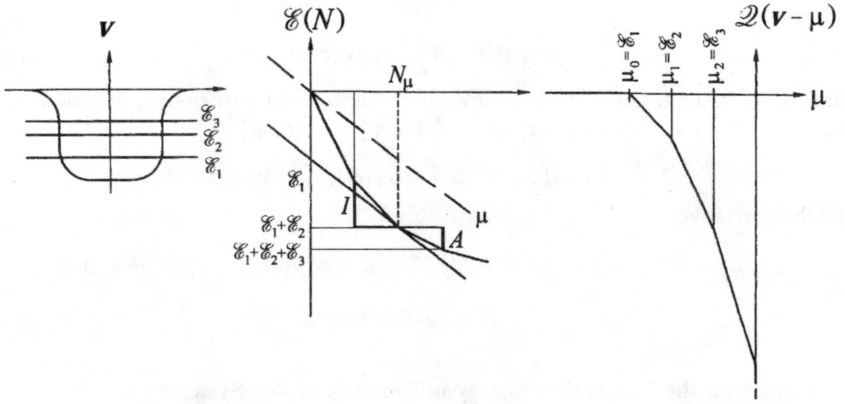

Figure 9. The ground-state energy $\mathscr{E}\,[N, v]$ for non-interacting electrons moving in the external potential v, as function of N for constant v, $\mathscr{E}\,(N)$, and the associated $\mathscr{Q}\,[v-\mu]$ potential (from Ref. [51]).

As long as the ionization potential I_N of the N electron state in the potential v is larger than $(-\mu)$ for a finite number of electrons $N \leq N_\mu$, $\mathscr{Q}[v-\mu]$ is finite and the infimum of Eq. 133 is a minimum at $N = N_\mu$.

The exact differential forms of the potentials $\mathscr{E}$, $\mathscr{Q}$, $\mathscr{F}$, and $\mathscr{R}$ are given by Eq. 129 and the following expressions:

$$d\mathscr{Q} = -N\,d\mu + \int \rho(\mathbf{r})\,dv(\mathbf{r})\,d\mathbf{r} = \int \rho(\mathbf{r})\,d[v(\mathbf{r}) - \mu]\,d\mathbf{r}\,, \tag{141}$$

$$d\mathscr{F} = \mu\,dN - \int v(\mathbf{r})\,d\rho(\mathbf{r})\,d\mathbf{r}\,, \tag{142}$$

$$d\mathscr{R} = -N\,d\mu - \int v(\mathbf{r})\,d\rho(\mathbf{r})\,d\mathbf{r}\,. \tag{143}$$

They give rise to a series of interesting Maxwell relations [21].

The $\mathscr{F}[N, \rho] = F[\rho]$ representation presents some difficulty since both state variables are linearly dependent. To remedy this Parr and Bartolotti [82] have introduced the "probability" function $\gamma(\mathbf{r})$,

$$\gamma(\mathbf{r}) = \rho(\mathbf{r})/N\,, \quad \int \gamma(\mathbf{r})\,d\mathbf{r} = 1\,, \tag{144}$$

in terms of which the density displacement

$$d\rho(\mathbf{r}) = N\,d\gamma(\mathbf{r}) + \gamma(\mathbf{r})\,dN \equiv d\rho_N(\mathbf{r}) + \gamma(\mathbf{r})\,dN\,, \tag{145}$$

can be partitioned into the internal (N- preserving) part $d\rho_N$, due to $d\gamma$, and the external part, due to dN. The exact differential of $\overline{\mathscr{F}}\,[N, \gamma] = \mathscr{F}[N, \rho]$ is given by the expression:

$$d\overline{\mathcal{F}} = (\mu - V_{ne}/N)dN - \int v(\mathbf{r})\,d\rho_N(\mathbf{r})\,d\mathbf{r}$$
$$\equiv \mu'\,dN - N\int v(\mathbf{r})\,d\gamma(\mathbf{r})\,d\mathbf{r}\,. \tag{146}$$

The above equation immediately follows from the familiar variational principle for the energy $\overline{\mathcal{E}}[N,\gamma] \equiv N\int \gamma(\mathbf{r})\ v(\mathbf{r})\,d\mathbf{r} + \overline{\mathcal{F}}[N,\gamma] = \mathcal{E}[N,\rho]$,

$$[\delta\overline{\mathcal{E}}/\delta\gamma(\mathbf{r})]_{N,v} \equiv [\delta\overline{\mathcal{F}}/\delta\gamma(\mathbf{r})]_{N,v} + N\,v(\mathbf{r}) = 0\,, \tag{147}$$

and the derivative

$$(\partial\overline{\mathcal{F}}/\partial N)_\gamma \equiv \mu' = (\partial\mathcal{F}/\partial N)_\rho + \int[\delta\mathcal{F}/\delta\rho(\mathbf{r})]_N\,[\partial\rho(\mathbf{r})/\partial N]_\gamma\,d\mathbf{r}$$
$$= \mu - \int v(\mathbf{r})\,\gamma(\mathbf{r})\,d\mathbf{r}\,. \tag{148}$$

8. More about the Interaction Energy in Donor-Acceptor Systems

In this section we reconsider in terms of the local charge sensitivities the quadratic interaction energy (see Section 4.2) in a general reactive system, $M = A \overset{Q}{-} B$, consisting of an electron donor [basic (B)] and an electron acceptor [acidic (A)] reactants at their current mutual separation and orientation, Q [12, 56, 57, 59, 61, 83, 84]. We follow the standard partitioning scheme of Morokuma [85], and previous DFT approaches to the EE and the corresponding definition of molecular fragments [11, 12, 57, 59, 61], e.g., the AIM. This approach considers the following stages of the reactant electron distribution in M:

(*i*) the separated (infinitely distant) reactants: $M^0(\infty) = A^0 + B^0$;

(*ii*) the polarized, mutually closed reactants at Q: $M^+(Q) = (A^+|B^+)$;

(*iii*) the mutually opened reactants at Q: $M^*(Q) = (A^*\,\vdots\,B^*)$.

The second stage corresponds to the *intra-reactant equilibrium* with $\mu_A^+ \neq \mu_B^+$, while the third stage represents the *global equilibrium* in M, with the fully equalized chemical potential (electronegativity), $\mu_A^* = \mu_B^* = \mu_M$, which settles after the inter-reactant B→A CT. One could also formally consider an additional stage (*iv*), intermediate between (*i*) and (*ii*), of "rigid", closed (interacting) reactants at Q, $M^0(Q) = (A^0|B^0)$ in which no EE within reactants has taken place [57].

The reactants in $M^*(Q)$ are displaced relative to those in $M^0(\infty)$ in both the external potential, $d\boldsymbol{v} = [dv_A, dv_B] \approx [\phi_B, \phi_A]$, and the electron populations, $d\boldsymbol{N} = [dN_A, dN_B] = [N_{CT}, -N_{CT}]$; here $\phi_X(\mathbf{r}) = v_X(\mathbf{r}) + \int d\mathbf{r}'\,\rho_X^0(\mathbf{r}')/|\mathbf{r}-\mathbf{r}'|$ is the electrostatic (Hartree) potential around X = A, B (defined with respect to the electron charge); here $v_X(\mathbf{r}) = -\sum_i^X Z_i/|\mathbf{r}-\boldsymbol{R}_i|$. The total interaction energy between A and B, $E_{AB} = E_{AB}^e + V_{AB}^{nn}$, consists of the corresponding electronic contribution, E_{AB}^e, and a trivial nuclear repulsion part, V_{AB}^{nn}. One observes that the quadratic approximation to E_{AB}^e [57],

$$E_{AB}^e = d\boldsymbol{N}(\boldsymbol{\mu}^0 + \frac{1}{2}d\boldsymbol{\mu})^\dagger + \int d\boldsymbol{v}(\mathbf{r})\,[\boldsymbol{\rho}^0(\mathbf{r}) + \frac{1}{2}d\boldsymbol{\rho}(\mathbf{r})]^\dagger\,d\mathbf{r}\,, \tag{149}$$

where $\boldsymbol{\rho}^0 = [\rho_A^0, \rho_B^0]$ and $\boldsymbol{\mu}^0 = [\mu_A^0, \mu_B^0]$ involves only the first-order responses $d\boldsymbol{\mu}$ and $d\boldsymbol{\rho}$, conjugated to the perturbations $d\boldsymbol{N}$ and $d\boldsymbol{v}$, respectively. These responses can be conveniently partitioned into the polarizational part (due to $d\boldsymbol{v}$, at $d\boldsymbol{N} = \boldsymbol{0}$) and the remaining CT contribution (due to N_{CT}, at $d\boldsymbol{v} \neq \boldsymbol{0}$):

$$d\boldsymbol{\mu} = d\boldsymbol{\mu}^+(d\boldsymbol{v}) + d\boldsymbol{\mu}^*(N_{CT}) = [d\mu_A, d\mu_B] , \tag{150}$$

$$d\boldsymbol{\rho} = d\boldsymbol{\rho}^+(d\boldsymbol{v}) + d\boldsymbol{\rho}^*(N_{CT}) = [d\rho_A, d\rho_B] . \tag{151}$$

First, let us consider the polarizational stage, $M^+(Q)$. The corresponding changes in $\boldsymbol{\mu}$ and $\boldsymbol{\rho}$ are:

$$d\boldsymbol{\mu}^+ = \int d\boldsymbol{v}(\mathbf{r}) \left(\frac{\delta \boldsymbol{\mu}}{\delta \boldsymbol{v}(\mathbf{r})} \right)_{\boldsymbol{N}} d\mathbf{r} \equiv \int d\boldsymbol{v}(\mathbf{r})\, \mathbf{f}^{\dagger}(\mathbf{r})\, d\mathbf{r} , \tag{152}$$

$$d\boldsymbol{\rho}^+(\mathbf{r}) = \int d\boldsymbol{v}(\mathbf{r}') \left(\frac{\delta \boldsymbol{\rho}(\mathbf{r})}{\delta \boldsymbol{v}(\mathbf{r}')} \right)_{\boldsymbol{N}} d\mathbf{r}' \equiv \int d\boldsymbol{v}(\mathbf{r}')\, \boldsymbol{\beta}(\mathbf{r}', \mathbf{r})\, d\mathbf{r}' . \tag{153}$$

The local FF matrix in Eq. 152,

$$\mathbf{f}(\mathbf{r}) = \left\{ f_{Y,X}(\mathbf{r}) = \frac{\delta \mu_Y}{\delta v_X(\mathbf{r})} = \frac{\partial \rho_X(\mathbf{r})}{\partial N_Y} \right\} , \quad (X,Y) = (A,B) , \tag{154}$$

includes the diagonal (CT) elements, $f_{X,X}(\mathbf{r}) = \partial \rho_X(\mathbf{r}) / \partial N_X$, normalized to $\int f_{X,X}(\mathbf{r})\, d\mathbf{r} = \partial N_X / \partial N_X = 1$, and the off-diagonal (relaxational) elements, $f_{X,Y}(\mathbf{r}) = \partial \rho_Y(\mathbf{r}) / \partial N_X$, normalized to $\int f_{X,Y}(\mathbf{r})\, d\mathbf{r} = (\partial N_Y / \partial N_X)_{N_Y} = 0$. For a given reactant, say A,

$$d\overset{+}{\mu}_A = \int [f_{A,A}(\mathbf{r})\, dv_A(\mathbf{r}) + f_{A,B}(\mathbf{r})\, dv_B(\mathbf{r})] d\mathbf{r} , \quad \text{etc.} \tag{155}$$

The density polarization matrix kernel of Eq. 153,

$$\boldsymbol{\beta}(\mathbf{r}', \mathbf{r}) = \left\{ \beta_{X,Y}(\mathbf{r}', \mathbf{r}) = \frac{\delta \rho_Y(\mathbf{r})}{\delta v_X(\mathbf{r}')} \right\} , \quad (X,Y) = (A,B) , \tag{156}$$

also involves the coupling (off-diagonal) response kernels, which contribute to the first-order change in the density of polarized reactants,

$$d\overset{+}{\rho}_A(\mathbf{r}) = \int [dv_A(\mathbf{r}')\, \beta_{A,A}(\mathbf{r}', \mathbf{r}) + dv_B(\mathbf{r}')\, \beta_{B,A}(\mathbf{r}', \mathbf{r})]\, d\mathbf{r}' , \quad \text{etc.} \tag{157}$$

The polarizational shifts in the reactant chemical potentials of Eqs. 152 and 155 determine $\overset{+}{\mu}_{CT} = \overset{0}{\mu}_{CT} + d\overset{+}{\mu}_A - d\overset{+}{\mu}_B$, which together with η_{CT} determines both N_{CT} and E_{CT} (Eqs. 100 and 101), as well as the CT displacements in $\boldsymbol{\mu}$ and $\boldsymbol{\rho}$:

$$d\mu_X^* = \frac{\partial \mu_X}{\partial N_{CT}} N_{CT} = \eta_X^M N_{CT}, \quad X = A, B ; \tag{158}$$

$$d\rho^*(\mathbf{r}) = \frac{\partial \rho(\mathbf{r})}{\partial N_{CT}} N_{CT} = f^{CT}(\mathbf{r})\, N_{CT}, \tag{159}$$

where the two components of $f^{CT}(\mathbf{r})$ are:

$$f_A^{CT}(\mathbf{r}) = f_{A,A}(\mathbf{r}) - f_{B,A}(\mathbf{r}), \qquad f_B^{CT}(\mathbf{r}) = f_{B,B}(\mathbf{r}) - f_{A,B}(\mathbf{r}). \tag{160}$$

In this approach the external potential displacements, responsible for a transition from stage (i) to stage (ii), create conditions for the subsequent CT effects, in the spirit of the Born-Oppenheimer approximation. Clearly, the consistent second-order Taylor expansion at $M^0(\infty)$ does not involve the coupling hardness $\eta_{A,B}$ and the off-diagonal response quantities of Eqs. 154 and 156, which vanish identically for infinitely separated reactants [12]. However, since the interaction at Q modifies both the chemical potential difference and the second-order charge sensitivities (non-vanishing off-diagonal elements) one should obtain a more realistic estimate of the interaction energy by considering the CT stage at the finite separation between reactants [57, 61]. Therefore, in this approach the overall interaction energy is obtained as the sum of the energy differences between the above three stages of interaction:

$$E_{AB}^e = [E_{AB}^e(ii) - E_{AB}^e(i)] + [E_{AB}^e(iii) - E_{AB}^e(ii)]$$

$$\equiv E_{AB}^+(d\boldsymbol{v}, N_{CT} = 0) + E_{AB}^*(d\boldsymbol{v}, N_{CT}) = (E_{ES} + E_P) + E_{CT}. \tag{161}$$

CHAPTER 3

CONCEPTS AND RELATIONS OF MOLECULAR CHARGE SENSITIVITY ANALYSIS

As we have seen in the previous Chapter the chemical potential (electronegativity) difference between reactants, μ^{+}_{CT}, and the hardness quantity η_{CT} determine the direction and amount of the inter-reactant CT. Recently, a variety of related potentials and charge sensitivities, both local, regional, and non-local has been defined at different levels of resolution, e.g., the hardnesses and softnesses of functional groups and reactants, hardness and softness matrices and kernels, the Fukui function indices (functions, vectors, matrices, and matrix kernels), response quantities involving external potential, etc. A systematic procedure for determining chemically interesting charge sensitivities and responses, called the *Charge Sensitivity Analysis* (CSA) [9, 32, 59], has been developed to probe trends in chemical reactivity; the *Electronegativity Equalization Method* (EEM) [10] represents a particular realization of the CSA. The main purpose of this Chapter is to survey basic CSA concepts and relations, with particular emphasis placed upon the quantities which can be used to probe reactivities of different local sites within a molecule and selectivities of alternative reaction paths.

The method can be formulated in alternative levels of resolution. Although most of the theoretical development of this Chapter will be presented in the AIM-resolution, sufficient for most chemical purposes, some vital formulas related to other levels will also be summarized. Each molecular fragment has a distinct molecular environment, which moderates the fragment responses to external perturbations. We shall examine in detail these relaxational contributions to charge sensitivities, and we shall comment upon their physical implications. The *in situ* quantities of reactants in the A---B systems will also be discussed. Finally, specific concepts for reactive systems, involving collective modes of charge displacements will be defined. They include the normal modes, which decouple the EE equations, the minimum energy coordinates, and alternative intermediate hardness decoupling modes.

In the previous Chapter we have argued that one can define various derivative properties related to local, regional (molecular fragment) or global descriptions, in which the relevant local density or electron populations quantities are treated as independent state variables. Depending upon the resolution involved, specified by a given partitioning of the system in the physical or function spaces, one considers the following charge distribution variables: $\rho(\mathbf{r})$ (*local,* L*-resolution*), the electron populations of constituent AIM, $\boldsymbol{N} = (N_1, ..., N_m)$ or AIM net charges (AIM*-resolution*), the overall populations attributed to larger molecular fragments (group of AIM), $\boldsymbol{N}_G = (N_X, N_Y, ..., N_Z)$ (*group,* G*-resolution*), or finally, the global number of electrons, $N = \int \rho(\mathbf{r})d\mathbf{r} = \Sigma_i N_i = \Sigma_X N_X$ (*global,* g*-resolution*). Of course, various intermediate levels, e.g., those defined in the function space spanned by the molecular orbitals (MO*-resolution*), can also be envisaged with the corresponding MO occupation number variables $\boldsymbol{n} = (n_1, ..., n_s)$. The system energy for a constant external potential is considered as the function (functional) of these electron population (density) variables: $E^{L}(\mathbf{r})$, $E^{AIM}(\boldsymbol{N})$, $E^{G}(\boldsymbol{N}_G)$, and $E^{g}(N)$. The populational gradient of the system energy defines the chemical potential variables while the populational (density) hessian represents the hardness matrix (kernel) for a given resolution. We therefore call the description based upon the populational variables *the hardness (interaction) representation.*

Each resolution implies its own case of the intra-system equilibrium. For example, the L-resolution corresponds to the totally constrained ("frozen") electron distribution, with all local, infinitesimal volume elements considered as being mutually closed, while the g-resolution represents the opposite extreme case of the totally relaxed electron distribution, with all local volume elements regarded as mutually opened. Similarly, the intermediate levels of resolution apply to the cases of partially constrained equilibrium, in which all fragments defining the partitioning of a molecular system M,

$$\mathrm{M} : (\mathbf{r}|\mathbf{r}'|...)^{\mathrm{L}},\ (\mathrm{i}|\mathrm{j}|...)^{\mathrm{AIM}},\ (\mathrm{X}|\mathrm{Y}|...)^{\mathrm{G}},$$

$$(\mathbf{r}\,\vdots\,\mathbf{r}'\,\vdots\,...)^{\mathrm{g}} = (\mathrm{i}\,\vdots\,\mathrm{j}\,\vdots\,...)^{\mathrm{g}} = (\mathrm{X}\,\vdots\,\mathrm{Y}\,\vdots\,...)^{\mathrm{g}},$$

are in their respective internal equilibrium states. As indicated above in the g-resolution all molecular fragments are free to exchange electrons. Thus, for the global equilibrium state of M the chemical potential (electronegativity) is equalized throughout the whole electron distribution:

$$\begin{aligned} \delta E^{\mathrm{L}}[\rho]/\delta\rho(\mathbf{r}) &\equiv \mu(\mathbf{r}) = \mu(\mathbf{r}') = ... \\ &= \partial E^{\mathrm{AIM}}(\boldsymbol{N})/\partial N_{\mathrm{i}} \equiv \mu_{\mathrm{i}} = \mu_{\mathrm{j}} = ... \\ &= \partial E^{\mathrm{G}}(\boldsymbol{N}_{\mathrm{G}})/\partial N_{\mathrm{X}} \equiv \mu_{\mathrm{X}} = \mu_{\mathrm{Y}} = ... \\ &= \partial E^{\mathrm{g}}(N)/\partial N \equiv \mu\ . \end{aligned} \tag{162}$$

The corresponding populational derivatives of the system energy define the respective conjugate variables, the so called "chemical potentials" (negative values of the corresponding electronegativities) for the resolution in question. For example, in the AIM picture one defines the AIM chemical potentials, while in the reactant resolution of the A---B systems one introduces the separate chemical potentials of the acidic and basic reactants. These conjugate quantities are the state variables in the inverse, *softness (response) representation.* The softness matrix (kernel), grouping the negative second partials of the relevant Legendre transform of the energy, is therefore the inverse of the system hardness matrix (kernel) for the resolution under consideration, representing the corresponding charge compliant matrix.

1. Hardness and Softness Quantities in the AIM-Resolution

Consider a given molecular system consisting of m atoms. In what follows we adopt the AIM-resolution to define the canonical AIM chemical potentials (electron population gradient), $\boldsymbol{\mu} = \partial E/\partial \boldsymbol{N} = (\mu_1, \mu_2, ..., \mu_m)$, and the corresponding AIM hardness matrix (electron population hessian): $\boldsymbol{\eta} = \partial^2 E/\partial \boldsymbol{N}\,\partial \boldsymbol{N} = \partial \boldsymbol{\mu}/\partial \boldsymbol{N} = \{\eta_{i,j}\}$; here all differentiations are carried out for the fixed external potential $v(\mathbf{r})$ represented by vector $\boldsymbol{v}$ in the AIM description. This canonical charge sensitivity information will be used to generate a variety of the system *charge sensitivities* (CS) probing the system responses to various populational and external potential perturbations at alternative constraints.

Let us first examine a general combination problem of determining CS of the system as a whole; we call them global or resultant quantities, since they imply the global equilibrium in the system under consideration. The complementary canonical compliant information is obtained by inverting $\boldsymbol{\eta}$,

$$\boldsymbol{\sigma} \equiv \partial \boldsymbol{N}/\partial \boldsymbol{\mu} = \boldsymbol{\eta}^{-1} = \{\sigma_{i,j} = \partial N_j/\partial \mu_i\}\ . \qquad (163)$$

This softness matrix can be used to generate the local (regional) and global softness quantities:

$$S_i \equiv \partial N_i/\partial \mu = \sum_j^M \frac{\partial \mu_j}{\partial \mu}\frac{\partial N_i}{\partial \mu_j} = \sum_j^M \sigma_{j,i}\ ,$$

$$S_X = \partial N_X/\partial \mu = \sum_j^X S_j\ ,$$

$$S = \frac{\partial N}{\partial \mu} = \sum_X^M S_X\ ; \qquad (164)$$

here Σ_X^M denotes the summation over all atomic fragments X in M. The global softness S is the inverse of the global hardness $\eta = \partial\mu/\partial N = 1/S$. In Eq. 164 we have used the EE Eq. 162.

The above softness parameters immediately give the corresponding FF indices, e.g.,

$$f_i = \partial N_i/\partial N = (\partial N_i/\partial \mu)/(\partial N/\partial \mu) = S_i/S\ ,$$

$$f_X^M = \partial N_G/\partial N = S_X/S\ , \quad \text{etc.}\ , \qquad (165)$$

which provide the "weighting" factors for combining the derivatives with respect to the electron population variables. For example, for the global chemical potential one obtains:

$$\mu = \sum_i^M \frac{\partial N_i}{\partial N}\frac{\partial E^{AIM}}{\partial N_i} = \sum_i^M f_i\,\mu_i\ . \qquad (166)$$

In terms of the AIM FF indices the global hardness can be expressed as the following quadratic form:

$$\eta = \sum_i^M \sum_j^M \frac{\partial N_i}{\partial N}\frac{\partial^2 E^{AIM}}{\partial N_i \partial N_j}\frac{\partial N_j}{\partial N} = \sum_i^M \sum_j^M f_i\,\eta_{i,j}\,f_j\ . \qquad (167)$$

It is also of interest to examine the mixed derivative characterizing the resultant local hardness of an atom i, $\eta_i = \partial^2 E^{AIM}/\partial N\,\partial N_i$,

$$\eta_i = \frac{\partial\mu_i}{\partial N} = \sum_j^M \frac{\partial N_j}{\partial N}\frac{\partial\mu_i}{\partial N_j} = \sum_j^M f_j\,\eta_{j,i}$$

$$= \frac{\partial\mu}{\partial N_i} = \frac{\partial N}{\partial N_i}\frac{\partial\mu}{\partial N} = \eta\,, \tag{168}$$

which is also equalized at the global hardness level. Notice, that Eqs. 167 and 168 imply that $\sum_i^M f_i = 1$, which is indeed satisfied by the FF normalization: $\sum_i^M f_i = \sum_i^M \partial N_i/\partial N = \partial N/\partial N = 1$.

We would like to observe that the general second-order expansion for a change in the system energy due to the $\mathrm{d}\boldsymbol{N}$ displacement,

$$\Delta E(\mathrm{d}\boldsymbol{N}) \approx \mathrm{d}\boldsymbol{N}\,\boldsymbol{\mu}^\dagger + \frac{1}{2}\mathrm{d}\boldsymbol{N}\,\boldsymbol{\eta}\,\mathrm{d}\boldsymbol{N}^\dagger\,, \tag{169}$$

simplifies for the global equilibrium displacement $\mathrm{d}\boldsymbol{N}^{eq}(\mathrm{d}N) \equiv \mathrm{d}N\,\boldsymbol{f}$, $\boldsymbol{f} = (f_1, f_2, \ldots, f_m)$ to the familiar expression in terms of the net change in the global number of electrons, $\mathrm{d}N$,

$$\Delta E[\mathrm{d}\boldsymbol{N}^{eq}(\mathrm{d}N)] = \mathrm{d}N\,(\boldsymbol{f}\,\boldsymbol{\mu}^\dagger) + \frac{1}{2}(\mathrm{d}N)^2\,\boldsymbol{f}\,\boldsymbol{\eta}\,\boldsymbol{f}^\dagger = \mathrm{d}N\,\mu + \frac{1}{2}(\mathrm{d}N)^2\,\eta\,. \tag{170}$$

Let us now consider the effective CS of fragments of M, specified by the corresponding collections of AIM. In Section 5 of the preceding Chapter a particular case has been examined of a partitioning of M into two complementary subsystems, M=(A|B), e.g., reactants in the donor-acceptor system. Of interest in this case are the elements of the condensed hardness matrix of Eq. 111, e.g., $\eta^M_{A,A} \equiv (\partial\mu^M_A/\partial N_A)$, where the closed subsystem B is free to relax when A undergoes a populational displacement $\mathrm{d}N_A$. Using the chain rule transformation gives:

$$\eta^M_{A,A} = \sum_j^M \frac{\partial N_j}{\partial N_A}\frac{\partial\mu_A}{\partial N_j} \equiv \sum_a^A f_a^{A,A}\,\eta_{a,A} + \sum_b^B f_b^{A,B}\,\eta_{b,A}$$

$$= \sum_a^A\sum_{a'}^A f_a^{A,A}\,\eta_{a,a'}\,f_{a'}^{A,A} + 2\sum_a^A\sum_b^B f_a^{A,A}\,\eta_{b,a}\,f_b^{A,B}$$

$$+ \sum_b^B\sum_{b'}^B f_b^{A,B}\,\eta_{b,b'}\,f_{b'}^{A,B}\,; \tag{171}$$

here, as in Section 8 of Chapter 2, the diagonal FF indices sum up to 1, e.g., $\sum_a^A f_a^{A,A} = \partial N_A/\partial N_A = 1$, while the off-diagonal (relaxational) FF indices sum up to 0, e.g., $\sum_b^B f_b^{A,B} = (\partial N_B/\partial N_A)_{N_B} = 0.$

A similar expression can be derived for the effective coupling hardness between both fragments in M:

$$\eta_{A,B}^{M} \equiv \partial\mu_B/\partial N_A = \sum_j^M \frac{\partial N_j}{\partial N_A}\frac{\partial \mu_B}{\partial N_j} = \sum_a^A f_a^{A,A}\,\eta_{a,B} + \sum_b^B f_b^{A,B}\,\eta_{b,B}$$

$$= \sum_j^M \sum_k^M \frac{\partial N_j}{\partial N_A}\frac{\partial^2 E^{AIM}}{\partial N_j \partial N_k}\frac{\partial N_k}{\partial N_B}$$

$$= \sum_a^A \sum_b^B f_a^{A,A}\,\eta_{a,b}\,f_b^{B,B} + \sum_b^B \sum_{b'}^B f_b^{A,B}\,\eta_{b,b'}\,f_{b'}^{B,B}$$

$$+ \sum_a^A \sum_{a'}^A f_a^{A,A}\,\eta_{a,a'}\,f_{a'}^{B,A} + \sum_b^B \sum_a^A f_b^{A,B}\,\eta_{b,a}\,f_a^{B,A}. \qquad (172)$$

In a similar way one can express any derivative involving a differentiation with respect to the overall fragment electron population. For example, for the group chemical potential one obtains:

$$\mu_A^M = \left(\frac{\partial E^{AIM}}{\partial N_A}\right)_{N_B} = \sum_j^M \frac{\partial N_j}{\partial N_A}\frac{\partial E^{AIM}}{\partial N_j} = \sum_a^A f_a^{A,A}\,\mu_a + \sum_b^B f_b^{A,B}\,\mu_b\,. \qquad (173)$$

The same chain rule transformation gives the system of linear equations for determining the relaxed FF indices appearing in Eqs. 171-173:

$$f_a^{A,A} = \left(\frac{\partial N_a}{\partial N_A}\right)_{N_B} = \sum_j^M \frac{\partial N_j}{\partial N_A}\frac{\partial N_a}{\partial N_j} = \left(\frac{\partial N_a}{\partial N_A}\right)_{N_B} + \sum_b^B f_b^{A,B}\,T_{b,a}^{(B|A)}$$

$$\equiv f_a^{rgd} + \delta f_a(M)\,, \qquad (174)$$

where the first term represents the *rigid* FF index, calculated for the frozen electron distribution in B, and the second term accounts for the electron relaxation in B determined by the relaxation matrix

$$T_{b,a}^{(B|A)} = \left(\frac{\partial N_a}{\partial N_b}\right)_{N_A}. \qquad (175)$$

As shown elsewhere [59] an explicit expression for this matrix element immediately follows from considering changes in the AIM chemical potentials in A due to dN_b, $\Delta\mu_A(dN_b)$, responsible for the subsequent relaxational flows in A, $dN_A(dN_b)$. Namely, in the quadratic approximation a given populational perturbation dN_b ($dN_{b' \neq b} = 0$) shifts the AIM chemical potentials in A by $\{\Delta\mu_a(dN_b) = dN_b\,\eta_{b,a}\}$. Thus, the displacements of the local chemical potentials relative to the new equilibrium level, $d\mu_A(dN_b) = dN_b\,\eta_{b,A}$, are:

$$d\boldsymbol{\mu}_A(dN_b) = \{\Delta\mu_a(dN_b) - d\mu_A(dN_b) = dN_b\,(\eta_{b,a} - \eta_{b,A})\} \,. \tag{176}$$

The populational displacements in A, relative to the new equilibrium AIM electron populations, $d\boldsymbol{N}_A(dN_b)$, the negative values of the spontaneous relaxational flows $\delta\boldsymbol{N}_A(dN_b)$ restoring the equilibrium, are related to $d\boldsymbol{\mu}_A(dN_b)$ by the EE matrix equation:

$$d\boldsymbol{\mu}_A = -\delta\boldsymbol{N}_A\,\boldsymbol{\eta}^{A,A} = d\boldsymbol{N}_A\,\boldsymbol{\eta}^{A,A} \qquad \text{or} \qquad \delta\boldsymbol{N}_A = -d\boldsymbol{N}_A = -d\boldsymbol{\mu}_A\,\boldsymbol{\sigma}^A \,, \tag{177}$$

where $\boldsymbol{\eta}^{A,A}$ is the diagonal (A, A) block of $\boldsymbol{\eta}$, and $\boldsymbol{\sigma}^A = (\boldsymbol{\eta}^{A,A})^{-1}$. Using Eqs. 176 and 177 gives the desired expression:

$$T^{(B|A)}_{b,a} = \eta_{b,A}\,S^A_a - \sum\nolimits^A_{a'} \eta_{b,a'}\,\sigma^A_{a',a} \,, \tag{178}$$

where $S^A_a = \Sigma^A_{a'}\,\sigma^A_{a',a}$.

The equation defining the relaxed off-diagonal FF indices is (compare with Eq. 174):

$$f^{A,B}_b = \left(\frac{\partial N_b}{\partial N_A}\right)_{N_B} = \sum\nolimits^M_j \frac{\partial N_j}{\partial N_A}\frac{\partial N_b}{\partial N_j} = \sum\nolimits^A_a \left(\frac{\partial N_a}{\partial N_A}\right)_{N_B}\left(\frac{\partial N_b}{\partial N_a}\right)_{N_B}$$

$$= \sum\nolimits^A_a f^{A,A}_a\,T^{(A|B)}_{a,b} \,. \tag{179}$$

The relaxational matrix also defines the relaxational corrections to the diagonal blocks $\boldsymbol{\eta}^{A,A}$ and $\boldsymbol{\eta}^{B,B}$ and the off-diagonal block $\boldsymbol{\eta}^{A,B} = (\boldsymbol{\eta}^{B,A})^\dagger$ in $\boldsymbol{\eta}$, e.g.,

$$\boldsymbol{\eta}^M_{A,A} \equiv \boldsymbol{\eta}^{A,A} + \delta\boldsymbol{\eta}_{A,A}(M) \,,$$

$$\boldsymbol{\eta}^M_{A,B} \equiv \boldsymbol{\eta}^{A,B} + \delta\boldsymbol{\eta}_{A,B}(M) \,. \tag{180}$$

For the explicit expressions of the relaxational corrections $\delta\boldsymbol{\eta}_{X,Y}(M)$ the reader is referred to the Appendix B.

The rigid FF indices of Eq. 174 characterize the uncoupled (non-interacting) molecular fragments. They can be obtained from the non-interacting fragment softnesses $\boldsymbol{\sigma}^X$, X = (A, B) and the associated local and global softnesses (see Eqs. 163-165). Approximating the relaxed diagonal FF indices $\boldsymbol{f}_A = \{f^{rgd}_a\}$, $\boldsymbol{f}_B = \{f^{rgd}_b\}$, and using them in the first terms of Eqs. 171 and 172 give the relevant *rigid* condensed hardness matrix elements,

$$\eta^{\rm rgd}_{\rm A,\,A} = \boldsymbol{f}_{\rm A}\, \boldsymbol{\eta}^{\rm A,\,A}\, \boldsymbol{f}^{\dagger}_{\rm A} \quad , \quad \eta^{\rm rgd}_{\rm A,\,B} = \boldsymbol{f}_{\rm A}\, \boldsymbol{\eta}^{\rm A,B}\, \boldsymbol{f}^{\dagger}_{\rm B} \,. \tag{181}$$

In a similar way one approaches the environmental effects in a general partitioning, M = (X|Y|Z|...), involving more than two mutually closed subsystems [34, 59]. For example, in a catalytic reaction involving two species A and B adsorbed on a crystal C, one explicitly considers the effective charge couplings in M = (A|B|C), characterized by the relevant 3×3 condensed hardness matrix. For such general partitionings into more than two fragments also the off-diagonal blocks, say $\boldsymbol{\eta}^{\rm A,\,B}$, are modified due to the charge relaxation of the remaining fragments, $\delta\boldsymbol{\eta}_{\rm A,\,B}(\rm M)$. In Section 4 of this Chapter we shall summarize the algorithm for calculating the molecular fragment charge sensitivities for a general partitioning M = (X|Y|Z|...).

2. Charge Sensitivities in the Local Description

We continue with a survey of the hardness and softness quantities in the local resolution [29, 74]. In this case the equilibrium (ground-state) density satisfies the HK variational principle of Eq. 138, which can be formulated in terms of the associated Euler equation:

$$\frac{\delta E^{\rm L}[\rho]}{\delta\rho(\mathbf{r})} = \mu(\mathbf{r}) = v(\mathbf{r}) + \frac{\delta F[\rho]}{\delta\rho(\mathbf{r})} = \mu \,, \tag{182}$$

where $F[\rho]$ is the universal HK density functional generating the sum of the electronic kinetic and repulsion energies. Differentiating this equation with respect to density gives the hardness kernel in terms of the relative external potential $u(\mathbf{r}) = v(\mathbf{r}) - \mu$:

$$\eta(\mathbf{r}', \mathbf{r}) = \frac{\delta^2 F[\rho]}{\delta\rho(\mathbf{r}')\,\delta\rho(\mathbf{r})} = -\frac{\delta u(\mathbf{r})}{\delta\rho(\mathbf{r}')} \,. \tag{183}$$

The softness kernel is the inverse of the above hardness kernel,

$$\int \eta(\mathbf{r}, \mathbf{r}')\,\sigma(\mathbf{r}', \mathbf{r}'')\,d\mathbf{r}' = \delta(\mathbf{r} - \mathbf{r}'') \,, \tag{184}$$

$$\sigma(\mathbf{r}', \mathbf{r}) = -\frac{\delta\rho(\mathbf{r})}{\delta u(\mathbf{r}')} \,. \tag{185}$$

The local softness,

$$s(\mathbf{r}) = \left(\frac{\partial\rho(\mathbf{r})}{\partial\mu}\right)_v = \int \frac{\partial u(\mathbf{r}')}{\partial\mu}\,\frac{\delta\rho(\mathbf{r})}{\delta u(\mathbf{r}')}\,d\mathbf{r}' = \int \sigma(\mathbf{r}', \mathbf{r})\,d\mathbf{r}' \,, \tag{186}$$

defines the global softness:

$$S = \partial N/\partial\mu = \int s(\mathbf{r})\,d\mathbf{r} = \eta^{-1} . \tag{187}$$

It follows from the Euler Eq. 182 that $\rho = \rho[u]$; therefore,

$$d\rho(\mathbf{r}) = \int du(\mathbf{r}')\frac{\delta\rho(\mathbf{r})}{\delta u(\mathbf{r}')}d\mathbf{r}' = -\int[dv(\mathbf{r}') - d\mu]\,\sigma(\mathbf{r}', \mathbf{r})\,d\mathbf{r}' . \tag{188}$$

Also, since $\mu = \mu[N, v]$,

$$d\mu = \eta dN + \int f(\mathbf{r})\,dv(\mathbf{r})\,d\mathbf{r} , \tag{189}$$

where, by the Maxwell relation,

$$f(\mathbf{r}) = \frac{\partial}{\partial N}\frac{\delta E[N, v]}{\delta v(\mathbf{r})} = \left(\frac{\delta\mu}{\delta v(\mathbf{r})}\right)_N = \left(\frac{\partial\rho(\mathbf{r})}{\partial N}\right)_v . \tag{190}$$

These equations give

$$d\rho(\mathbf{r}) = s(\mathbf{r})\,\eta\,dN + \int dv(\mathbf{r}')[-\sigma(\mathbf{r}', \mathbf{r}) + f(\mathbf{r}')\,s(\mathbf{r})]\,d\mathbf{r}' . \tag{191}$$

On the other hand $d\rho = d\rho[N, v]$ reads:

$$d\rho(\mathbf{r}) = f(\mathbf{r})\,dN + \int dv(\mathbf{r}')\,\beta(\mathbf{r}', \mathbf{r})\,d\mathbf{r}' \tag{192}$$

where the linear response function [27],

$$\beta(\mathbf{r}', \mathbf{r}) = \left(\frac{\delta\rho(\mathbf{r})}{\delta v(\mathbf{r}')}\right)_N . \tag{193}$$

A comparison between Eqs. 191 and 192 shows that [27]

$$\beta(\mathbf{r}', \mathbf{r}) = -\sigma(\mathbf{r}', \mathbf{r}) + f(\mathbf{r}')\,s(\mathbf{r}) , \tag{194}$$

$$f(\mathbf{r}) = s(\mathbf{r})/S . \tag{195}$$

In the local resolution one derives the following combination formulas for the global chemical potential and hardness, which are analogous to the corresponding formulas of the AIM resolution (Eqs. 166, 167):

$$\mu = \left(\frac{\partial E^{g}}{\partial N}\right)_v = \int\left(\frac{\partial\rho(\mathbf{r})}{\partial N}\right)_v\left(\frac{\delta E^{L}}{\delta\rho(\mathbf{r})}\right)_v d\mathbf{r} = \int f(\mathbf{r})\,\mu(\mathbf{r})\,d\mathbf{r} , \tag{196}$$

$$\eta = \left(\frac{\partial^2 E^g}{\partial N^2}\right)_v = \left(\frac{\partial \mu}{\partial N}\right)_v = \iint \left(\frac{\partial \rho(\mathbf{r})}{\partial N}\right)_v \left(\frac{\delta^2 E^L}{\delta \rho(\mathbf{r})\, \delta \rho(\mathbf{r}')}\right)_v \left(\frac{\partial \rho(\mathbf{r}')}{\partial N}\right)_v d\mathbf{r}\, d\mathbf{r}'$$

$$= \int f(\mathbf{r})\, \eta(\mathbf{r}, \mathbf{r}')\, f(\mathbf{r}')\, d\mathbf{r}\, d\mathbf{r}' \,. \qquad (197)$$

In the global equilibrium state (see the Euler Eq. 182) one obtains the local hardness equalization at the global hardness level:

$$\eta(\mathbf{r}) = \left(\frac{\partial \mu(\mathbf{r})}{\partial N}\right)_v = -\left(\frac{\partial u(\mathbf{r})}{\partial N}\right)_v = -\int \left(\frac{\partial \rho(\mathbf{r}')}{\partial N}\right)_v \left(\frac{\delta u(\mathbf{r})}{\delta \rho(\mathbf{r}')}\right) d\mathbf{r}'$$

$$= \int f(\mathbf{r}')\, \eta(\mathbf{r}', \mathbf{r})\, d\mathbf{r}' = \left(\frac{\partial \mu}{\partial N}\right)_v = \eta \,. \qquad (198)$$

It should be observed that the last two equations are consistent with each other since $\int f(\mathbf{r})\, d\mathbf{r} = 1$. One also notices that the local softness and hardness satisfy the following reciprocity relation

$$\int s(\mathbf{r})\, \eta(\mathbf{r})\, d\mathbf{r} = S\eta \int f(\mathbf{r})\, d\mathbf{r} = 1 \,. \qquad (199)$$

The above local g.s. development can be summarized in terms of the following matrix, integral equations, which link the vectors of the g.s. perturbations, $[dN, dv(\mathbf{r})]$, with the g.s. response vector $[d\mu, d\rho(\mathbf{r})]$:

$$[d\mu,\, d\rho(\mathbf{r})] = \left[dN,\, \int d\mathbf{r}'\, dv(\mathbf{r}')\right] \begin{bmatrix} \eta & f(\mathbf{r}) \\ f(\mathbf{r}') & \beta(\mathbf{r}', \mathbf{r}) \end{bmatrix},$$

$$[dN,\, dv(\mathbf{r})] = \left[d\mu,\, \int d\mathbf{r}'\, d\rho(\mathbf{r}')\right] \begin{bmatrix} 0 & 1 \\ 1 & -\eta(\mathbf{r}', \mathbf{r}) \end{bmatrix}.$$

The local hardness $\eta(\mathbf{r})$ can be interpreted as the mixed derivative of the energy, including the differentiation with respect to the global number of electrons and electron density,

$$\eta(\mathbf{r}) = \left(\frac{\partial}{\partial N} \frac{\delta E^L}{\delta \rho(\mathbf{r})}\right)_v = \left(\frac{\partial \mu(\mathbf{r})}{\partial N}\right)_v = \left(\frac{\delta}{\delta \rho(\mathbf{r})} \frac{\partial E^g}{\partial N}\right)_v = \left(\frac{\delta \mu}{\delta \rho(\mathbf{r})}\right)_v . \qquad (200)$$

In Eq. 198 we have used the first part of the above Maxwell relation. Clearly, the same result, of the local hardness equalization, follows from the second part of this identity. To demonstrate this one may use either the obvious chain rule in the $[N, v]$ representation:

$$\left(\frac{\delta \mu}{\delta \rho(\mathbf{r})}\right)_v = \frac{\delta N}{\delta \rho(\mathbf{r})} \left(\frac{\partial \mu}{\partial N}\right)_v = \eta \,, \qquad (201)$$

or one can adopt the partitioning of Eq. 144:

$$[\delta\mu/\delta\rho(\mathbf{r})]_v = \int [\delta\gamma(\mathbf{r}')/\delta\rho(\mathbf{r})]\,[(\delta\bar{\mathscr{E}}/\delta\gamma(\mathbf{r}')]\,d\mathbf{r}' + [\delta N/\delta\rho(\mathbf{r})](\partial E^g/\partial N) = \eta\,, \tag{202}$$

where we have used the variational principle of Eq. 147.

One can interpret the quantity of Eq. 198 as an averaging over $\mathbf{r}'$ with $f(\mathbf{r}')$ providing the "weighting" factor; it should be stressed, however, that $f(\mathbf{r})$ exhibits both negative and positive values. Clearly, the derivative of Eq. 198 does not differentiate between local sites in a molecule, so that it cannot be used as a measure of local differences in hardness of the equilibrium electron distribution. Parr et al. [26] have suggested another weighting average with the probability distribution $\gamma(\mathbf{r})$ [Eq. 144] replacing the FF $f(\mathbf{r})$:

$$\bar{\eta}(\mathbf{r}) = \int \gamma(\mathbf{r}')\,\eta(\mathbf{r}', \mathbf{r})\,d\mathbf{r}'\,. \tag{203}$$

This quantity, not equalized throughout the space, does not have a clear interpretation as a specific derivative, so its possible utility in probing the properties of the equilibrium electron distribution still remains to be justified on a more physical basis.

Obviously, any "weighting" factor, which integrates to 1, defines its own "local hardness" distribution. In our opinion this arbitrariness of the averaged quantities of Eq. 203 puts in doubt the very purpose of searching for a local hardness quantity of this type.

It is the hardness kernel which monitors local responses in $\mu(\mathbf{r}')$ $[-u(\mathbf{r}')]$ per unit density displacement at any given point $\mathbf{r}$. The averaging process of Eq. 203 focuses on probability of *finding* an electron at point $\mathbf{r}'$, before any CT, thus providing a "static" measure of an average charge coupling between an added/removed electron at $\mathbf{r}$ and its environment in a molecule. A different emphasis is made in Eq. 198, where the "weighting" factor reflects local *changes* in density due to CT, thus generating a "dynamic" response of the system to the external perturbation dN.

3. Relative Internal Development

Here we would like to comment upon the *internal* hardness/softness quantities [86]. The preceding section has been devoted to derivative properties involving a net change in the global number of electrons, $dN \neq 0$, i.e., an exchange of electrons between M and a hypothetical external reservoir (the system environment). As such these quantities can be called *external*. However, it is also of interest to examine the *internal* hardness/softness type derivatives, corresponding to $dN = 0$, i.e., the intra-system polarization [9, 32, 86]. An example of such internal treatment has already been given in Section 4.2 of Chapter 2, where the isoelectronic CT in the reactive system A---B has been examined. Another typical example is the charge adjustment due to a given displacement of nuclei, which changes the external potential $v(\mathbf{r})$ at constant N.

One way to incorporate the $dN = 0$ constraint is to consider electron population of one atom, say m-th, N_m, as dependent on electron populations of the remaining AIM $\boldsymbol{N}' = (N_1, \ldots, N_{m-1})$: $N_m = N_m(\boldsymbol{N}')$ [32, 86]. The closed-system energy in the AIM resolution can therefore be expressed in this relative formulation as a function of the $m-1$ independent variables $\boldsymbol{N}'$:

$$E^{AIM}(\boldsymbol{N})_N = E^{AIM}[\boldsymbol{N}', N_m(\boldsymbol{N}')] \equiv E_m^{AIM}(\boldsymbol{N}') \,. \tag{204}$$

where, for simplicity, we have dropped the external potential vector in the list of arguments. The closure relation,

$$dN_m = -\sum_{i=1}^{m-1} dN_i \,, \tag{205}$$

gives the trivial implicit derivatives of N_m with respect to $\boldsymbol{N}'$, $(\partial N_m \, \partial N_i)_N = -1$, $i < m$. The corresponding electron population derivatives of E_m^{AIM}, which take into account this dependence, define the (relative) *internal chemical potentials* of AIM,

$$\mu_i^m \equiv \frac{\partial E^{AIM}}{\partial N_i} + \frac{\partial N_m}{\partial N_i}\frac{\partial E^{AIM}}{\partial N_m} = \mu_i - \mu_m \,, \quad i < m \,, \tag{206}$$

and the corresponding hardness matrix,

$$H_{j,i}^m \equiv \frac{\partial \mu_i^m}{\partial N_j} + \frac{\partial N_m}{\partial N_j}\frac{\partial \mu_i^m}{\partial N_m} = \eta_{j,i} - \eta_{j,m} - \eta_{m,i} + \eta_{m,m} \,, \quad (i, j) < m \,. \tag{207}$$

In terms of these quantities the quadratic energy expression for the closed system becomes:

$$\Delta E_m^{AIM}(\boldsymbol{N}') \cong d\boldsymbol{N}' (\boldsymbol{\mu}^m)^\dagger + \frac{1}{2} d\boldsymbol{N}' \, \mathbf{H}^m \, (d\boldsymbol{N}')^\dagger \,, \tag{208}$$

where $(\boldsymbol{\mu}^m) = (\mu_1^m, \ldots, \mu_{m-1}^m) = \partial E_m^{AIM} / \partial \boldsymbol{N}'$ and the $(m-1)\times(m-1)$ matrix $\mathbf{H}^m = \{H_{i,j}^m\} = \partial^2 E_m^{AIM} / \partial \boldsymbol{N}' \partial \boldsymbol{N}'$. The EE equations in this N- constrained approach are:

$$\partial E_m^{AIM} / \partial \boldsymbol{N}' \big|_{\boldsymbol{N}_e'} \equiv \boldsymbol{\mu}_e^m = \boldsymbol{\mu}^m + d\boldsymbol{N}_e' \, \mathbf{H}^m = \boldsymbol{0}' \,; \tag{209}$$

where $\boldsymbol{N}_e'$ denotes the equilibrium electron populations of independent atoms; hence the electron population shifts

$$d\boldsymbol{N}_e' = \boldsymbol{N}_e' - \boldsymbol{N}' = -\boldsymbol{\mu}^m \, \mathbf{S}^m \,, \tag{210}$$

where the *internal* (relative) *softness matrix* $\mathbf{S}^m = \partial \boldsymbol{N}' / \partial \boldsymbol{\mu}^m = (\mathbf{H}^m)^{-1}$. The dN_m then immediately follows from the closure constraint of Eq. 205.

Although both $\boldsymbol{\mu}^m$ and $\mathbf{H}^m$ ($\mathbf{S}^m$) are dependent upon the choice of the atom m, the predicted responses $d\boldsymbol{N}$ are independent of such a choice; the same is true in the case of other response quantities, e.g., the *internal softness matrix, including all* m *atoms*, $\mathbf{S} = (\partial \boldsymbol{N}/\partial \boldsymbol{\mu})_N$ [86]. This can be explicitly demonstrated using the following chain rule transformation

$$S_{i,j} = \left(\frac{\partial N_j}{\partial \mu_i}\right)_N = \sum_{k=1}^{m-1}\left(\frac{\partial \mu_k^m}{\partial \mu_i}\right)_N \frac{\partial N_j}{\partial \mu_k^m} \equiv \sum_{k=1}^{m-1} \tau_{i,k}^m S_{k,j}^m , \quad (i, j) = 1, \ldots, m, \tag{211}$$

where

$$\tau_{i,k}^m = \begin{cases} \delta_{i,k}, & i < m, \\ & k < m. \\ -1, & i = m, \end{cases} \tag{212}$$

This immediately gives $S_{i,j} = S_{i,j}^m$ for $(i,j) < m$, $S_{m,j} = S_{j,m} = -\Sigma_{k=1}^{m-1} S_{k,j}^m$ for $j < m$, and $S_{m,m}$ follows from the closure relation of the last column (row) of $\mathbf{S}$, $\Sigma_{i=1}^m S_{i,m} = \Sigma_{j=1}^m S_{m,j} = 0$, $S_{m,m} = \Sigma_{i=1}^{m-1}\Sigma_{j=1}^{m-1} S_{i,j}^m$. Thus, the m-independent block of $\mathbf{S}$ is identical with $\mathbf{S}^m$, while the remaining m-dependent elements of $\mathbf{S}$ immediately follow from the relevant closure relations as simple functions of $\mathbf{S}^m$.

Obviously, in symmetrical systems all atoms which are symmetry related to the atom m should also be excluded from $\boldsymbol{N}'$, if the symmetry is to be preserved in the charge displacement process.

The difference between the respective elements of the external (N-unrestricted) and internal (N-*restricted*) softness matrices, $\sigma_{i,j} = \partial N_j/\partial \mu_i$ and $S_{i,j} = (\partial N_j/\partial \mu_i)_N$, respectively, is due to the external CT component:

$$\sigma_{i,j} = \left(\frac{\partial N_j}{\partial \mu_i}\right)_N + \frac{\partial \mu}{\partial \mu_i}\frac{\partial N}{\partial \mu}\frac{\partial N_j}{\partial N} = S_{i,j} + f_i\, S\, f_j\,. \tag{213}$$

This equation represents the AIM equivalent of the Berkowitz and Parr relation [27] of Eq. 194 for the local (diagonal) response quantities of M as a whole:

$$\sigma(\mathbf{r}', \mathbf{r}) = -\beta(\mathbf{r}', \mathbf{r}) + f(\mathbf{r}')\, S\, f(\mathbf{r})\,. \tag{214}$$

A comparison between Eq. 213, $\boldsymbol{\sigma} = \mathbf{S} + \boldsymbol{f}^\dagger S \boldsymbol{f}$, and the AIM equivalent of Eq. 214, $\boldsymbol{\sigma} = -\boldsymbol{\beta} + \boldsymbol{f}^\dagger S \boldsymbol{f}$, which we derive below, identifies the linear response function matrix:

$$\boldsymbol{\beta} \equiv \{\beta_{i,j} \equiv (\partial N_j/\partial v_i)_N = -S_{i,j}\}\,, \tag{215}$$

which represents the local kernel $\beta(\mathbf{r}', \mathbf{r})$ in the AIM resolution. This interpretation of the internal softness $\mathbf{S}$ shows that it provides the AIM-resolved linear response tensor probing the system responses to shifts in the effective level of the external potential at atomic positions. Therefore, the polarization changes in the electron distribution (responses to the external potential displacements) can be determined from the external softness properties calculated for the fixed nuclear geometry (external potential). This very property is used in determining the mapping relations between the collective modes of the electron populations and alternative modes of displacements in nuclear positions [86], which we shall discussed later in the book.

We conclude this section with a comment on the nature of the AIM discretization. It can be viewed as resulting from the phenomenological energy expression,

$$E(\boldsymbol{N}[N, v], \boldsymbol{v}[N, v]) \equiv E^{\mathrm{AIM}}(\boldsymbol{N}, \boldsymbol{v}) = \textstyle\sum_i^m N_i v_i + F^{\mathrm{AIM}}(\boldsymbol{N}) , \tag{216}$$

representing the AIM equivalent of the familiar local density functional $E_v[\rho] = \int \rho v \,\mathrm{dr} + F[\rho]$ (Eq. 4). The effective potential vector $\boldsymbol{v}$ groups the external potential parameters associated with constituent atoms, $\boldsymbol{v} = (v_1, \ldots, v_m)$. The AIM chemical potential equalization equations can be interpreted as the Euler equations (see Eq. 182),

$$\frac{\partial E^{\mathrm{AIM}}(\boldsymbol{N}, \boldsymbol{v})}{\partial N_i} = \mu_i = v_i + \frac{\partial F^{\mathrm{AIM}}}{\partial N_i} = \mu , \tag{217}$$

resulting from the variational principle:

$$\min_{\boldsymbol{N}} \{ E^{\mathrm{AIM}}(\boldsymbol{N}, \boldsymbol{v}) - \mu(N(\boldsymbol{N}) - N^0) \} . \tag{218}$$

Continuing this analogy to the local description one could introduce the relative AIM external potential parameters $\{u_i \equiv v_i - \mu\} = \boldsymbol{u}$, which uniquely define the equilibrium AIM electron populations: $\boldsymbol{N} = \boldsymbol{N}(\boldsymbol{u})$. In terms of $\boldsymbol{u}$ (see Eqs. 183 and 185)

$$\boldsymbol{\eta} = \frac{\partial^2 F^{\mathrm{AIM}}}{\partial \boldsymbol{N} \partial \boldsymbol{N}} = -\frac{\partial \boldsymbol{u}}{\partial \boldsymbol{N}} , \quad \boldsymbol{\sigma} = -\frac{\partial \boldsymbol{N}}{\partial \boldsymbol{u}} , \quad \boldsymbol{f} = \left(\frac{\partial \mu}{\partial \boldsymbol{v}}\right)_N = \left(\frac{\partial \boldsymbol{N}}{\partial N}\right)_{\boldsymbol{v}} . \tag{219}$$

We can now repeat the local derivations of Eqs. 188, 191 and 192 in the AIM resolution by comparing

$$\begin{aligned} \mathrm{d}\boldsymbol{N}(\boldsymbol{u}) &= \mathrm{d}\boldsymbol{u} \frac{\partial \boldsymbol{N}}{\partial \boldsymbol{u}} = -[\mathrm{d}\boldsymbol{v} - \mathrm{d}\mu \, \boldsymbol{1}]\, \boldsymbol{\sigma} = \eta \, \mathrm{d}N \boldsymbol{S} + \mathrm{d}\boldsymbol{v}[-\boldsymbol{\sigma} + \boldsymbol{f}^{\dagger} S \, \boldsymbol{f}] \\ &= \mathrm{d}\boldsymbol{N}[N, v] = \mathrm{d}N \boldsymbol{f} + \mathrm{d}\boldsymbol{v} \boldsymbol{\beta} , \end{aligned} \tag{220}$$

which gives Eqs. 165, 213 and 215. Also, by writting the chain rule of Eq. 213, in terms of $\boldsymbol{u}$, one can directly express $\sigma_{i,j}$ in terms of $\beta_{i,j}$ itself:

$$\sigma_{i,j} = \frac{\partial N_j}{\partial \mu_i} = -\frac{\partial N_j}{\partial u_i} = -\left(\frac{\partial N_j}{\partial u_i}\right)_N - \frac{\partial \mu}{\partial u_i}\frac{\partial N}{\partial \mu}\frac{\partial N_j}{\partial N}$$

$$= -\left(\frac{\partial N_j}{\partial v_i}\right)_N + f_i\, S\, f_j = -\beta_{i,j} + f_i\, S\, f_j\,. \tag{221}$$

4. Determining Charge Sensitivities from Canonical AIM Hardnesses

In this section we would like to present a simple algorithm for calculating charge sensitivities of the preceding sections, organized as a single matrix inversion step [60]. Suppose that the molecular system M was shifted from the initial, global equilibrium ($\mu_1^0 = \mu_2^0 = \ldots = \mu_m^0 = \mu^0$) by an external perturbation (dN, $d\boldsymbol{v}$); as a result the system reaches the displaced equilibrium state corresponding to a new level of the global chemical potential: $\mu_1 = \mu_2 = \ldots = \mu_m = \mu(dN, d\boldsymbol{v})$. In accordance with our previous development we first determine the displaced AIM chemical potentials, $\boldsymbol{\mu}^+ = \mu^0 \boldsymbol{1} + d\boldsymbol{v}$, using the chemical potential displacements $d\boldsymbol{\mu}(d\boldsymbol{v}) = (\boldsymbol{\mu}^+ - \mu \boldsymbol{1})$ as the forces behind subsequent net changes in the AIM electron populations, due to both dN and $d\boldsymbol{v}$, $\delta\boldsymbol{N} = (N_1 - N_1^0, \ldots, N_m - N_m^0)$. The corresponding EE equations linking the initial and final chemical potentials, incorporating the obvious relation $\Sigma_i^M dN_i = dN$, may be written in the following matrix form:

$$(-\mu,\ \delta\boldsymbol{N})\begin{pmatrix} 0 & \boldsymbol{1} \\ \boldsymbol{1}^\dagger & \boldsymbol{\eta} \end{pmatrix} \equiv \mathfrak{R}\,\tilde{\boldsymbol{\eta}} = (dN,\ -\boldsymbol{\mu}^+) \equiv \mathfrak{P}\,, \tag{222}$$

where $\mathfrak{R}$ and $\mathfrak{P}$ represent the generalized response and perturbation vectors, respectively. One can solve this equation by inverting the generalized hardness matrix $\tilde{\boldsymbol{\eta}}$,

$$\mathfrak{R} = (-\mu,\ \delta\boldsymbol{N}) = \mathfrak{P}\,\tilde{\boldsymbol{\eta}}^{-1} = \mathfrak{P}\,\tilde{\boldsymbol{\sigma}} = (dN,\ -\boldsymbol{\mu}^+)\begin{pmatrix} -\eta & \boldsymbol{f} \\ \boldsymbol{f}^\dagger & -\boldsymbol{\beta} \end{pmatrix}, \tag{223}$$

where $\tilde{\boldsymbol{\sigma}}$ denotes the generalized softness matrix. Indeed, the first component of this equation reads: $\mu - \mu^0 \equiv d\mu(dN, d\boldsymbol{v}) = dN\,\eta + d\boldsymbol{v}\,\boldsymbol{f}^\dagger$, while the second component expresses $\delta\boldsymbol{N}(dN, d\boldsymbol{v}) = dN\,\boldsymbol{f} + \boldsymbol{\mu}^+\boldsymbol{\beta} = dN\,\boldsymbol{f} + \mu^0\,\boldsymbol{1}\,\boldsymbol{\beta} + d\boldsymbol{v}\boldsymbol{\beta} = dN\,\boldsymbol{f} + d\boldsymbol{v}\boldsymbol{\beta}$, since $\Sigma_i^M \beta_{i,j} = (\partial N/\partial v_i)_N = 0$.

In an analogous way one can determine the fragment responses corresponding to the constrained equilibrium in $M^G = (X|Y|\ldots|Z)$. In this case the system is divided into mutually closed fragments, each exhibiting its own equilibrium chemical potential: $\boldsymbol{\mu}_X = \mu_X^M \boldsymbol{1}_X, \ldots, \boldsymbol{\mu}_Z = \mu_Z^M \boldsymbol{1}_Z$. The constrained equilibrium EE equations, which describe the intra-fragment rearrangements, together with the obvious relations $\{\Sigma_i^X dN_i^X = dN_X\}$, may be written in the following matrix form [60],

$$\begin{pmatrix} 0 & 0 & \cdots & 0 & \boldsymbol{1}_X & \boldsymbol{0}_Y & \cdots & \boldsymbol{0}_Z \\ 0 & 0 & \cdots & 0 & \boldsymbol{0}_X & \boldsymbol{1}_Y & \cdots & \boldsymbol{0}_Z \\ \cdots & \cdots & \cdots & \cdots & \cdots & \cdots & \cdots & \cdots \\ 0 & 0 & \cdots & 0 & \boldsymbol{0}_X & \boldsymbol{0}_Y & \cdots & \boldsymbol{1}_Z \\ \boldsymbol{1}_X^\dagger & \boldsymbol{0}_X^\dagger & \cdots & \boldsymbol{0}_X^\dagger & \boldsymbol{\eta}^{X,X} & \boldsymbol{\eta}^{X,Y} & \cdots & \boldsymbol{\eta}^{X,Z} \\ \boldsymbol{0}_Y^\dagger & \boldsymbol{1}_Y^\dagger & \cdots & \boldsymbol{0}_Y^\dagger & \boldsymbol{\eta}^{Y,X} & \boldsymbol{\eta}^{Y,Y} & \cdots & \boldsymbol{\eta}^{Y,Z} \\ \cdots & \cdots & \cdots & \cdots & \cdots & \cdots & \cdots & \cdots \\ \boldsymbol{0}_Z^\dagger & \boldsymbol{0}_Z^\dagger & \cdots & \boldsymbol{1}_Z^\dagger & \boldsymbol{\eta}^{Z,X} & \boldsymbol{\eta}^{Z,Y} & \cdots & \boldsymbol{\eta}^{Z,Z} \end{pmatrix} \begin{pmatrix} -\mu_X^M \\ -\mu_Y^M \\ \cdots \\ -\mu_Z^M \\ \delta \boldsymbol{N}_X^\dagger \\ \delta \boldsymbol{N}_Y^\dagger \\ \cdots \\ \delta \boldsymbol{N}_Z^\dagger \end{pmatrix} = \begin{pmatrix} dN_X^M \\ dN_Y^M \\ \cdots \\ dN_Z^M \\ -(\overset{+}{\boldsymbol{\mu}}_X)^\dagger \\ -(\overset{+}{\boldsymbol{\mu}}_Y)^\dagger \\ \cdots \\ -(\overset{+}{\boldsymbol{\mu}}_Z)^\dagger \end{pmatrix},$$

$$\text{or in short:} \quad \tilde{\boldsymbol{\eta}}_G \, \boldsymbol{\Re}_G^\dagger = \boldsymbol{\rho}_G^\dagger, \tag{224}$$

where $\boldsymbol{\Re}_G$ and $\boldsymbol{\rho}_G$ stand for the generalized response and perturbation vectors in M^G, respectively. As before for the global equilibrium case, the vectors $\overset{+}{\boldsymbol{\mu}}_A = \mu^0 \boldsymbol{1}_A + d\boldsymbol{v}_A$ (A = X, Y, ..., Z), group the initially displaced chemical potentials of constituent atoms, which define the intra-fragment forces, $d\overset{+}{\boldsymbol{\mu}}_A = \overset{+}{\boldsymbol{\mu}}_A - \mu_A^M \boldsymbol{1}_A$ (A = X, Y, ..., Z), determining the changes in the electron populations of atoms, $\delta \boldsymbol{N}_A = \boldsymbol{N}_A - \boldsymbol{N}_A^0$, etc., allowed by the partitioning M^G. Again, the solution of this equation can be found by the inversion of the generalized hardness matrix $\tilde{\boldsymbol{\eta}}^G$:

$$\begin{pmatrix} -\mu_X^M \\ -\mu_Y^M \\ \cdots \\ -\mu_Z^M \\ \delta \boldsymbol{N}_X^\dagger \\ \delta \boldsymbol{N}_Y^\dagger \\ \cdots \\ \delta \boldsymbol{N}_Z^\dagger \end{pmatrix} = \begin{pmatrix} -\eta_{X,X}^M & -\eta_{X,Y}^M & \cdots & -\eta_{X,Z}^M & \boldsymbol{f}^{X,X} & \boldsymbol{f}^{X,Y} & \cdots & \boldsymbol{f}^{X,Z} \\ -\eta_{Y,X}^M & -\eta_{Y,Y}^M & \cdots & -\eta_{Y,Z}^M & \boldsymbol{f}^{Y,X} & \boldsymbol{f}^{Y,Y} & \cdots & \boldsymbol{f}^{Y,Z} \\ \cdots & \cdots & \cdots & \cdots & \cdots & \cdots & \cdots & \cdots \\ -\eta_{Z,X}^M & -\eta_{Z,Y}^M & \cdots & -\eta_{Z,Z}^M & \boldsymbol{f}^{Z,X} & \boldsymbol{f}^{Z,Y} & \cdots & \boldsymbol{f}^{Z,Z} \\ (\boldsymbol{f}^{X,X})^\dagger & (\boldsymbol{f}^{Y,X})^\dagger & \cdots & (\boldsymbol{f}^{X,Z})^\dagger & -\boldsymbol{\beta}^{X,X} & -\boldsymbol{\beta}^{X,Y} & \cdots & -\boldsymbol{\beta}^{X,Z} \\ (\boldsymbol{f}^{X,Y})^\dagger & (\boldsymbol{f}^{Y,Y})^\dagger & \cdots & (\boldsymbol{f}^{Y,Z})^\dagger & -\boldsymbol{\beta}^{Y,X} & -\boldsymbol{\beta}^{Y,Y} & \cdots & -\boldsymbol{\beta}^{Y,Z} \\ \cdots & \cdots & \cdots & \cdots & \cdots & \cdots & \cdots & \cdots \\ (\boldsymbol{f}^{X,Z})^\dagger & (\boldsymbol{f}^{Y,Z})^\dagger & \cdots & (\boldsymbol{f}^{Z,Z})^\dagger & -\boldsymbol{\beta}^{Z,X} & -\boldsymbol{\beta}^{Z,Y} & \cdots & -\boldsymbol{\beta}^{Z,Z} \end{pmatrix} \begin{pmatrix} dN_X^M \\ dN_Y^M \\ \cdots \\ dN_Z^M \\ -(\overset{+}{\boldsymbol{\mu}}_X)^\dagger \\ -(\overset{+}{\boldsymbol{\mu}}_Y)^\dagger \\ \cdots \\ -(\overset{+}{\boldsymbol{\mu}}_Z)^\dagger \end{pmatrix},$$

$$\boldsymbol{\Re}_G^\dagger = \tilde{\boldsymbol{\sigma}}^G \boldsymbol{\rho}_G^\dagger, \tag{225}$$

where $\tilde{\sigma}^{G} = (\tilde{\eta}^{G})^{-1}$ is the molecular fragment generalized softness matrix. It can be partitioned into two main diagonal blocks, representing the condensed hardness matrix elements and those of the linear response function, respectively, with the off-diagonal blocks giving the relaxed FF indices (diagonal and off-diagonal). For each fragment one obtains the familiar equation $d\mu_A^M = \Sigma_B^M dN_B^M \eta_{A,B}^M + \Sigma_B^M dv_B (f^{A,B})^\dagger$, which generates the shift in the equilibrium chemical potential of fragment A , and the second equation $\delta N_A = \Sigma_B^M dN_B^M f^{B,A} + \Sigma_B^M dv_B^M \beta^{B,A}$ (A, B = X, Y, ...Z), for the relevant fragment charge reorganization. Here, the matrix blocks of the FF and linear response matrices are defined as follows: $f^{X,Y} \equiv \partial N_Y/\partial N_X = \partial \mu_X/\partial v_Y$, $\beta^{X,Y} \equiv \partial N_Y/\partial v_X$.

Using the chain rule transformation one can generalize Eqs. 174 and 179 for this general constrained equilibrium case,

$$f^{A,B} = \frac{\partial N_B}{\partial N_A} = \sum_C^M \frac{\partial N_C}{\partial N_A}\frac{\partial N_B}{\partial N_C} \equiv \sum_C^M f^{A,C}\,\mathbf{T}^{(C|B)}, \tag{226}$$

where $\mathbf{T}^{(C|B)}$ is a block of the relaxational matrix of Eq. 175, generalized for the partitioning of a molecule into molecular fragments. Knowledge of the FF indices allows one to derive the combination formulas for the condensed quantities of Eq. 225, i.e., the fragment chemical potentials and condensed hardness matrix elements, in terms of the relevant AIM charge sensitivities:

$$\mu_A^M = \left(\frac{\partial E^G}{\partial N_A}\right)_{N_{B\neq A}} = \left(\frac{\partial \mathbf{N}}{\partial N_A}\right)\left(\frac{\partial E^{AIM}}{\partial \mathbf{N}}\right) = \sum_B^M \left(\frac{\partial \mathbf{N}_B}{\partial N_A}\right)\left(\frac{\partial E}{\partial \mathbf{N}_B}\right) =$$
$$= \sum_B^M f^{A,B}(\boldsymbol{\mu}_B)^\dagger, \quad A, B = (X, Y, \ldots, Z), \tag{227}$$

$$\eta_{A,B}^M = \frac{\partial^2 E^G}{\partial N_A\,\partial N_B} = \frac{\partial \mu_A^M}{\partial N_B} = \frac{\partial \mu_B^M}{\partial N_A} = \sum_{C,D}^M \frac{\partial \mathbf{N}_C}{\partial N_A}\frac{\partial^2 E}{\partial \mathbf{N}_C\,\partial \mathbf{N}_D}\frac{\partial \mathbf{N}_D}{\partial N_B}$$
$$= \sum_{C,D}^M f^{A,C}\,\boldsymbol{\eta}^{C,D}(f^{B,D})^\dagger, \quad A, B, C, D = (X, Y, \ldots, Z). \tag{228}$$

One can generalize the Berkowitz Parr formula of Eqs. 214 and 221 into the form valid for the constrained equilibrium in M^G :

$$\boldsymbol{\sigma}^{A,B} = -\boldsymbol{\beta}^{A,B} + \sum_{C,D}^M (f^{D,A})^\dagger \sigma_{D,C}^M f^{C,B}. \tag{229}$$

where the condensed softness matrix element,

$$\sigma_{A,B}^M = \frac{\partial N_B}{\partial \mu_A} = \mathbf{1}_A\,\boldsymbol{\sigma}^{A,B}\,\mathbf{1}_B^\dagger \equiv \mathbf{S}^{A,B}\,\mathbf{1}_B^\dagger. \tag{230}$$

5. *In situ* Quantities in the Donor–Acceptor Systems

In this section we briefly summarize the *in situ* electron population/chemical potential derivatives along the N-restricted CT process, $dN = 0$ or $dN_A = -dN_B = N_{CT}$, in a general system M = A---B consisting of an acidic (A) and basic (B) reactants [61]. We shall adopt the AIM resolution of the preceding sections. The *in situ* derivatives involve the differentiation with respect to the amount of CT, N_{CT} , taking into account all environmental effects due to the presence of the other reactant. Such derivatives have already appeared in our previous discussion of the quadratic interaction energy in such systems (see Section 4.2 and 8 of Chapter 2); they are simple functions of the condensed or the AIM resolved molecular fragment (reactant) quantities defined in the preceding sections.

As we have already observed before the closed M constraint implies trivial dependencies of $f_A = \partial N_A/\partial N_{CT} = 1$ and $f_B = \partial N_B/\partial N_{CT} = -1$, which enter the *in situ* derivatives. In the hypothetical state $M^+ = (A|B)$, when both reactants are mutually closed but, internally, fully relaxed, one obtains the *in situ* chemical potential difference,

$$\mu_{CT}^{+} = \frac{\partial E(A|B)}{\partial N_A} f_A + \frac{\partial E(A|B)}{\partial N_B} f_B = \mu_A^{+} - \mu_B^{+}, \tag{231}$$

while the effective *in situ* hardness is also given by Eq. 89:

$$\eta_{CT} = \eta_A^M + \eta_B^M, \tag{232}$$

is the sum of the *in situ* hardnesses of reactants, $\eta_A^M = \eta_{A,A}^M - \eta_{B,A}^M$ and $\eta_B^M = \eta_{B,B}^M - \eta_{A,B}^M$. The condition of the vanishing inter-reactant CT gradient, $g(N_{CT}) = 0$ (Eq. 99), gives the equilibrium value of N_{CT}, which equalizes the chemical potentials of the mutually open reactants in $M^* = (A \mid B)$ (Eq. 100).

Consider next the *in situ* CT softnesses, which involve derivatives with respect to the fragment chemical potentials, and related quantities, e.g., the FF indices. The global CT softness is immediately given by the inverse of η_{CT},

$$S_{CT} = \partial N_{CT}/\partial \mu_{CT} = 1/\eta_{CT}. \tag{233}$$

The condensed reactant FF indices f_A and f_B also represent the *in situ* quantities, and as such must sum up to zero: $f_A + f_B = 0$, in accordance with the fixed global number of electrons in the system.

For the mixed derivatives representing the *in situ* AIM hardnesses in AB, e.g., for constituent atoms in A, one obtains:

$$\eta_A^{CT} = \frac{\partial \mu_{CT}}{\partial N_A} = \frac{\partial \boldsymbol{\mu}_A}{\partial N_{CT}} = f_A\left(\frac{\partial \boldsymbol{\mu}_A}{\partial \boldsymbol{N}_A}\right)_{N_B} + f_B\left(\frac{\partial \boldsymbol{\mu}_A}{\partial \boldsymbol{N}_B}\right)_{N_A}$$

$$= \left(\frac{\partial \boldsymbol{N}}{\partial \boldsymbol{N}_A}\right)_{N_B}\left(\frac{\partial \boldsymbol{\mu}_A}{\partial \boldsymbol{N}}\right)_{N_B} + \left(\frac{\partial \boldsymbol{N}}{\partial \boldsymbol{N}_B}\right)_{N_B}\left(\frac{\partial \boldsymbol{\mu}_A}{\partial \boldsymbol{N}}\right)_{N_A}$$

$$\equiv (\boldsymbol{f}^{A,A} - \boldsymbol{f}^{B,A})\,\boldsymbol{\eta}^{M}_{A,A} - (\boldsymbol{f}^{B,B} - \boldsymbol{f}^{A,B})\,\boldsymbol{\eta}^{M}_{B,A} \equiv \boldsymbol{f}_A^{CT}\,\boldsymbol{\eta}^{M}_{A,A} + \boldsymbol{f}_B^{CT}\,\boldsymbol{\eta}^{M}_{B,A}\,, \quad (234)$$

where $\mathbf{f}^{(A|B)} = \{\boldsymbol{f}^{X,Y}\}$ and $\boldsymbol{\eta}^{(A|B)} = \{\boldsymbol{\eta}^{M}_{X,Y}\}$, $(X, Y) = (A, B)$ represent the relaxed AIM FF indices and hardnesses defined by Eqs. 174, 179, and 180.

Finally, the global *in situ* hardness can also be related to the above AIM quantities by the chain rule:

$$\eta_{CT} = f_A\left(\frac{\partial \mu_{CT}}{\partial \boldsymbol{N}_A}\right)_{N_B} + f_B\left(\frac{\partial \mu_{CT}}{\partial \boldsymbol{N}_B}\right)_{N_A} = \left(\frac{\partial \boldsymbol{N}}{\partial \boldsymbol{N}_A}\right)_{N_B}\left(\frac{\partial \mu_{CT}}{\partial \boldsymbol{N}}\right) - \left(\frac{\partial \boldsymbol{N}}{\partial \boldsymbol{N}_B}\right)_{N_A}\left(\frac{\partial \mu_{CT}}{\partial \boldsymbol{N}}\right)$$

$$= \boldsymbol{f}_A^{CT}\,(\boldsymbol{\eta}_A^{CT})^{\dagger} + \boldsymbol{f}_B^{CT}\,(\boldsymbol{\eta}_B^{CT})^{\dagger}\,. \quad (235)$$

The *in situ* AIM-resolved FF vectors of reactants in AB are defined in Eq. 234:

$$\boldsymbol{f}_A^{CT} = \left(\frac{\partial \boldsymbol{N}_A}{\partial N_{CT}}\right)_N = f_A\left(\frac{\partial \boldsymbol{N}_A}{\partial \boldsymbol{N}_A}\right)_{N_B} + f_B\left(\frac{\partial \boldsymbol{N}_A}{\partial \boldsymbol{N}_B}\right)_{N_A} = \boldsymbol{f}^{A,A} - \boldsymbol{f}^{B,A}\,,$$

$$\boldsymbol{f}_B^{CT} = \left(\frac{\partial \boldsymbol{N}_B}{\partial N_{CT}}\right)_N = f_A\left(\frac{\partial \boldsymbol{N}_B}{\partial \boldsymbol{N}_A}\right)_{N_B} + f_B\left(\frac{\partial \boldsymbol{N}_B}{\partial \boldsymbol{N}_B}\right)_{N_A} = \boldsymbol{f}^{A,B} - \boldsymbol{f}^{B,B}\,; \quad (236)$$

they are given by differences between the corresponding diagonal and off-diagonal FF indices. The above FF indices for the isoelectronic CT between reactants define the partitioning of the global CT softness into the corresponding atomic contributions, e.g.,

$$\boldsymbol{S}_A^{CT} = \frac{\partial N_{CT}}{\partial \boldsymbol{\mu}_A} = \frac{\partial \boldsymbol{N}_A}{\partial \mu_{CT}} = \frac{\partial N_{CT}}{\partial \mu_{CT}}\left(\frac{\partial \boldsymbol{N}_A}{\partial N_{CT}}\right)_N = S^{CT}\,\boldsymbol{f}_A^{CT}\,. \quad (237)$$

The final expression in Eq. 231 also follows from the combination formula for $\overset{+}{\mu}_{CT}$ in terms of the relaxed chemical potentials of atoms in a single reactant, $\boldsymbol{\mu}_A^{M}$ or $\boldsymbol{\mu}_B^{M}$, since the intra-reactant equilibrium AIM populations of one reactant are unique functions of the AIM populations of the other reactant: $E(A|B) = E[\boldsymbol{N}_A, \boldsymbol{N}_B(\boldsymbol{N}_A)] = E[\boldsymbol{N}_A(\boldsymbol{N}_B), \boldsymbol{N}_B]$,

$$\overset{+}{\mu}_{CT} = \frac{\partial N_A}{\partial N_{CT}}\left(\frac{\partial E(A|B)}{\partial N_A}\right)_{N_B} \equiv f_A^{CT}\,(\mu_A^M)^{\dagger} = \frac{\partial N_B}{\partial N_{CT}}\left(\frac{\partial E(A|B)}{\partial N_B}\right)_{N_A} \equiv f_B^{CT}\,(\mu_B^M)^{\dagger}\,, \quad (238)$$

where, e.g., the relaxed chemical potentials of atoms in A are:

$$\mu_A^M = \left(\frac{\partial E(A|B)}{\partial N_A}\right)_{N_B} = \left(\frac{\partial E(A|B)}{\partial N_A}\right)_{N_B} + \left(\frac{\partial E(A|B)}{\partial N_B}\right)_{N_A}\frac{\partial N_B}{\partial N_A}$$

$$= \mu_A^{rgd} + \mu_B^{rgd}\,\mathbf{T}^{(A|B)} \equiv \mu_A^{rgd} + \delta\mu_A(M)\,. \quad (239)$$

The *in situ* condensed hardnesses of Eq. 232 can also be expanded in terms of η_X^{CT} and f_X^{CT}, $X = A, B$. Namely, taking A as an example one finds:

$$\eta_A^M = \frac{\partial^2 E(A|B)}{\partial N_A\,\partial N_{CT}} = \frac{\partial \mu_A}{\partial N_{CT}} = \left(\frac{\partial N_A}{\partial N_{CT}}\right)\left(\frac{\partial \mu_A}{\partial N_A}\right)_{N_B} = f_A^{CT}\,\eta_A^M$$

$$= \frac{\partial \mu_{CT}}{\partial N_A} = \left(\frac{\partial N_A}{\partial N_A}\right)_{N_B}\left(\frac{\partial \mu_{CT}}{\partial N_A}\right)_{N_B} = f^{A,A}\,\eta_A^{CT}\,. \quad (240)$$

6. Collective Charge Displacements

The AIM electron population displacements, $d\mathbf{N}$, are strongly coupled through the off-diagonal hardness matrix elements $\eta_{i,j} \approx \gamma_{i,j}$. Thus, a given shift dN_k strongly affects the chemical potentials of all AIM. This representation considers all AIM populational parameters as independent variables, which can be interpreted as projections of the populational vector $\vec{\mathcal{N}} = (N_1\vec{e}_1 + N_2\vec{e}_2 + \ldots N_m\vec{e}_m)$ onto the orthogonal system of populational axes associated with the constituent atoms, AIM populational basis of vectors $\vec{\mathbf{e}}_M = (\vec{e}_1, \ldots, \vec{e}_m)$. In what follows the product of vectors are scalar products.

A more compact description of chemical reactivity can be obtained in representations involving collective charge displacements [20, 32, 33, 57, 59, 61, 87, 88], $d\mathbf{p} = d\mathbf{N}\mathbf{T}$, defined by the transformation matrix $\mathbf{T}$; they also represent a degree of the mode decoupling. In this section we shall summarize the most important sets of such collective (delocalized) modes of potential interest in the theory of chemical reactivity. We start with the total decoupling, which is obtained in the representation of the hardness matrix eigenvectors, called *populational normal modes* (PNM) [32, 33, 57]. They are analogous to the normal modes of nuclear motions defined in the nuclear position space. Next, we consider the (non-orthogonal) set of the *minimum energy coordinates* (MEC) [32, 57, 87], defined in the softness representation, again in direct analogy to the compliant formalism of nuclear motions [41, 87-91]. We conclude this section with a brief summary of alternative intermediate decoupling schemes, focusing our attention on those of interest in the theory of chemical reactivity. These collective modes include the internally and externally decoupled modes of reactants [88, 92].

6.1. Populational Normal Modes

The totally decoupled representation involves displacements in populations of PNM,

$$\mathrm{d}\boldsymbol{n} = \mathrm{d}\boldsymbol{N}\,\mathbf{U} \quad , \qquad \mathrm{d}\vec{\mathcal{N}} = \vec{\boldsymbol{U}}_{\mathrm{M}}\,\mathrm{d}\boldsymbol{n}^{\dagger} , \tag{241}$$

defined by the columns of the transformation matrix $\mathbf{T} = \mathbf{U} \equiv (\boldsymbol{U}_1 \,|\, \boldsymbol{U}_2 \,|\ldots|\, \boldsymbol{U}_m)$, which diagonalizes the AIM hardness matrix, i.e., "rotates" the AIM axes $\vec{\boldsymbol{e}}_{\mathrm{M}}$ to the (orthogonal) principal axes of the hardness tensor (see Fig. 10), $\vec{\boldsymbol{U}}_{\mathrm{M}} = \vec{\boldsymbol{e}}_{\mathrm{M}}\mathbf{U}$:

$$\mathbf{U}^{\dagger}\,\boldsymbol{\eta}\,\mathbf{U} = \mathbf{h} = \{\mathrm{h}_{\alpha}\,\delta_{\alpha\beta}\} \ , \quad \mathbf{U}^{\dagger}\,\mathbf{U} = \mathbf{I} , \tag{242}$$

where $\mathrm{h}_{\alpha} = \partial^2 E^{\mathrm{PNM}}/\partial n_{\alpha}^2 = \partial\xi_{\alpha}/\partial n_{\alpha}$ is the principal hardness of mode α, and $\xi_{\alpha} = \boldsymbol{\mu}\,\boldsymbol{U}_{\alpha}$ is the mode chemical potential. Obviously, the quadratic form of the second differential of the system energy with respect to the electron population variables becomes diagonal in the $\vec{\boldsymbol{U}}_{\mathrm{M}}$ reference frame:

$$\mathrm{d}^2 E^{\mathrm{AIM}} \equiv \frac{1}{2}\,\mathrm{d}\boldsymbol{N}\,\boldsymbol{\eta}\,\mathrm{d}\boldsymbol{N}^{\dagger} = \frac{1}{2}\sum_{\alpha}^{\mathrm{M}} \mathrm{h}_{\alpha}\,(\mathrm{d}n_{\alpha})^2 \equiv \mathrm{d}^2 E^{\mathrm{PNM}} . \tag{243}$$

Also decoupled are the corresponding EE equations (see Eq. 177) in the PNM representation,

$$\mathrm{d}\xi_{\alpha} = \mathrm{d}n_{\alpha}\mathrm{h}_{\alpha} \quad \text{or} \quad \mathrm{d}\boldsymbol{\xi} = \mathrm{d}\boldsymbol{n}\,\mathbf{h} , \tag{244}$$

where $\mathrm{d}\boldsymbol{\xi} = (\mathrm{d}\xi_1, \ldots, \mathrm{d}\xi_m)$ and $\mathrm{d}\boldsymbol{n} = (\mathrm{d}n_1, \ldots, \mathrm{d}n_m)$ are the displacements from the corresponding global equilibrium values. The global chemical potential and hardness can now be formally expressed in terms of the relevant PNM quantities,

$$\mu = \frac{\partial \boldsymbol{n}}{\partial N}\,\frac{\partial E^{\mathrm{PNM}}}{\partial \boldsymbol{n}} \equiv \boldsymbol{F}\,\boldsymbol{\xi}^{\dagger} , \tag{245}$$

$$\eta = \boldsymbol{f}\,\boldsymbol{\eta}\,\boldsymbol{f}^{\dagger} = \sum_{\alpha}^{\mathrm{M}} F_{\alpha}^2\,\mathrm{h}_{\alpha} , \tag{246}$$

where the normal FF indices $F_{\alpha} = \partial n_{\alpha}/\partial N = \boldsymbol{f}\,\boldsymbol{U}_{\alpha}$.

The phases of the CT-active PNM are fixed by the convention [93]:

$$\varphi_{\alpha} \equiv \frac{\partial N}{\partial n_{\alpha}} = \frac{\partial \boldsymbol{N}}{\partial n_{\alpha}}\,\frac{\partial N}{\partial \boldsymbol{N}} = \boldsymbol{1}\,\boldsymbol{U}_{\alpha} > 0 , \tag{247}$$

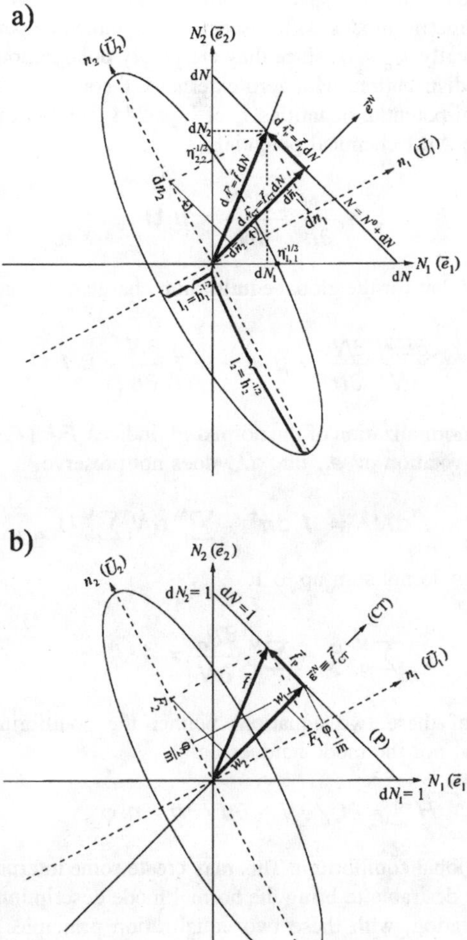

Figure 10. The PNM (principal axes), $\boldsymbol{U}_M$, in the system with two AIM populational degrees of freedom, $\boldsymbol{e}_M$, the associated charge sensitivities, and a separation of the pure CT and P components, $d\vec{\mathcal{N}}_{CT}$ and $d\vec{\mathcal{N}}_P$, of the equilibrium charge displacement $d\vec{\mathcal{N}} = \vec{f}dN$, along the FF vector $\vec{f} = \boldsymbol{e}_M\,\boldsymbol{f}^\dagger = \boldsymbol{U}_M\,\boldsymbol{F}^\dagger$. Panel (a) refers to a general displacement $dN > 0$, while panel (b) corresponds to $dN = 1$. The hardness tensor ellipse with semiaxes of length $\{l_\alpha\}$, is defined by the quadratic surface: $d\boldsymbol{N}\,\boldsymbol{\eta}\,d\boldsymbol{N}^\dagger = \Sigma_\alpha^M h_\alpha n_\alpha^2 \equiv \Sigma_\alpha^M n_\alpha^2 / l_\alpha^2 = 1$. As explicitly shown in panel (a), the shifts of the modified PNM population variables, $d\bar{\boldsymbol{n}}$, are obtained by projecting the relevant PNM variables, $d\boldsymbol{n}$, onto the pure-CT direction $\vec{e}^N = \boldsymbol{1}\,\boldsymbol{e}_M^\dagger$. One similarly obtains a resolution of $d\vec{\mathcal{N}}$ into $d\vec{\mathcal{N}}_{CT} = d\vec{\mathcal{N}} \cdot \vec{e}^N$ and $d\vec{\mathcal{N}}_P = d\vec{\mathcal{N}} - d\vec{\mathcal{N}}_{CT}$ [panel (a)], and of $\vec{f}$ into $\vec{f}_{CT} = \vec{f} \cdot \vec{e}^N$ and $\vec{f}_P = \vec{f} - \vec{f}_{CT}$ [panel (b)].

analogous to the equivalent AIM space equation: $\partial N/\partial N_i = 1$. We would like to point out, that the antisymmetric modes with respect to a symmetry operation of a system in question, have identically $\varphi_\alpha = 0$, since they are purely polarizational, representing only an internal charge redistribution, with zero external CT component.

The PNM chemical potential quantities $\boldsymbol{\xi}$ of Eqs. 244 and 245 can be obtained by the transformation of the AIM chemical potentials:

$$\boldsymbol{\xi} = \frac{\partial \boldsymbol{N}}{\partial \boldsymbol{n}} \frac{\partial E^{AIM}}{\partial \boldsymbol{N}} = \boldsymbol{\mu}\,\mathbf{U}\,. \tag{248}$$

It should be observed that for the global equilibrium charge distribution, where $\boldsymbol{\mu} = \mu\,\boldsymbol{1}$,

$$\boldsymbol{\xi}^{eq} = \frac{\partial E^{g}}{\partial N} \frac{\partial N}{\partial \boldsymbol{n}} \equiv \mu\,\boldsymbol{\varphi} = \mu\,\boldsymbol{1}\,\frac{\partial \boldsymbol{N}}{\partial \boldsymbol{n}} = \mu\,\boldsymbol{1}\,\mathbf{U}\,. \tag{249}$$

Consider now the normalization of the normal FF indices $\boldsymbol{F} = (F_1, \ldots, F_m)$ of Eq. 245. Obviously, since the rotation of $\vec{\boldsymbol{e}}_M$ into $\vec{\boldsymbol{U}}_M$ does not preserve

$$dN = \boldsymbol{1}\,d\boldsymbol{N}^{\dagger} \neq \boldsymbol{1}\,d\boldsymbol{n}^{\dagger} = \sum_i^M dN_i \sum_\alpha^M U_{i,\alpha}\,, \tag{250}$$

the normal FF indices do not sum up to 1,

$$\sum_\alpha^M F_\alpha = \sum_\alpha^M \frac{\partial n_\alpha}{\partial N} \neq 1\,. \tag{251}$$

As a consequence of these two equations neither the equilibrium normal chemical potentials, $\boldsymbol{\xi}^{eq} = \mu\,\boldsymbol{\varphi}$, nor the mode hardnesses

$$\boldsymbol{H}^{eq} = \partial\boldsymbol{\xi}/\partial N = \partial\mu/\partial\boldsymbol{n} = \eta\,\boldsymbol{\varphi}\,, \tag{252}$$

are equalized at the global equilibrium. This may create some interpretational difficulties.

It would be highly desirable to bring the normal mode description as close as possible to the AIM interpretation, with these two equalization principles being automatically satisfied and the appropriate FF indices providing the real mode contributions to the unit change in the global number of electrons. In order to reach this goal one has to redefine the mode population variables [9, 92], by projecting the $d\boldsymbol{n}$ displacements of Eq. 241 onto the pure CT-direction represented by the vector (not normalized to 1),

$$\vec{e}^{N} = \sum_i^M \left(\frac{\partial N}{\partial N_i}\right) \vec{e}_i = \boldsymbol{1}\,\vec{\boldsymbol{e}}_M^{\dagger}\,, \tag{253}$$

which determines the shift in the global number of electrons, $dN = \vec{e}^{N} \cdot d\vec{\mathcal{N}} = \boldsymbol{1}\,d\boldsymbol{N}^{\dagger}$. This procedure is illustrated in Fig. 10. The CT-projected shifts in mode populations,

$$\mathrm{d}\bar{n}_\alpha = \mathrm{d}n_\alpha \frac{\partial N}{\partial n_\alpha} = \mathrm{d}n_\alpha\, \varphi_\alpha\,, \quad \alpha = 1, \ldots, \mathrm{m}\,, \tag{254}$$

are phase-independent and correspond to the non-unitary transformation of the AIM population displacements:

$$\mathrm{d}\bar{\boldsymbol{n}} = \mathrm{d}\boldsymbol{N}\,\bar{\mathbf{U}}\,; \quad \bar{\mathbf{U}} = (\varphi_1 \boldsymbol{U}_1 \,|\, \varphi_2 \boldsymbol{U}_2 \,|\ldots|\, \varphi_\mathrm{m} \boldsymbol{U}_\mathrm{m})\,. \tag{255}$$

The inverse transformation is:

$$\mathrm{d}\boldsymbol{N} = \mathrm{d}\bar{\boldsymbol{n}}\,\bar{\mathbf{U}}^{-1}\,; \quad (\bar{\mathbf{U}}^{-1})^\dagger = (\varphi_1^{-1} \boldsymbol{U}_1 \,|\, \varphi_2^{-1} \boldsymbol{U}_2 \,|\ldots|\, \varphi_\mathrm{m}^{-1} \boldsymbol{U}_\mathrm{m})\,. \tag{256}$$

It can be easily verified that

$$\textstyle\sum_\alpha^\mathrm{M} \mathrm{d}\bar{n}_\alpha = \mathrm{d}N\,, \tag{257}$$

so that the associated CT-projected FF indices,

$$\boldsymbol{w} = \frac{\partial \bar{\boldsymbol{n}}}{\partial N} = (\varphi_1 F_1,\, \varphi_2 F_2, \ldots\, \varphi_\mathrm{m} F_\mathrm{m})\,, \tag{258}$$

exhibit the desired "normalization":

$$\textstyle\sum_\alpha^\mathrm{M} w_\alpha = \frac{\partial N}{\partial N} = 1\,. \tag{259}$$

Moreover, since

$$\bar{\boldsymbol{\varphi}} = \frac{\partial N}{\partial \bar{\boldsymbol{n}}} = \boldsymbol{1}\,, \tag{260}$$

one obtains the desired AIM-like equalization of the respective mode chemical potentials,

$$\bar{\boldsymbol{\xi}}^{\mathrm{eq}} = \frac{\partial \bar{E}^{\mathrm{PNM}}(\bar{\boldsymbol{n}})}{\partial \bar{\boldsymbol{n}}} = \frac{\partial N}{\partial \bar{\boldsymbol{n}}} \frac{\partial E^{\mathrm{g}}}{\partial N} = \mu\, \boldsymbol{1}\,, \tag{261}$$

and the associated mode hardnesses,

$$\bar{\boldsymbol{H}}^{\mathrm{eq}} = \frac{\partial \bar{\boldsymbol{\xi}}}{\partial N} = \frac{\partial \mu}{\partial \bar{\boldsymbol{n}}} = \frac{\partial N}{\partial \bar{\boldsymbol{n}}} \frac{\partial \mu}{\partial N} = \eta\, \boldsymbol{1}\,. \tag{262}$$

We again emphasize that for a non-equilibrium charge distribution, e.g., that obtained from independent SCF calculations, $\bar{\boldsymbol{\xi}} = \boldsymbol{\mu}\,\bar{\mathbf{U}}$, may not be equalized. It should be observed that the AIM components of w_α indices,

$$w_\alpha = F_\alpha \sum_i^M \frac{\partial N_i}{\partial n_\alpha} = \sum_i^M F_\alpha U_{i,\alpha} \equiv \sum_i^M w_{i,\alpha}, \tag{263}$$

also provide a resolution of the AIM FF indices into the PNM contributions:

$$f_i = \frac{\partial N_i}{\partial N} = \frac{\partial \boldsymbol{n}}{\partial N}\frac{\partial N_i}{\partial \boldsymbol{n}} = \sum_\alpha^M F_\alpha U_{i,\alpha} = \sum_\alpha^M w_{i,\alpha}. \tag{264}$$

In terms of the CT-projected population variables of Eq. 254 the hardness tensor is given by the diagonal matrix

$$\bar{\mathbf{h}} \equiv \frac{\partial^2 \bar{E}^{PNM}(\bar{\boldsymbol{n}})}{\partial \bar{\boldsymbol{n}}\, \partial \bar{\boldsymbol{n}}} = \frac{\partial \bar{\xi}}{\partial \bar{\boldsymbol{n}}} = \left\{ \frac{1}{\varphi_\alpha^2} h_\alpha \delta_{\alpha\beta} \equiv \bar{h}_\alpha \delta_{\alpha,\beta} \right\}; \tag{265}$$

the corresponding expression for the PNM softness matrix is :

$$\bar{\mathcal{S}} = \bar{\mathbf{h}}^{-1} = \frac{\partial \bar{\boldsymbol{n}}}{\partial \bar{\xi}} = \{\varphi_\alpha^2 \mathcal{S}_\alpha \delta_{\alpha\beta} \equiv \bar{\mathcal{S}}_\alpha \delta_{\alpha\beta}\}, \tag{266}$$

where $\mathcal{S}_\alpha = h_\alpha^{-1}$. Therefore, the purely polarizational ($\varphi_\alpha = 0$) modes (internal redistribution channels) have infinite modified mode hardness (zero modified softness), $\bar{h}_\alpha = \infty$ ($\bar{\mathcal{S}}_\alpha = 0$).

Consider now a unit displacement in the global number of electrons $dN = 1$ (see Fig. 10b). The equilibrium AIM electron population displacements are then given by $\boldsymbol{f} = \boldsymbol{e}_M \boldsymbol{f}^\dagger$; together with its CT component, $\boldsymbol{f}_{CT} = (1/m)\, \boldsymbol{e}_M^\dagger \equiv \bar{\boldsymbol{e}}^N \boldsymbol{e}_M^\dagger$, it defines the CT induced polarization vector: $\boldsymbol{f}_P = \boldsymbol{f} - \boldsymbol{f}_{CT}$. The $\boldsymbol{f}_{CT}$ represents the scaled CT vector $\boldsymbol{e}^N = (\partial N/\partial \boldsymbol{N})\, \boldsymbol{e}_M^\dagger = \boldsymbol{\varphi}^{AIM} \boldsymbol{e}_M^\dagger = \boldsymbol{1}\, \boldsymbol{e}_M^\dagger$, appropriately "renormalized" to represent a single electron displacement. Obviously, the external CT to/from the system, involving an electron reservoir, will also induce the associated internal polarization (P) in a molecule, so that the two vectors $\boldsymbol{f}$ and $\boldsymbol{f}_{CT}$ do not coincide, due to a non-vanishing P-component. It is of interest in the theory of chemical reactivity to separate these two components, and provide their AIM, PNM and (AIM + PNM) resolutions, respectively (see Fig. 10).

When considering the PNM representation, one should remember that the corresponding FF quantities include the overall (P + CT)-displacement, while the renormalized $\boldsymbol{\varphi}$-derivatives represent the CT part only. Thus, in the PNM framework $\boldsymbol{F}$-components represent the resultant (P + CT) vector $\boldsymbol{f}$, the corresponding $\boldsymbol{f}_{CT}$-components are given by vector $\boldsymbol{\varepsilon}^N \equiv (1/m)(\partial N/\partial \boldsymbol{n}) = (1/m)\, \boldsymbol{\varphi}$, and the $\boldsymbol{f}_P$-components are defined by the vector $(\boldsymbol{F} - \boldsymbol{\varepsilon}^N) = \{F_\alpha - \varphi_\alpha/m\}$. Of great interest also is the (AIM + PNM) - resolution. As we have observed above, $f_i = \Sigma_\alpha^M w_{i,\alpha}$; also,

$$\frac{1}{m}\varphi_i^{AIM} = \frac{1}{m}\frac{\partial N}{\partial N_i} = \frac{1}{m}\sum_{\alpha}^{M}\frac{\partial N}{\partial n_\alpha}\frac{\partial n_\alpha}{\partial N_i} = \frac{1}{m}\sum_{\alpha}^{M}\varphi_\alpha U_{i,\alpha} = \frac{1}{m}\sum_{\alpha}^{M}\overline{U}_{i,\alpha}\,. \quad (267)$$

Therefore, the quantities $\overline{U}_{i,\alpha}/m$ and $(w_{i,\alpha} - \overline{U}_{i,\alpha}/m)$ provide the contributions to the CT and P components, respectively, from the mode α at position of the i -th AIM , of the unit equilibrium charge displacement $dN = 1$ in M.

The transformation $\overline{\mathbf{U}} = \partial\overline{\boldsymbol{n}}/\partial\boldsymbol{N}$ of Eq. 255 measures the CT-projected shifts in the mode populations, per unit test displacements in the AIM populations $\boldsymbol{N}$. Thus the elements of these columns in $\overline{\mathbf{U}}$, which represent purely P-modes, must vanish identically ($\varphi_\alpha = 0$). The associated rows of the inverse transformation $\overline{\mathbf{U}}^{-1} = \partial\boldsymbol{N}/\partial\overline{\boldsymbol{n}}$ of Eq. 256, providing similar measures of responses in the AIM populations per unit test displacements in the mode populations $\overline{\boldsymbol{n}}$, must therefore be infinite. Indeed, since they may be considered as corresponding to $\overline{n}_\alpha \to 0$, a finite $\Delta\overline{n}_\alpha = 1$ has to correspond to infinite AIM displacements.

In order to illustrate the above normal mode concepts and quantities we have generated the relevant charge sensitivities for toluene, using the MNDO [94] AIM charges and the corresponding semiempirically interpolated hardness tensor of Eqs. 23, 25, 67 and 68. The results are displayed in Table 4 and in Figs. 11 and 12, with modes ordered in accordance

TABLE 4. Basic second-order quantities (a.u.) characterizing the PNM's of toluene.[3] The global quantities are: η = 0.190, S = 5.255, and μ = -0.254.[4]

PNM	h_α	F_α	φ_α	$\overline{h}_\alpha$	w_α
1	0.026	0.057	0.008	436.4	0.004
2	0.037	0.089	0.017	122.8	0.002
3	0.042	0	0	∞	0
4	0.054	0.082	0.023	99.6	0.002
5	0.090	0	0	∞	0
6	0.107	0.010	0.006	3496	0.000
7	0.182	0.101	0.096	19.62	0.010
8	0.249	0	0	∞	0
9	0.269	0.002	0.003	25980	0.000
10	0.315	0.024	0.040	197.5	0.001
11	0.374	0	0	∞	0
12	0.466	0.040	0.099	47.86	0.004
13	0.695	0	0	∞	0
14	0.954	0.017	0.083	140.1	0.001
15	2.908	0.253	3.869	0.1942	0.980

[3] For the mode contours and numbering see Fig. 11.

[4] The MNDO charges for the optimized geometry have been used.

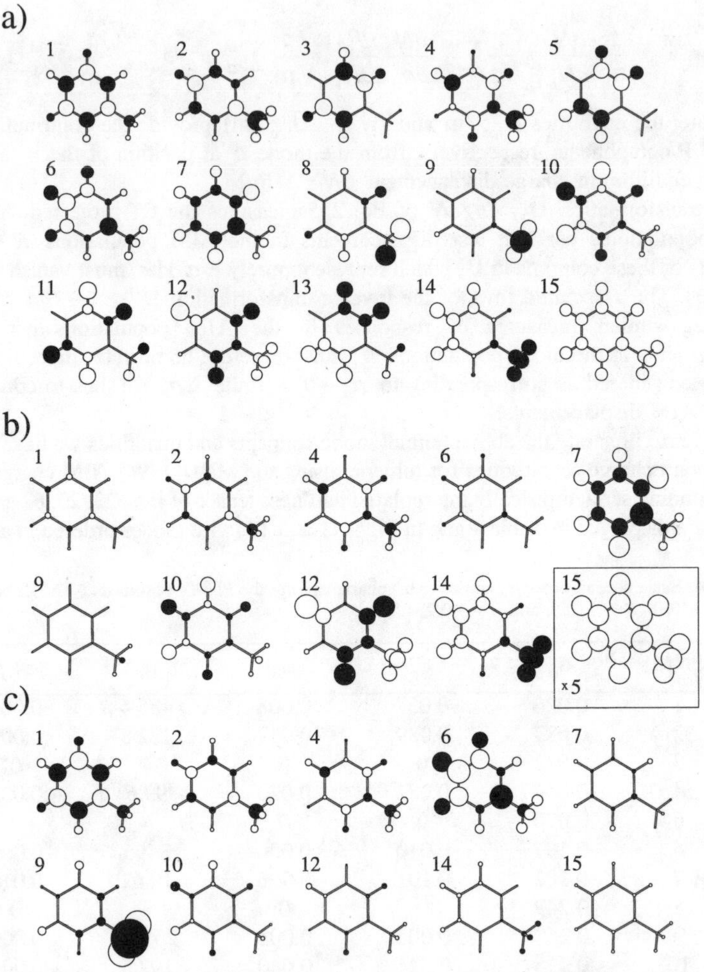

Figure 11. Schematic diagrams of the PNM in toluene, with the shaded areas denoting negative values. In panel (a) the area within a circle on atom i in the mode α is proportional to $|U_{i,\alpha}|$, with all diagrams being plotted using the same scaling factor, and the CT-active mode phases being determined by the convention $\varphi_\alpha > 0$ (Eq. 247). The panels (b) and (c), both limited only to the ten CT-active ($\varphi_\alpha \neq 0$) modes, represent the corresponding diagrams illustrating the distribution of the $\bar{U}_{i\alpha} = \varphi_\alpha U_{i\alpha}$ and $\bar{U}^{-1}_{i\alpha} = U_{i\alpha} / \varphi_\alpha$ coefficients, which define $d\bar{\boldsymbol{n}} = d\boldsymbol{N}\,\bar{\mathbf{U}}$ and $d\boldsymbol{N} = d\bar{\boldsymbol{n}}\,\bar{\mathbf{U}}^{-1}$, populational shifts, respectively; in panel (b) the last mode corresponding to the largest principal hardness, h_α, has different scaling factor from the remaining modes, for demonstrative purposes; the actual AIM displacements in this mode are obtained by multiplying the panel displacements by factor 5, as indicated in the figure. The mode ordering is in accordance with increasing h_α values. Selected (second-order) charge sensitivities of these PNM are listed in Table 4. The same diagram convention is used throughout the book.

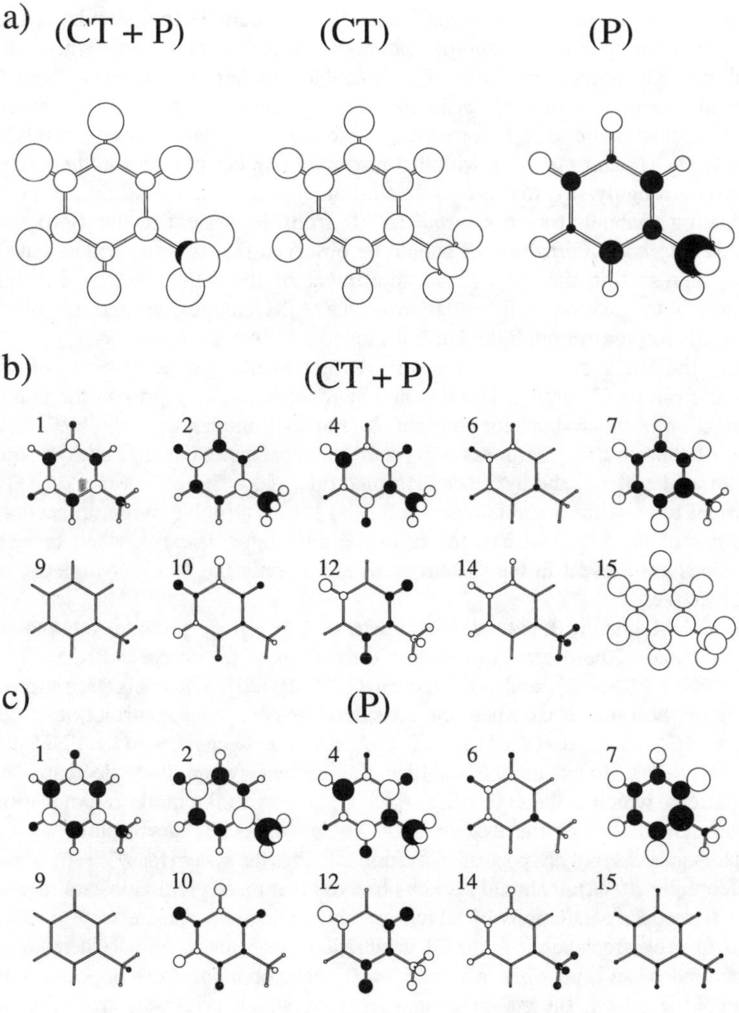

Figure 12. The AIM- (part a) and (AIM + PNM)- (parts b and c) resolutions of the (P + CT), CT, and P electron population displacements for $dN = 1$, represented by the vectors, $\vec{f} = \boldsymbol{f}\,\boldsymbol{e}_{M}^{\dagger} = \boldsymbol{F}\,\boldsymbol{U}_{M}^{\dagger}$, $\overline{e}^{N} = \overline{e}^{N}\,\boldsymbol{e}_{M}^{\dagger}$ and $(\vec{f} - \overline{e}^{N})$, respectively. The circle areas in three diagrams of part a are proportional to $|f_i|$, $1/m$, and $|f_i - 1/m|$, respectively; in parts b and c, where only the CT-active ($\varphi_\alpha > 0$) modes are shown, the two groups of diagrams represent $w_{i,\alpha}$ and $(w_{i,\alpha} - \overline{U}_{i,\alpha}/m)$ quantities, respectively.

with the increasing principal hardness, h_α. The highest value of h_α (lowest value of $\bar{h}_\alpha$), corresponds to the non-selective mode $\alpha = 15$, in which electron populations of all AIM are shifted "in phase" by approximately the same amount, i.e., when all of them simultaneously accept (or loose) a comparable number of electrons; thus, this mode practically exhibits a pure CT-character ($w_\alpha \approx 1$), and zero nodes in the mode contour. The h_α-softest mode, $\alpha = 1$, corresponding to the lowest eigenvalue, exhibits practically a purely P-character ($w_\alpha \approx 0$), with the maximum number of nodes in the mode diagram. One could roughly say, that the h_α-hardest and -softest modes exhibit, respectively, the dominating channels for an *external CT* (to/from the system in question) and the *CT-induced internal polarization*. It should be observed that it is the softest (site-selective) mode, representing the most facile adjustment of the AIM electron distribution, that determines the internal differentiation of the AIM charges, as a result of the charge donated (accepted) through the hardest mode to (from) the reservoir.

Only the ten symmetric modes with respect to the symmetry plane of the toluene molecule can be CT-active. The five anti-symmetric modes represent the purely internal polarization of the system, for constant N. For such modes $\varphi_\alpha = w_\alpha = F_\alpha = \xi_\alpha = 0$. In terms of the modified hardnesses $\bar{h}_\alpha$ and chemical potential $\bar{\xi}_\alpha$, such P-modes have infinite values of $\bar{h}_\alpha$, and by the de l'Hospital rule, $\bar{\xi}_\alpha = \xi_\alpha/\varphi_\alpha = (\partial\xi_\alpha/\partial n_\alpha)/(\partial\varphi_\alpha/\partial n_\alpha)$ are equal to $(+\infty)$ for a stable mode and $(-\infty)$ for an unstable mode, in accordance with the signs of the principal hardnesses in the numerator. Clearly, when more symmetry operations are present in the system point group, only the totally symmetric modes can be CT-active.

The PNM chemical potential parameters, $\boldsymbol{\xi}^{eq} = \mu\,\boldsymbol{\varphi}$, provide yet another natural classification of the normal modes into three groups: (a) *acceptor* PNM ($\xi_\alpha < 0$), (b) *donor* PNM ($\xi_\alpha > 0$), and (c) P-modes ($\xi_\alpha = 0$) [20]. The energetic rationale for this distinction becomes clear when one considers the first-order contributions to the system energy, $dE_M^{(1)} \equiv \boldsymbol{\xi}\,[d\boldsymbol{n}(dN)]^\dagger = \Sigma_\alpha^M \xi_\alpha F_\alpha dN$, due to changes in the PNM populations, $d\boldsymbol{n}(dN)$. The donor modes stabilize the system, when they decrease their mode populations, which is the case when $F_\alpha > 0$, i.e., when the mode α population changes in phase with that of the system as a whole ($dN < 0$, electrophilic attack). In the acceptor mode case only its population increase stabilizes the system for $F_\alpha > 0$, when $dN > 0$ (nucleophilic attack). It should also be observed that these predictions are reversed for the $F_\alpha < 0$ modes, populations of which are shifted in the opposite way to dN. In other words, in an electrophilic attack the CT stabilization energy results from the dominating $F_\alpha > 0$ donor modes and, possibly, a few $F_\alpha < 0$ acceptor modes; the opposite is true in the nucleophilic attack, the stabilization energy of which originates from the dominating $F_\alpha > 0$ acceptor modes and the $F_\alpha < 0$ donor modes; clearly, all remaining CT-active modes increase the system energy.

The dominant role of the hardest mode $\alpha = 15$ (Fig. 11) immediately follows from its w_α value (Table 4) and the $\bar{U}_{15}$ vector coefficients (see Fig. 11b). The same conclusion, therefore, must follow from the relevant resolution of the pure CT-vector, given by the uniformly scaled $\bar{\mathbf{U}}$ transformation, by the factor $1/m$. This feature of the $\alpha = 15$ mode is also seen in the (AIM + PNM) resolution of the FF vector, shown in Fig. 12b, where all but last CT-active modes describe mainly the CT-induced polarization (explicitly extracted in Fig. 12c).

The h_α - hierarchy of PNM, used as the ordering criterion, identifies the softest mode as the one with the largest number of nodes in the mode diagram. This is opposite to the familiar relation between the nodal structure and energy of the canonical molecular orbitals (MO), for which the lowest orbital energy corresponds to the most delocalized occupied MO with the least number of nodes, while the highest orbital energy identifies the antibonding MO with the largest number of nodes. A reference to the $\bar{h}_\alpha$ column of Table 4 shows, that this modified mode hardness brings the PNM description closer to the above MO analogy. Namely, the CT-dominant mode is now identified by the *lowest* value of $\bar{h}_\alpha$. Moreover, the mode $\alpha = 7$, which in toluene dominates the CT-induced polarization (second largest w_α, compare also the mode $\alpha = 7$ contours of Figs. 11a and 12c, with the AIM resolved polarization part of $\boldsymbol{f}$ in Fig. 12a), appears as the second $\bar{h}_\alpha$ - softest mode. Thus, the $\bar{h}_\alpha$ values do indeed reflect the PNM participation in the CT-related charge rearrangement, with the $\bar{h}_\alpha$ - soft modes participating the most. A reference to Fig. 11a also shows that with increasing h_α larger groups of neighbouring atoms change their electron population in phase, thus giving rise to a long range polarization; similarly, the h_α - soft modes are seen to be responsible for the short-range polarization between neighbouring atoms.

It follows from Figs. 12b, c that the CT-induced polarization involves mostly modes $\alpha = 1, 2, 4, 7, 10$, and 12, each exhibiting a small CT component measured by the $w_\alpha \neq 0$ indices. We would like to remark at this point that the non-selective mode 15 is of little interest in the theory of chemical reactivity, since it practically uniformly affects all AIM. The CT-active polarizational modes differentiate net changes on AIM as a result of external CT, and as such they influence the reactivity/selectivity trends of open molecular systems. In toluene this CT-induced polarization component (last diagram in Fig. 12a) is mainly a polarization between the carbons and surrounding hydrogens, with $dN > 0$ (an inflow of electrons, e.g., due to an attack by a nucleophile) partly reversing the C $\leftarrow$ H polarization in an isolated toluene; the reverse charge rearrangement enhancing this initial polarization is predicted for $dN < 0$ (an outflow of electrons, e.g., due to an attack by an electrophile). As also seen in Fig. 12a, the methyl carbon exhibits a negative FF index, thus changing its electron population in a way opposite to the system as a whole.

As we have mentioned earlier in this section, the $\bar{\mathbf{U}}^{-1}$ transformation, shown schematically in Fig. 11c, reflects relative local responses in the AIM populations $\boldsymbol{N}$ per unit displacements in the CT-projected PNM populations, $\bar{\boldsymbol{n}}$. It follows from the figure that the most efficient in shifting the AIM charges are modes $\alpha = 1$, 6 and 9. The last mode causes mainly the charge shift between methyl hydrogens while the remaining modes additionally involve a charge differentiation between the ring atoms.

6.2. Minimum Energy Coordinates

Following the compliance formalism of nuclear displacements [89, 90] one can define the corresponding concepts of the populational *minimum energy coordinates* (MEC) and the associated compliance constants in the space of the AIM electron populations [32, 57, 87, 91]. They can be defined for both open ($dN \neq 0$) and closed ($dN = 0$) systems;

however, it has been shown numerically [32] that the MEC characteristics remain practically the same in both the *external* and *internal* cases, respectively. Therefore, in an illustrative application to toluene here we report only the open system MEC and their hardness parameters. The MEC concept can also be given the relaxational interpretation as the minimum energy path of the electron redistribution in the molecular environment of the primarily displaced atom. We shall call such collective charge displacements in the molecular fragment resolution the *relaxational coordinates* (REC). Since all these concepts represent the system responses to localized charge displacements (test oxidations or reductions) they obviously are of great interest in the theory of chemical reactivity [32, 57, 87, 91] where such very questions are often asked.

Let us first define the external MEC in M consisting of m atoms. Consider the global equilibrium of M in contact with a hypothetical electron reservoir (r): $\boldsymbol{\mu}_0 = \mu \boldsymbol{1}$ and $\mu = \mu_r$. Let $\boldsymbol{z} = \boldsymbol{N} - \boldsymbol{N}^0 \equiv \mathrm{d}\boldsymbol{N}$ denotes the vector of a hypothetical AIM electron population displacement from the equilibrium distribution $\boldsymbol{N}^0$. Since $\mathrm{d}N = -\mathrm{d}N_r$, the assumed equilibrium removes the first-order contribution to the associated change in the energy $E = E_M + E_r$ of the combined (closed) system (M¦r), due to $\boldsymbol{z}$; moreover, taking into account the infinitely soft character of a macroscopic reservoir, the only contribution to the energy change in the quadratic approximation is:

$$E(\boldsymbol{z}) - E^0 \approx \frac{1}{2} \boldsymbol{z} \boldsymbol{\eta} \boldsymbol{z}^{\dagger} = \mathrm{d}^2 E(\boldsymbol{z}) , \qquad (268)$$

where $E^0 = E(\boldsymbol{0})$ and $\boldsymbol{\eta} = (\partial^2 E / \partial \boldsymbol{z} \partial \boldsymbol{z})_0$ is the AIM hardness matrix of M. The equilibrium restoring force is defined by the AIM electronegativities, relative to that of the reservoir:

$$\bar{\boldsymbol{\chi}} \equiv -\frac{\partial E}{\partial \boldsymbol{z}} = -\frac{\partial E_M}{\partial \boldsymbol{z}} + \frac{\partial N}{\partial \boldsymbol{z}} \frac{\partial E_r}{\partial N} = \boldsymbol{\chi} - \chi_r \boldsymbol{1}$$

$$= (\boldsymbol{\chi}^0 - \boldsymbol{z}\boldsymbol{\eta}) - \chi_r \boldsymbol{1} = -\boldsymbol{z}\boldsymbol{\eta} , \qquad (269)$$

where $\chi_r = -\mu_r = \chi^0 = \chi_i^0$. The last equation allows one to express population variables $\boldsymbol{z}$ in terms of the restoring force, $\boldsymbol{z} = -\bar{\boldsymbol{\chi}} \boldsymbol{\eta}^{-1} \equiv \bar{\boldsymbol{\mu}} \boldsymbol{\sigma}$, which identifies the softness matrix $\boldsymbol{\sigma}$ as the populational compliance matrix, $\boldsymbol{\sigma} = \partial \boldsymbol{z} / \partial \bar{\boldsymbol{\chi}}$. Expressing the energy change in terms of force $\bar{\boldsymbol{\chi}}$ gives

$$E(\bar{\boldsymbol{\chi}}) - E^0 = \frac{1}{2} \bar{\boldsymbol{\chi}} \boldsymbol{\sigma} \bar{\boldsymbol{\chi}}^{\dagger} . \qquad (270)$$

One can use this equation to define the k-th MEC,

$$\vec{\mathcal{N}}_k \equiv \sum_i^M (i)^k \vec{e}_k \equiv \vec{\boldsymbol{e}}_M \, (\mathcal{N}^{(k)})^{\dagger} , \qquad (271)$$

where the k-th vector of population *interaction constants* $\mathcal{N}^{(k)} = \{(i)^k\}$, is defined by the minimum energy criterion:

$$(i)^k \equiv (\partial z_i/\partial z_k)_{E=\min} . \tag{272}$$

The explicit expression for these quantities immediately follows from Eq. 270, since the definition of Eq. 272 implies $\bar{\chi}_{i\neq k} = 0$. Thus,

$$(i)^k = (\partial z_i/\partial \bar{\chi}_k)/(\partial z_k/\partial \bar{\chi}_k) = \sigma_{i,k}/\sigma_{k,k} . \tag{273}$$

The associated hardness (compliance constant) is defined by a similar minimum energy criterion:

$$\Gamma_k \equiv (\partial^2 E/\partial z_k^2)_{E=\min} = (\partial z_k/\partial \bar{\chi}_k)^{-1} = \sigma_{k,k}^{-1} . \tag{274}$$

It represents the curvature of the section of the energy function (Eq. 270) along $\vec{\mathcal{N}}_k$. This construction of the MEC is illustrated in Fig. 13, where one makes a unit populational shift along one AIM direction and searches for the minimum in the other AIM direction, identified by the point, where the direction line is tangent to the corresponding ellipse contour of the force field. The figure shows that the MEC are not orthogonal, both being dominated by the softest PNM and the internal polarization components. The sum of the AIM displacements,

$$dN^{(k)} \equiv \sum_i^M (i)^k = (\partial N/\partial N_k)_{E=\min} = \sigma_{k,k}^{-1} \sum_i^M \sigma_{i,k} , \tag{275}$$

measures the global shift dN per unit displacement in N_k, when all remaining AIM are opened both internally (relative to the other AIM) and externally (relative to the reservoir).

Consider now the internal MEC. In this case we examine the closed M (N = const.), with all AIM being free to exchange electrons among themselves. Let us again assume the initial global equilibrium $\boldsymbol{\mu}^0 = \mu^0 \mathbf{1}$, and envisage the hypothetical polarizational displacement from $\boldsymbol{N}^0$, $\bar{\boldsymbol{z}} = (d\boldsymbol{N})_N$. The relevant compliants are now given by the internal softness matrix $\mathbf{S} = (\partial \boldsymbol{N}/\partial \boldsymbol{\mu})_N = -\boldsymbol{\beta}$, which defines the corresponding internal MEC and their characteristics:

$$\vec{n}_k \equiv \sum_i^M (\bar{i})^k \vec{e}_k \equiv \vec{\boldsymbol{e}}_M (\boldsymbol{n}^{(k)})^\dagger , \tag{276}$$

$$(\bar{i})^k \equiv (\partial z_i/\partial z_k)_{N, E_M=\min} = (\partial \bar{z}_i/\partial \bar{z}_k)_{E_M=\min} = S_{i,k}/S_{k,k} = \beta_{i,k}/\beta_{k,k} , \tag{277}$$

$$\gamma_k \equiv (\partial^2 E/\partial z_k^2)_{N, E_M=\min} = (\partial^2 E/\partial \bar{z}_k^2)_{E_M=\min} = S_{k,k}^{-1} = -\beta_{k,k}^{-1} . \tag{278}$$

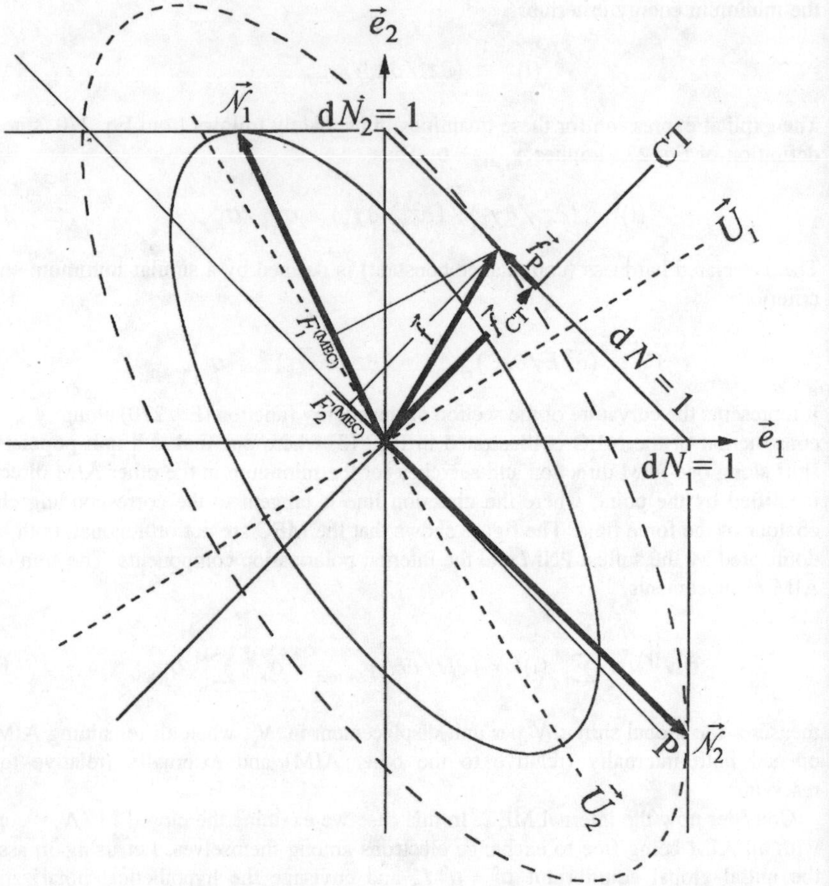

Figure 13. The external MEC {$\vec{\mathcal{N}}_k$, k = 1, 2} in the system exhibiting two AIM populational degrees of freedom. Both MEC are seen to be dominated by the components from the soft PNM, $\vec{U}_2$, or the corresponding projection onto the internal polarization axis, P, thus representing only minor changes in the overall number of electrons (projections onto the CT direction).

We would like to emphasize, that due to the closure constraint there are only $(m-1)$ linearly independent internal MEC. Thus, the m vectors defined by Eq. 276 in reality span the $(m-1)$-dimensional space of the internal MEC. In order to remove this linear dependence one could adopt the relative internal approach of Section 3 in this Chapter. Namely, one would then select the electron population of one atom in the system as dependent upon populations of all remaining atoms, and discard the MEC associated with that atom. All remaining MEC can be also directly constructed from the corresponding internal relative softness matrix. Although the sets of independent internal MEC for alternative choices of dependent atom will differ from one another, they must span the same $(m-1)$-dimensional linear space of independent internal MEC. For example, in the two-AIM system of Fig. 13 there is only one independent internal MEC direction along the P-line.

It has been shown numerically that the above two sets of the MEC, external and internal, are almost indistinguishable exhibiting practically identical compliance constants of Eqs. 274 and 278. Therefore, in the illustrative MEC plots for toluene (Fig. 14a) we have displayed only the external MEC contours and listed their corresponding hardness parameters.

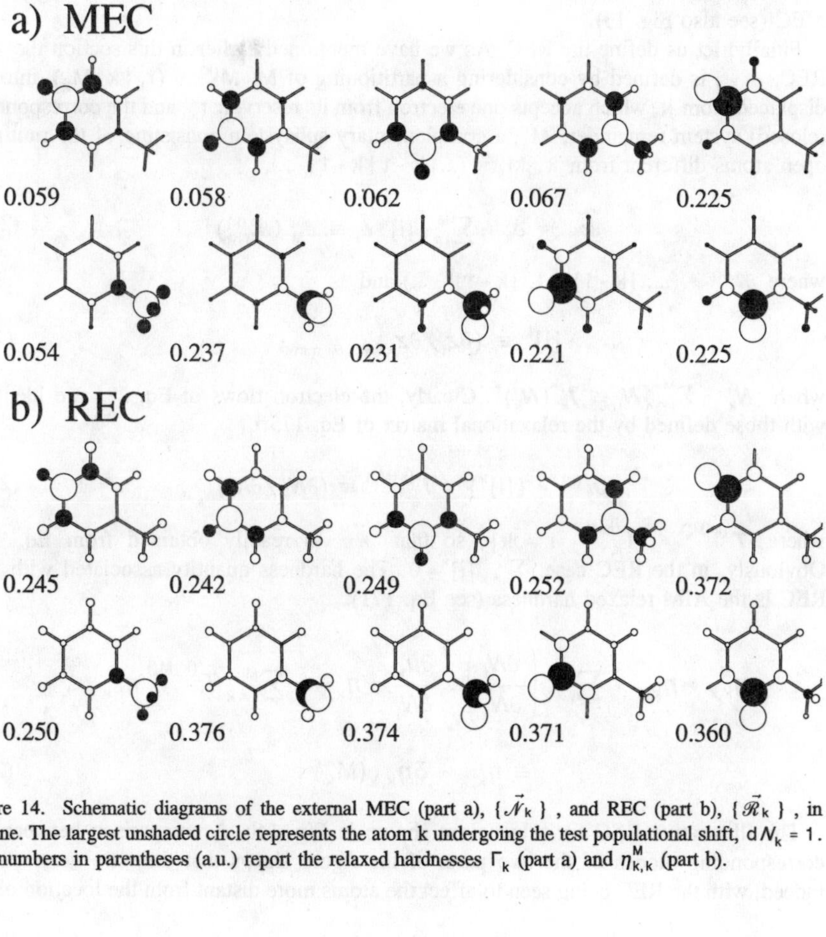

Figure 14. Schematic diagrams of the external MEC (part a), { $\vec{\mathcal{N}}_k$ } , and REC (part b), { $\vec{\mathcal{R}}_k$ } , in toluene. The largest unshaded circle represents the atom k undergoing the test populational shift, $dN_k = 1$. The numbers in parentheses (a.u.) report the relaxed hardnesses Γ_k (part a) and $\eta^{M}_{k,k}$ (part b).

Clearly, due to the closure constraint,

$$d\bar{N}^{(k)} \equiv \sum_i^M (\bar{i})^k = (\partial N/\partial N_k)_{N,\,E=\min} = 0\ , \tag{279}$$

so that $\Sigma_{i\neq k}^M (\bar{i})^k = -(\bar{k})^k = -1$.

An inspection of diagrams shown in Fig. 14a reveals that the primary (unit) positive displacement $(k)^k = 1$, identified by the largest open circle in each contour, is mainly at the expense of the negative populational shifts on neighbouring atoms, which effectively screen the primary chemical reduction of atom k, thus generating oscillating responses on remaining atoms, with decaying magnitude for increasing distance from the perturbed atom. This explains why the $\vec{\mathcal{N}}_k$ and $\vec{n}_k$ MEC are almost identical. It is also worth noting that all contours exhibit the maximum sign alternation around the primary displaced atom. This indicates, that the softest PNM will always contribute strongly to all MEC (see also Fig. 13).

Finally, let us define the REC. As we have mentioned earlier in this section the k-th REC, $\vec{\mathcal{R}}_k$, is defined by considering a partitioning of M, $M^{(k)} = (r_k \mid k \mid M'_k)$, into the displaced atom k, which accepts one electron from its reservoir r_k, and the corresponding (closed) system remainder, M'_k, a complementary subsystem consisting of the mutually open atoms different from k, $M'_k = (\ldots \mid k-1 \mid k+1 \mid \ldots)$,

$$\vec{\mathcal{R}}_k = \vec{e}_k + \sum_{i\neq k}^M [i]^k\, \vec{e}_i \equiv \vec{\boldsymbol{e}}_M\, (\boldsymbol{\mathcal{R}}^{(k)})^\dagger\ , \tag{280}$$

where $\boldsymbol{\mathcal{R}}^{(k)} = (\ldots,\ [k-1]^k,\ 1,\ [k+1]^k,\ \ldots)$ and

$$[i]^k \equiv (\partial z_i/\partial z_k)_{N'_k,\,E=\min}\ , \tag{281}$$

where $N'_k = \Sigma_{i\neq k}^M N_i \equiv \boldsymbol{1}'_k\,(\boldsymbol{N}'_k)^\dagger$. Clearly, the electron flows of Eq. 281 are identical with those defined by the relaxational matrix of Eq. 175:

$$\boldsymbol{\mathcal{R}}^{(k)} = \{[i]^k\} = \boldsymbol{T}^{(k|M'_k)} \equiv (\partial \boldsymbol{N}'_k/\partial N_k)_{N'_k}\ , \tag{282}$$

where $\boldsymbol{T}^{(k|M'_k)} = \{T_{k,i}^{(k|M'_k)},\ i\neq k\}$, so that $\vec{\mathcal{R}}_k$ is readily obtained from Eq. 178. Obviously, in the REC case: $\Sigma_{i\neq k}^M [i]^k = 0$. The hardness quantity associated with k-th REC is the AIM relaxed hardness (see Eq. 171):

$$\eta_{k,k}^M = \eta_{k,k} + \sum_{i\neq k}^M \left(\frac{\partial N_i}{\partial N_k}\right)_{N'_k} \frac{\partial \mu_k}{\partial N_i} = \eta_{k,k} + \sum_{i\neq k}^M T_i^{(k|M'_k)}\,\eta_{i,k}$$

$$\equiv \eta_{k,k} + \delta\eta_{k,k}(M'_k)\ . \tag{283}$$

The illustrative REC for toluene are shown in Fig. 14b. A comparison between the corresponding modes of the two parts of the figure shows that they are very similar indeed, with the REC being seen to affect the atoms more distant from the location of the primary displacement; in its vicinity, where the amplitudes are the largest, both sets of contours are practically indistinguishable.

An examination of the MEC hardnesses of Fig. 14a shows that the test oxidations/reductions on carbon atoms give rise to comparable MEC hardnesses, the softest MEC in the system; the *para* (*p*)- and *meta* (*m*)- carbons in the ring are seen to be slightly softer in this respect, than the *ortho* (*o*)- carbons. The softest population displacement is on the methyl carbon. The hard category of MEC involves test oxidations/reductions on hydrogen atoms, with a slightly harder changes being predicted for the methyl hydrogens. A similar picture emerges, when one considers the relaxed (diagonal) atomic hardnesses reported below the REC diagrams of Fig. 14b. Namely, the hardness differentiation between the *(o-m-p)*-carbons remains the same, but the methyl carbon appears almost as hard as the carbon chemically bonded to the methyl group. The softest hydrogen is now at the *o*-position, in contrast to the MEC hardness, which is predicted to be the lowest at the *m*-location. Again, the methyl hydrogens are seen to be the hardest atoms in terms of the relaxed AIM hardness parameters.

Let us denote the MEC/REC directions by the representative vector $\vec{r}_M = (\vec{\mathcal{N}}_M, \vec{n}_M, \vec{\mathcal{R}}_M)$, where $\vec{\mathcal{N}}_M = (\vec{\mathcal{N}}_1, \ldots, \vec{\mathcal{N}}_m)$, etc., and define the corresponding transformation matrix $\mathbf{W} = (\mathcal{N}, \boldsymbol{n}, \mathcal{R})$ between the basis set of the orthonormal AIM vectors $\vec{e}_M$ $(\vec{e}_M^{\dagger}\vec{e}_M = \mathbf{I})$ and the corresponding MEC/REC directions:

$$\vec{r}_M = \vec{e}_M\, \mathbf{W}. \tag{284}$$

The three transformation matrices $\mathbf{W}$ combine the MEC/REC components:

$$\mathcal{N} = (\mathcal{N}^{(1)} \,|\, \mathcal{N}^{(2)} \,|\, \ldots \,|\, \mathcal{N}^{(m)}) = \partial \vec{\mathcal{N}}_M / \partial \vec{e}_M\,, \tag{285}$$

$$\boldsymbol{n} = (\boldsymbol{n}^{(1)} \,|\, \boldsymbol{n}^{(2)} \,|\, \ldots \,|\, \boldsymbol{n}^{(m)}) = \partial \vec{\boldsymbol{n}}_M / \partial \vec{e}_M\,, \tag{286}$$

$$\mathcal{R} = (\mathcal{R}^{(1)} \,|\, \mathcal{R}^{(2)} \,|\, \ldots \,|\, \mathcal{R}^{(m)}) = \partial \vec{\mathcal{R}}_M / \partial \vec{e}_M\,. \tag{287}$$

It should be observed that, contrary to the ortho-normal character of the AIM basis vectors, the MEC/REC vectors $\vec{r}_M$ are not orthogonal, giving rise to the metric

$$\mathcal{M} = \vec{r}_M^{\dagger}\, \vec{r}_M = \mathbf{W}^{\dagger}\, \vec{e}_M^{\dagger}\, \vec{e}_M\, \mathbf{W} = \mathbf{W}^{\dagger}\, \mathbf{W}. \tag{288}$$

Clearly, the transformation $\mathbf{W}$ also defines the projections of the population displacement vector $\vec{z} = \sum_i^M dN_i\, \vec{e}_i$ onto $\vec{r}_M = (\vec{r}_1, \ldots, \vec{r}_m)$:

$$z_k^{(r)} \equiv \vec{z}\cdot\vec{r}_k = \sum_i^M z_i\, W_{i,k} = \sum_i^M dN_i\, W_{i,k}\,, \qquad \boldsymbol{z}^{(r)} = \boldsymbol{z}\,\mathbf{W} = d\boldsymbol{N}\,\mathbf{W}. \tag{289}$$

Thus, the corresponding expression for the energy change in the quadratic approximation becomes:

$$\Delta E = \mu\, z^\dagger + \frac{1}{2} z\, \eta\, z^\dagger$$

$$= [\mu\, (\mathbf{W}^{-1})^\dagger]\, (z^{(r)})^\dagger + \frac{1}{2} z^{(r)}\, [\mathbf{W}^{-1} \eta\, (\mathbf{W}^{-1})^\dagger]\, (z^{(r)})^\dagger$$

$$\equiv \mu^{(r)}\, (z^{(r)})^\dagger + \frac{1}{2} z^{(r)}\, \eta^{(r)}\, (z^{(r)})^\dagger \; . \tag{290}$$

This equation defines the chemical potential vector $[\mu^{(r)}]$ and the hardness matrix $[\eta^{(r)}]$ for the MEC/REC reference frame. Similarly,

$$F^{(r)} \equiv \frac{\partial z^{(r)}}{\partial N} = \frac{\partial z}{\partial N} \frac{\partial z^{(r)}}{\partial z} = f\, \mathbf{W} \; , \tag{291}$$

defines the relevant MEC/REC FF indices. Moreover, since $\vec{e}^{\,N} = \vec{e}_M\, 1^\dagger = \vec{r}_M\, (\mathbf{W}^{-1}\, 1^\dagger)$, the complementary derivatives are:

$$(\varphi^{(r)})^\dagger = \frac{\partial \vec{e}^{\,N}}{\partial \vec{r}_M} = \frac{\partial N}{\partial z^{(r)}} = \mathbf{W}^{-1}\, 1^\dagger \; . \tag{292}$$

These quantities are identically vanishing for the internal MEC, $\varphi^{(\mathscr{M})} = 0$, and they are exactly equal to 1 in the REC case, $\varphi^{(\mathscr{R})} = 1$, since then only the primary (unit) population displacement contributes to dN.

We further observe that the FF indices of Eq. 291 do not sum up to 1. However, following the PNM development one could again define the modified, CT-projected indices,

$$w_k^{(r)} = F_k^{(r)}\, \varphi_k^{(r)} \; , \qquad k = 1, \ldots, m \; , \tag{293}$$

which satisfy the desired "normalization":

$$F^{(r)}\, (\varphi^{(r)})^\dagger = f\, \mathbf{W}\, \mathbf{W}^{-1}\, 1^\dagger = f\, 1^\dagger = 1 \; . \tag{294}$$

The above observation calls for the corresponding CT-projected redefinition (see Eqs. 254, 255) of the external MEC population variables,

$$\bar{z}^{(\mathscr{N})} = z\, \bar{\mathscr{N}} \; , \quad \bar{\mathscr{N}} = (\mathscr{N}^{(1)}\, \varphi_1^{(\mathscr{N})}\, |\, \mathscr{N}^{(2)}\, \varphi_2^{(\mathscr{N})}\, |\, \ldots\, |\, \mathscr{N}^{(m)}\, \varphi_m^{(\mathscr{N})}) \; , \tag{295}$$

and the associated derivatives (see Eq. 290):

$$\bar{\mu}^{(\mathscr{N})} = \mu\, (\bar{\mathscr{N}}^{-1})^\dagger, \quad \bar{\eta}^{(\mathscr{N})} = \bar{\mathscr{N}}^{-1}\, \eta\, (\bar{\mathscr{N}}^{-1})^\dagger \; , \text{ etc.} \tag{296}$$

It can be easily verified using Eqs. 295 and 296 that $\bar{\boldsymbol{\mu}}^{(\mathcal{N})} = \mu\, \boldsymbol{1}$ for the global equilibrium state, when $\boldsymbol{\mu} = \mu\, \boldsymbol{1}$. This is because

$$\bar{\varphi}_k^{(\mathcal{N})} \equiv \partial N / \partial \bar{z}_k^{(\mathcal{N})} = (\partial z_k^{(\mathcal{N})} / \partial \bar{z}_k^{(\mathcal{N})})(\partial N / \partial z_k^{(\mathcal{N})})$$
$$= \varphi_k^{(\mathcal{N})} / \varphi_k^{(\mathcal{N})} = 1\,, \qquad k = 1, 2, ..., m\,. \tag{297}$$

In the internal MEC case all populational derivatives $\boldsymbol{\mu}^{(n)} = \boldsymbol{0}$ for the initial global equilibrium. Therefore, also in the REC description for a given partitioning $M^{(k)} = (k|M_k') = (k|i \neq k)$ $\boldsymbol{\mu}^{(\mathcal{R})} = \boldsymbol{0}$, if the initial electron distribution corresponds to the global equilibrium; then also $\mu_k^{(\mathcal{R})} = \partial E / \partial z_k^{(\mathcal{R})} = \varphi_k^{(\mathcal{R})} \mu = \mu$. Thus, the representative REC chemical potential $\mu_k^{(\mathcal{R})}$ for the $M^{(k)}$ partitioning, representing the relaxed chemical potential of atom k, is also at the initial global chemical potential level.

The REC coefficient $[i]^k$ of Eq. 281 can be also expressed as the ratio of the corresponding softness quantities (see Eqs. 273 and 277). Since the forces (internal electronegativities) associated with the relaxational degrees of freedom must exactly vanish by the minimum energy criterion, the following chain rule applies:

$$[i]^k = \left(\frac{\partial N_{i\neq k}}{\partial N_k}\right)_{N_k'} = \left(\frac{\partial N_{i\neq k}}{\partial \mu_k}\right)_{N_k'} \Bigg/ \left(\frac{\partial N_k}{\partial \mu_k}\right)_{N_k'}, \tag{298}$$

where we have removed the E = min condition from the list of constraints, to simplify notation. The derivative in the numerator of the last expression is the off-diagonal softness for the $M^{(k)} = (k|M_k')$ partitioning of M:

$$S_{k,i}^{(k)} = \left(\frac{\partial N_k}{\partial \mu_k}\right)_{N_k'} \left(\frac{\partial N_{i\neq k}}{\partial N_k}\right)_{N_k'} = \frac{f_i^{k,M_k'}}{\eta_{k,k}^{M}}, \tag{299}$$

defined by the ratio of the corresponding off-diagonal FF index and the relaxed hardness of the displaced atom k in M, the inverse of which also appears in the denominator of Eq. 298. Using the last two equations gives Eq. (281).

6.3. Collective Modes of Molecular Fragments

The concepts of PNM and MEC/REC of the preceding section may be extended into those representing molecular fragments corresponding to a given partitioning of the system in question into mutually closed, interacting subsystems. As we have already argued before, such an extension could be of great interest in the theory of chemical reactivity. For example, in the gas-phase reactions one is interested in the role played by the charge displacement modes of the A and B reactants in $M^+ = (A|B)$, before the B→A CT; of interest also are the changes in the PNM/MEC/REC of reactants in M^+, due to the presence of the other reactant, relative to the modes in the separated reactants limit.

In the M^+ case complementary fragments (reactants) are considered as mutually closed subsystems. One encounters a similar type of a partitioning in the physisorption catalytic system consisting of the catalyst ($\mathscr{C}$) and the adsorbate ($\mathscr{A}$): ($\mathscr{A}|\mathscr{C}$). However, in the chemisorption systems ($\mathscr{A}\,¦\,\mathscr{C}$), both fragments, adsorbate and substrate are considered as mutually opened. When two reactants $\mathscr{A}_1$ and $\mathscr{A}_2$ are chemisorbed on a surface, they are both externally open subsystems, in contact with their respective active sites $\mathscr{C}_1$ and $\mathscr{C}_2$. Both adsorbates may be considered as mutually closed, ($\mathscr{C}_1\,¦\,\mathscr{A}_1|\mathscr{A}_2\,¦\,\mathscr{C}_2$), to refer to their hypothetical state before the direct CT: ($\mathscr{C}_1\,¦\,\mathscr{A}_1$) → ($\mathscr{A}_2\,¦\,\mathscr{C}_2$). Clearly, instead of mentally separating the charges of the two active sites from the surface remainder, one may also consider them as parts of $\mathscr{C}$, one opened with respect to the other, thus envisaging the two adsorbates to be able to exchange electrons indirectly, through the catalyst: ($\mathscr{A}_1\,¦\,\mathscr{C}\,¦\,\mathscr{A}_2$).

Yet another important case of partitioning of model catalytic systems involves the cluster (C) representations of $\mathscr{C}$, C = (C_{as}|D), where C_{as} is a *small* cluster representing an *active site* (as) of the surface cluster C, while D is a *large* cluster reminder [87] which electronically conditions the adsorption/rearrangement surface processes and provides an environment, which moderates the $\mathscr{A}$ - C_{as} CT. Clearly, all alternative divisions of a general catalytic system involving $\mathscr{A}$ (or $\mathscr{A}_1$ and $\mathscr{A}_2$), C and an infinite system remainder, R, may be examined, in order to characterize different aspects (hypothetical intermediate stages) of such elementary adsorption/desorption/rearrangement processes, e.g., ($\mathscr{A}|C_{as}|D|R$), ($\mathscr{A}|C_{as}\,¦\,D|R$), ($\mathscr{A}\,¦\,C_{as}\,¦\,D|R$), etc. The last of the above partitionings, with $\mathscr{A}$ and C in equilibrium, could then be used to probe the CT trends of the ($\mathscr{A}\,¦\,C$) system as a whole towards an external agent, e.g., the second adsorbate.

Let us first consider the M^+ = (A|B) system consisting of an acidic (A) and basic (B) reactants. As explicitly shown in Eq.180 the presence of the other reactant modifies the respective blocks $\boldsymbol{\eta}^{M}_{X,Y}$, of the hardness matrix, due to the relaxational contributions. The relaxed hardness matrix, as we shall demonstrate in the Appendix B, corresponds to the collective relaxational modes $\boldsymbol{N}^{rel} = \boldsymbol{N}\,\mathbf{t}$, $\boldsymbol{\eta}^{rel} = \mathbf{t}^{-1}\,\boldsymbol{\eta}\,\mathbf{t}$, where the transformation matrix **t** is defined by the relaxational matrices $\mathbf{T}^{(A|B)}$ and $\mathbf{T}^{(B|A)}$ (Eqs. 175 and 178):

$$\mathbf{t} = \begin{pmatrix} \mathbf{I}_A & \mathbf{T}^{(A|B)} \\ \mathbf{T}^{(B|A)} & \mathbf{I}_B \end{pmatrix},$$

where the diagonal reactant blocks are unit (identity) matrices. This transformation and the eigenvectors of the diagonal blocks $\boldsymbol{\eta}^{M}_{A,A}$ and $\boldsymbol{\eta}^{M}_{B,B}$,

$$(\mathbf{U}^{M}_{X})^{\dagger}\,\boldsymbol{\eta}^{M}_{X,X}\,\mathbf{U}^{M}_{X} = \mathbf{h}^{M}_{X}\,(\text{diagonal}), \qquad (\mathbf{U}^{M}_{X})^{\dagger}\,\mathbf{U}^{M}_{X} = \mathbf{I}_X\,, \qquad X = A, B\,, \tag{300}$$

define new sets of the PNM of reactants. Namely, by combining the $\mathbf{U}^{M}_{X}$ blocks into the corresponding intra-fragment decoupling transformation,

$$\tilde{\mathbf{U}}^{rel}_{int} = \begin{pmatrix} \mathbf{U}^{M}_{A} & \mathbf{0}_{A,B} \\ \mathbf{0}_{B,A} & \mathbf{U}^{M}_{B} \end{pmatrix}, \tag{301}$$

one obtains the relaxed PNM of interacting reactants in terms of the AIM populations:

$$d\boldsymbol{n}^{rel}_{int} = d\boldsymbol{N}\,\mathbf{t}\,\tilde{\mathbf{U}}^{rel}_{int} \equiv d\boldsymbol{N}\,\mathbf{U}^{rel}_{int}\,, \tag{302}$$

which should differ from those of isolated reactants. The latter are defined by the matrix

$$\mathbf{U}_{int} = \begin{pmatrix} \mathbf{U}_{A} & \mathbf{0}_{A,B} \\ \mathbf{0}_{B,A} & \mathbf{U}_{B} \end{pmatrix}, \tag{303}$$

where the unitary block $\mathbf{U}_X$ diagonalizes the corresponding diagonal block $\boldsymbol{\eta}^{X,X}$ of the rigid AIM hardness matrix.

Obviously, one may also probe M^+ via the test charge displacements of AIM in one reactant, say on atom a in A. Such a forced, localized displacement $\Delta N_a = 1$ creates, via the minimum energy criterion for the assumed charge separation between reactants, the corresponding collective charge displacement on both reactants, consisting of the internal MEC part in A and the associated REC component in B:

$$\vec{\imath}^{\,M+}_{a} = \vec{n}^{M+}_{A(a)} + \vec{\mathscr{R}}^{M+}_{B(a)}\,, \tag{304}$$

where:

$$\vec{n}^{M+}_{A(a)} = \sum\nolimits_{a'}^{A} (\bar{a}')^{M^+}_{a}\,\vec{e}_{a'}\,, \qquad (\bar{a}')^{M^+}_{a} \equiv \left(\frac{\partial N_{a'}}{\partial N_a}\right)_{N_A, N_B} = \frac{S^{M^+}_{a',a}}{S^{M^+}_{a,a}}, \tag{305}$$

$$\vec{\mathscr{R}}^{M+}_{B(a)} = \sum\nolimits_{b}^{B} [b]^{M^+}_{a}\,\vec{e}_{b}\,, \qquad [b]^{M^+}_{a} \equiv \left(\frac{\partial N_b}{\partial N_a}\right)_{N_A, N_B}; \tag{306}$$

here $\mathbf{S}^{M^+}_{A,A} = (\partial \boldsymbol{N}_A/\partial \boldsymbol{\mu}_A)^{M^+}_{N_A, N_B}$ is the (relaxed) internal softness matrix of A in M^+ including both the reactant-diagonal and reactant off-diagonal contributions. The REC components are defined by the relevant relaxational matrix $\mathbf{T}^{(A|B)}$ (Eqs. 175 and 178) and the perturbing MEC displacement $\vec{n}^{M+}_{A(a)}$:

$$[b]_a^{M^+} = \sum_{a'}^{A} (\bar{a}')_a^{M^+} T_{a',b}^{(A|B)} . \quad (307)$$

In yet another hypothetical case of the primary displaced fragment being in equilibrium with an external reservoir (r), $M_A^r = (r \mid A|B)$, one has to replace the internal MEC, $\vec{n}_{A(a)}^{M^+}$ in Eqs. 304 and 305, by the corresponding external MEC, $\vec{\mathcal{N}}_{A(\hat{a})}^{M_A^r} = \Sigma_{a'}^{A} (a')_a^{M_A^r} \dot{e}_{a'}$, where $(a')_a^{M_A^r} \equiv (\partial N_{a'}/\partial N_a)_{N_B}^{M^r} = \sigma_{a',a}^{M_A^r}/\sigma_{a,a}^{M_A^r}$, and $\sigma_{A,A}^{M_A^r} = (\partial \boldsymbol{N}_A/\partial \boldsymbol{\mu}_A)_{N_B}^{M^r}$ is the (relaxed) external softness matrix of A in M_A^r. The resulting expression for the accompanying relaxational adjustment $\vec{\mathcal{R}}_{B(\hat{a})}^{M_A^r}$ in

$$\vec{\imath}_a^{M_A^r} = \vec{\mathcal{N}}_{A(\hat{a})}^{M_A^r} + \vec{\mathcal{R}}_{B(\hat{a})}^{M_A^r} \quad (308)$$

is:

$$\vec{\mathcal{R}}_{B(\hat{a})}^{M_A^r} = \sum_b^B [b]_a^{M_A^r} \vec{e}_b , \quad [b]_a^{M_A^r} = \left(\frac{\partial N_b}{\partial N_a}\right)_{N_B}^{M^r} = \sum_{a'}^{A} (a')_a^{M_A^r} T_{a',b}^{(A|B)} . \quad (309)$$

The above MEC/REC generalizations can also be extended to a general M = $(A \mid B|C|D| \ldots) \equiv (M_{open}|M_{closed})$ partitioning including the mutually open fragments in M_{open}, and the mutually closed fragments in M_{closed}. Namely, any unit displacement on an atom in M_{open} will generate the relevant MEC part in this subsystem, and the conjugated REC parts in each (internally open) molecular fragment of M_{closed}.

CHAPTER 4

CONCEPTS FOR CHEMICAL REACTIVITY

One of goals of theoretical chemistry is to identify various electronic and geometrical factors influencing chemical reactions, in order to probe chemical reactivity trends of large molecular systems, and to formulate the desired, favourable *"matching relations"* between reactants in terms of the appropriate *reactivity criteria*. The charge sensitivities discussed in the preceding chapters provide a novel, sound basis for formulating such *reactivity indices*, rooted in the corresponding expressions for the interaction energy, which can be used in both explaining and predicting trends in chemical reactivity, as well as in rationalizing and formulating general *reactivity rules* for various types of chemical processes. Since charge sensitivities carry the information about the system responses to both external potential and/or the electron population perturbations, it is natural to expect that the appropriately chosen set of sensitivity data should be adequate to characterize both the CT(dN)- and P(dv)-dominated processes.

When considering behaviour of a single molecule or a family of chemically similar molecules, in a given type of a reaction, e.g., during the electrophilic, nucleophilic, or radical attack by a small agent, various *single-reactant reactivity concepts* have proven their utility [12, 32, 96-104]. They are based upon the underlying notion of an inherent chemical reactivity of a molecule or relative reactivity of its parts, for a fixed *reaction stimulus*, defined by the assumed constant character of the other reactant. This notion, implying that the way a molecule reacts is somehow predetermined by its own structure, is very approximate one and relatively crude. Obviously, a more subtle *two-reactant description* is required to probe various arrangements of two large reactants, with the relevant reactivity criteria including charge sensitivities of both reaction partners, preferably accounting for their mutual influence at a given stage of their approach.

It should be emphasized, that the sensitivity data are parametrically dependent upon the geometric structure of the whole reactive system. Therefore, one also has to examine the geometry relaxation due to the coupling between the geometrical (nuclear position) and electronic (electron population) molecular degrees of freedom, in accordance with the classical *electron following* or *electron preceding perspectives* on nuclear motions [95].

Sensitivity criteria proposed in the literature fall mainly into the single-molecule criteria [12, 32], although some two-reactant concepts have also been reported [9, 57, 59, 60, 83, 91, 92, 105-109]. The molecular responses to the (dN, dv)-perturbations are to a large extent determined by the properties of the frontier molecular orbitals, i.e., the *lowest unoccupied* (LUMO) and the *highest occupied* (HOMO) orbitals of both reactants. Therefore, it should not be surprising that many sensitivity criteria are closely related to the frontier orbital theory [12, 39, 40, 100, 101].

A main one-reactant reactivity criterion in CSA, used to predict the system preferences towards localized attacks by small electrophilic, nucleophilic, or radical agents, is based upon the AIM FF indices. The corresponding reactivity trends immediately follow from the criterion of the maximum (minimum) FF index in the electrophilic (nucleophilic) attack, which locates the atomic site exhibiting the strongest local basic (acidic) character [9, 59]. This is because the sign of the FF index of a local site in a molecular system directly reflects upon the site donor/acceptor behaviour in a given type of the global charge displacement, dN. For an electrophilic attack a molecule acts as a *global base* ($dN < 0$), so that the negative FF index implies that the local site acts as a *local acid*, and the positive index identifies the *local basic site*. The reverse interpretation applies to the *nucleophilic attack*, when the system acts as a *global acid* ($dN > 0$): a negative FF value now corresponds to the *local basic site*, while a positive FF index identifies the *local acidic site* in the attacked molecule. Thus, the nucleophilic (B) reactant prefers the site in a molecule (A) with the lowest (preferably negative) FF index, since then the B→A coordination is locally stronger. Similarly, the electrophilic (A) reactant prefers the site in a molecule (B) with the highest (positive) FF value, since then the A←B bond is also locally strengthened.

The FF indices partition the amount of CT into regional (AIM, group) or collective (PNM) contributions. Therefore, they directly reflect the corresponding parts of the CT-energy. Also, as originally stressed by Parr and Yang [12, 39, 40], the local FF measures the chemical potential response to the local external potential perturbation, e.g., due to the presence of the other reactant. This has led to a postulate that the chemical reactions are preferred from the direction which produces the maximum initial chemical potential response of a reactant, which implies that the local FF values measure reactivity. A similar character has the maximum spin polarization criterion of Gázquez et al. [42], which gives an interesting spin polarization description of the formation of a covalent ($dN = 0$) bond.

Fukui function indices represent the "renormalized" softness information; thus they basically carry the same physical content as local softnesses. Therefore, the latter can also serve as important measures of chemical reactivity, e.g., in metals. As shown by Yang and Parr [40] in their ensemble approach, the local softnesses also reflect the local fluctuations in the electronic density. This has direct implications for catalytic activity on transition metals, for which such fluctuations in charge were shown to be very important [110].

Since the interaction energy E_{AB} is a truly two-reactant concept, all criteria directly linked to it should also be classified as belonging to that category. We also would like to stress that all changes in the sensitivity characteristics of reactants due to the presence of the other reactant, e.g., polarization or relaxation contributions, have also the two-reactant character. Such terms are expected to be large when a small reactant is coupled to a large reaction partner, e.g., a surface of a catalyst.

The PNM correspond to the total decoupling of the molecular AIM hardness matrix $\boldsymbol{\eta}$ in the representation of its eigenvectors $\vec{U}_M = \{\vec{U}_\alpha = \vec{e}_M U_\alpha\}$: $\boldsymbol{\eta}\mathbf{U} = \mathbf{h}\mathbf{U}$. Of interest in the theory of chemical reactivity also are alternative intermediate hardness decoupling modes, which we shall survey in this chapter. We again consider a general reactive system M = A–B of reactant A and B consisting of n and m–n AIM, respectively. We assume that all AIM of A are grouped as the first n atoms in $\boldsymbol{\eta}$, thus generating the block structure of the AIM rigid hardness matrix:

$$\boldsymbol{\eta} = \begin{pmatrix} \boldsymbol{\eta}^{A,A} & \boldsymbol{\eta}^{A,B} \\ \boldsymbol{\eta}^{B,A} & \boldsymbol{\eta}^{B,B} \end{pmatrix} . \tag{310}$$

As we have already demonstrated in previous chapter and in the Appendix B, the presence of the other reactant in $M^+ = (A|B)$ modifies $\boldsymbol{\eta}$ due to the relaxational contributions $\delta\boldsymbol{\eta}(M)$, generating the relaxed hardness matrix

$$\boldsymbol{\eta}^{rel} = \boldsymbol{\eta} + \delta\boldsymbol{\eta}(M) = \begin{pmatrix} \boldsymbol{\eta}^{M}_{A,A} & \boldsymbol{\eta}^{M}_{A,B} \\ \boldsymbol{\eta}^{M}_{B,A} & \boldsymbol{\eta}^{M}_{B,B} \end{pmatrix} . \tag{311}$$

Finally, in the reactant resolution, $M^+ = (A^+|B^+)$, the overall charge couplings are phenomenologically quantified by the condensed hardness matrix $\boldsymbol{\eta}_{(A|B)} = \{\eta^{M}_{X,Y}\}$, representing derivatives of the reactant chemical potentials with respect to their corresponding electron populations.

These hardness (interaction) parameters can be considered as the canonical input data in modelling the associated softness and FF (response) properties. Together they provide a basis for a variety of the reactivity concepts, which we shall survey in this chapter. After a brief summary of the partitioning of the FF data in reactive systems we shall consider alternative, partly decoupled collective populational modes of reactants, and a perturbational approach to the PNM based upon the matrix partitioning technique, which provides a basis for a truly two-reactant, mode-to-mode description of the CT/P phenomena. Next, the mapping relations between the electron population and nuclear position displacements, as well as their implications for the theory of electronic structure and chemical reactivity will be discussed. This will be followed by elementary charge stability considerations in various representations, and by schemes for partitioning the interaction energy.

1. Partitioning of the Fukui Function Indices in Reactive Systems

The basic charge response quantities of a given open molecular system M, the AIM FF indices, $f_i = \partial N_i / \partial N$, can be partitioned into the respective collective mode contributions (Eq. 264). In the reactive system $M^* = (A^* \,|\, B^*)$ one is additionally interested in the partitioning of the collective mode terms into the reactant contributions.

Consider general collective electron population displacements $\mathrm{d}\boldsymbol{p}$ (p-modes),

$$\mathrm{d}\boldsymbol{p} = \mathrm{d}\boldsymbol{N}\mathbf{C} , \tag{312}$$

defined by the respective columns of the linear transformation matrix $\mathbf{C} = (\boldsymbol{C}_1, \ldots, \boldsymbol{C}_\alpha, \ldots, \boldsymbol{C}_m) = \partial\boldsymbol{p}/\partial\boldsymbol{N}$. Let a and b denote representative constituent AIM of A and B, respectively.

A straightforward chain rule transformation of the AIM FF index f_i gives:

$$f_i = \sum_\alpha \frac{\partial p_\alpha}{\partial N}\frac{\partial N_i}{\partial p_\alpha} \equiv \sum_\alpha F_\alpha^{(p)} C_{\alpha,i}^{-1} \equiv \sum_\alpha w_{i,\alpha}^{(p)}$$

$$= \sum_\alpha \left(\frac{\partial N_A}{\partial N}\frac{\partial p_\alpha}{\partial N_A} + \frac{\partial N_B}{\partial N}\frac{\partial p_\alpha}{\partial N_B} \right) C_{\alpha,i}^{-1} \equiv \sum_\alpha \left(f_A^M F_\alpha^{A(p)} + f_B^M F_\alpha^{B(p)} \right) C_{\alpha,i}^{-1}$$

$$\equiv \sum_\alpha \left(f_A^M w_{i,\alpha}^{A(p)} + f_B^M w_{i,\alpha}^{B(p)} \right) \equiv \sum_\alpha \left(W_{i,\alpha}^{A(p)} + W_{i,\alpha}^{B(p)} \right), \qquad (313)$$

where the condensed reactant FF indices f_A^M and f_B^M determine the overall distribution of the external CT among reactants. The above partitioning first divides f_i into the p-mode contributions, $\{w_{i,\alpha}^{(p)}\}$, satisfying the usual normalization,

$$\sum_\alpha \left(\sum_i w_{i,\alpha}^{(p)} \right) \equiv \sum_\alpha w_\alpha^{(p)} = 1, \qquad (314)$$

and then additionally separates the components reflecting the mode participation in changing the overall electron populations of reactants. These reactant-resolved contributions, the f_X^M weighted $\{W_{i,\alpha}^{X(p)}\}$ and unweighted $\{w_{i,\alpha}^{X(p)}\}$, obey the following global summation rules:

$$\sum_\alpha \sum_i w_{i,\alpha}^{X(p)} = \sum_\alpha \frac{\partial p_\alpha}{\partial N_X}\frac{\partial N}{\partial p_\alpha} = \frac{\partial N}{\partial N_X} = 1, \qquad (315)$$

$$\sum_\alpha \sum_i W_{i,\alpha}^{X(p)} = f_X^M \frac{\partial N}{\partial N_X} = f_X^M, \quad X = (A, B). \qquad (316)$$

It is also of interest to divide the overall $w_\alpha^{(p)}$ quantities of Eq. 314 into the reactant contributions defined by the corresponding partial summations over constituent atoms of each reactant:

$$w_\alpha^{(p)} = \sum_a^A w_{a,\alpha}^{(p)} + \sum_b^B w_{b,\alpha}^{(p)} = w_\alpha^{A(p)} + w_\alpha^{B(p)}. \qquad (317)$$

The collective mode FF index,

$$F_\alpha^{A(p)} = \frac{\partial p_\alpha}{\partial N_A} = \sum_i \frac{\partial N_i}{\partial N_A}\frac{\partial p_\alpha}{\partial N_i} = \sum_a^A f_a^{A,A} C_{a,\alpha} + \sum_b^B f_b^{A,B} C_{b,\alpha}$$

$$\equiv F_\alpha^{A,A(p)} + F_\alpha^{A,B(p)} , \tag{318}$$

includes both the diagonal, $f_a^{A,A} \equiv \partial N_a/\partial N_A$, and off-diagonal, $f_b^{A,B} \equiv \partial N_b/\partial N_A$ contributions (see Eqs. 174, 179, 224, 225). Thus, the $F_\alpha^{A(p)}$ can be partitioned into the diagonal reactant contribution, $F_\alpha^{X,X(p)} = \Sigma_x^X(\partial N_x/\partial N_X)(\partial p_\alpha/\partial N_x)$, accounting for the charge reorganization inside X, and the associated off-diagonal part, $F_\alpha^{X,Y(p)} = \Sigma_y^Y(\partial N_y/\partial N_X)(\partial p_\alpha/\partial N_y)$, $Y \neq X$, due to the equilibrium changes in AIM populations in the other reactant.

The above partitioning applies when M in M^r = (r¦A¦B) changes the global number of electrons, $dN = dN_A + dN_B \neq 0$, e.g., in catalytic system, when A = adsorbate, B = active site, and the reservoir r = surface remainder. When N is constrained in M^* = (A¦B), i.e., when $dN = dN_A = -dN_B = N_{CT}$, the reactant FF indices are fixed: $(f_A^M)_N = f_A = 1$ and $(f_B^M)_N = f_B = -1$. For such isoelectronic (internal) CT processes the AIM FF indices are replaced by the corresponding *in situ* indices of Eq. 236. Their collective mode partitioning gives:

$$f_x^{CT} = \left(\frac{\partial N_x}{\partial N_{CT}}\right)_N = f_x^{X,X} - f_x^{Y,X} = \sum_\alpha \left[\frac{\partial p_\alpha}{\partial N_X} - \frac{\partial p_\alpha}{\partial N_Y}\right]\frac{\partial N_x}{\partial p_\alpha}$$

$$= \sum_\alpha \left[F_\alpha^{X(p)} - F_\alpha^{Y(p)}\right] C_{\alpha,x}^{-1} = \sum_\alpha \left(w_{x,\alpha}^{X(p)} - w_{x,\alpha}^{Y(p)}\right) \equiv \sum_\alpha \omega_{x,\alpha}^{CT(p)} ,$$

$$(X \neq Y) \in (A, B) , \quad x \in X . \tag{319}$$

This equation separates the reactant diagonal p-mode CT contributions , $\{w_{a,\alpha}^{A(p)}\}$ and $\{w_{b,\alpha}^{B(p)}\}$, from the reactant off-diagonal ones, $\{w_{b,\alpha}^{A(p)}\}$ and $\{w_{a,\alpha}^{B(p)}\}$, thus facilitating a discussion of a relative importance of the relaxation promotion effects reflected by the off-diagonal components. It also partitions the *in situ* AIM FF index into its p-mode contributions, analogs of the $w_{i,\alpha}^{(p)}$ components of Eq. 313, which sum up to zero:

$$\sum_i \sum_\alpha \omega_{i,\alpha}^{CT(p)} \equiv \sum_\alpha \omega_\alpha^{CT} = \sum_i f_i^{CT} = \sum_a^A f_a^{CT} + \sum_b^A f_b^{CT} = 0 . \tag{320}$$

The partial sum

$$\omega_\alpha^{CT(p)} = \frac{\partial p_\alpha}{\partial N_{CT}} \frac{\partial N}{\partial p_\alpha} \equiv F_\alpha^{CT(p)} \varphi_\alpha^{(p)} , \tag{321}$$

represents the CT analog of the $w_\alpha^{(p)}$ quantity of Eq. 314. It can be divided into the corresponding reactant contributions (see Eq. 317):

$$\omega_\alpha^{\mathrm{CT(p)}} = \sum_{X=A,B} \frac{\partial p_\alpha}{\partial N_{\mathrm{CT}}} \frac{\partial N_X}{\partial p_\alpha} \equiv \sum_{X=A,B} F_\alpha^{\mathrm{CT(p)}} \varphi_\alpha^{X(p)} \equiv \sum_{X=A,B} \Omega_{X,\alpha}^{\mathrm{CT(p)}} . \tag{322}$$

Therefore, the overall mode quantity $\omega_\alpha^{\mathrm{CT(p)}}$ represents the net change in the global number of electrons in M through the p-channel α, per unit CT between reactants; its reactant resolved components monitor the corresponding changes in the overall numbers of electrons in A and B in such a process.

2. Internal Hardness Decoupling Modes

Consider the *intra-reactant decoupling*, in which we rotate the AIM reference frame of M, $\vec{e}_M$ into the PNM of the mutually closed reactants, called the *internally decoupled modes* (IDM). Should one ignore the relaxational influence of one reactant upon another, one readily obtains the rigid IDM by diagonalizing the reactant blocks of $\boldsymbol{\eta}$:

$$\mathbf{U}_X^\dagger \boldsymbol{\eta}^{X,X} \mathbf{U}_X = \mathbf{h}_X \text{ (diagonal)} \qquad \mathbf{U}_X^\dagger \mathbf{U}_X = \mathbf{I}_X , \qquad X = A, B . \tag{323}$$

The columns of the resulting overall intra-reactant decoupling transformation matrix (see Eq. 303) define the *rigid IDM*. The orthogonal axes $\vec{\boldsymbol{U}}^{\mathrm{int}} = \vec{\boldsymbol{e}}_M \mathbf{U}_{\mathrm{int}}$ represent the new basis vectors in this representation. The transformed hardness matrix assumes the partly decoupled form:

$$\boldsymbol{\eta}_{\mathrm{int}} \equiv \mathbf{U}_{\mathrm{int}}^\dagger \boldsymbol{\eta} \, \mathbf{U}_{\mathrm{int}} = \begin{pmatrix} \mathbf{h}_A & \boldsymbol{\eta}_{\mathrm{int}}^{A,B} \\ \boldsymbol{\eta}_{\mathrm{int}}^{B,A} & \mathbf{h}_B \end{pmatrix} , \tag{324}$$

with $\boldsymbol{\eta}_{\mathrm{int}}^{A,B} = \mathbf{U}_A^\dagger \boldsymbol{\eta}^{A,B} \mathbf{U}_B = (\boldsymbol{\eta}_{\mathrm{int}}^{B,A})^\dagger$. Obviously, the assumed zero off-diagonal blocks in $\mathbf{U}_{\mathrm{int}}$ reflect the rigid electron distribution in the environment of the primary displaced reactant.

These rigid IDM in $M^+(Q) = (A^+|B^+)$ resemble the PNM of the separated reactants in $M^0(\infty)$. As we have already argued in Chapter 3 (Section 6.3) one can include the effect of the charge relaxation in the reaction partner by diagonalizing the reactant blocks $\boldsymbol{\eta}_{X,X}^{M}$ of $\boldsymbol{\eta}^{\mathrm{rel}}$. The modified internal transformation $\tilde{\mathbf{U}}_{\mathrm{int}}^{\mathrm{rel}}$ defines the *relaxed IDM*, $\vec{\boldsymbol{U}}_{\mathrm{rel}}^{\mathrm{int}} = \vec{\boldsymbol{e}}_M \tilde{\mathbf{U}}_{\mathrm{int}}^{\mathrm{rel}}$. By examining the differences between these two sets of IDM of reactants one can identify the effect of the *relaxation promotion* of the reactant PNM in a molecule. Similarly, by comparing the isolated reactant FF indices in M^0 with those corresponding to the reactant fragments in M, one can monitor the changes due to the charge coupling with the other reactant.

In Fig. 15 the relaxed IDM of toluene in the chemisorption complex M = (toluene$|V_2O_5$), involving the two-pyramid cluster of the vanadium pentoxide, are shown together with the AIM relaxed FF diagram and those representing its resolution into the CT- and P-components. These relaxed charge distribution patterns should be compared

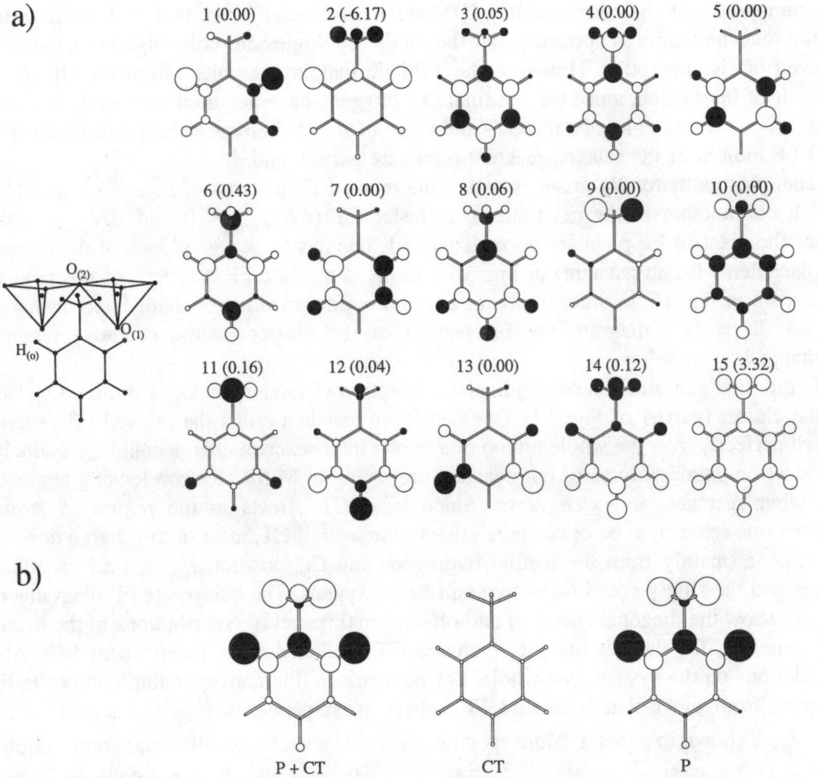

Figure 15. Diagrams of the relaxed IDM (part a) and the relaxed AIM FF indices (part b) of toluene in the indicated surface complex M with the bipyramidal cluster of the vanadium oxide surface. The modes are arranged in accordance with increasing hardnesses h_{α}^{M}; the hardness tensor reflects the isolated reactant AIM charges from MNDO and scaled-INDO calculations on toluene and cluster, respectively. Numbers in parentheses report the w_{α}^{M} values. In part b the three diagrams display the AIM FF distribution of toluene in M, calculated from the toluene block of η^{rel}, and its resolution into the CT- and P- components, respectively.

with the corresponding separated reactant analogs shown in Figs. 11 a and 12 a. Such a comparison reveals a substantial toluene promotion effect due to the interaction with the oxide cluster. For example, three new positions exhibiting negative f_i^{M} indices are observed when toluene interacts with the surface, also including the ring carbon which coordinates the methyl group and the ortho-hydrogens, $H_{(o)}$, which strongly interact with the vanadyl oxygens $O_{(1)}$ of the surface.

The effect of the relaxational promotion of the vanadium oxide cluster, due to the interaction with toluene, is illustrated in Fig. 16, where the unrelaxed PNM and FF data are compared with the corresponding IDM and the relaxed FF results. It follows from the figure that the hardness spectrum and the mode topologies are only slightly changed as a result of chemisorption. However, the AIM FF diagram exhibits a dramatic change as a result of interaction, since the terminal $O_{(1)}$ oxygens have acquired the negative values in M. Again, there is less of the CT- and more of the P-component in the resultant (P + CT) FF indices of the relaxed reactant (compare parts b and d).

The question naturally arises, whether the relaxed FF patterns of Figs. 15 b and 16 d match one another in the net toluene → cluster CT ($dN_{toluene} < 0$ and $dN_{cluster} > 0$). Since the reactant FF plots are normalized to 1, one has to change phases of the toluene displacements (basic reactant) of Fig. 15 b in the composite FF diagrams of reactants in M, shown in Fig. 17, in order to obtain a common interpretation of shaded and unshaded circles; in such a diagram the FF pattern of the cluster (acidic reactant) remains unchanged.

Let us first examine the perpendicular adsorption of toluene on $O_{(2)}$ from the $O_{(1)}$ side of the cluster (part A of Fig. 17). One detects in panels a-c that the relaxed FF patterns match perfectly over the whole region of a strong inter-reactant charge coupling. Namely, the electron gaining atoms of one reactant are facing in M the electron loosing atoms of the other reactant, and *vice versa*. Since local CT effects in the region of strong interactions are seen to be opposite relative to the overall CT, most of the charge donated by toluene (mainly from the methyl hydrogens and $C_{(o)}$ positions) goes into vanadium atoms and the singly coordinated pyramid-base oxygens. The composite FF diagrams of Fig. 17 show the diagonal (panel a) and off-diagonal (panel b) contributions to the *in situ* FF (panel c), for the net toluene → cluster CT predicted from independent SCF MO calculations on the system as a whole. For comparison, the corresponding composite FF diagram from the isolated reactant FF indices (first panels in Figs. 12 a and 16 b), $f_A - f_B$, is shown in panel d. More specifically, panel a shows the diagonal contributions, $f^{A,A} - f^{B,B} \equiv (\partial N_{cluster}/\partial N_{cluster}) - (\partial N_{toluene}/\partial N_{toluene})$, the off-diagonal displacements, $f^{A,B} - f^{B,A} \equiv (\partial N_{toluene}/\partial N_{cluster}) - (\partial N_{cluster}/\partial N_{toluene})$, are reported in panel b, while their sum, $f_A^{CT} + f_B^{CT} \equiv (f^{A,A} - f^{B,A}) - (f^{B,B} - f^{A,B}) = (\partial N_{cluster}/\partial N_{CT}) + (\partial N_{toluene}/\partial N_{CT})$, is shown in panel c.

One immediately observes that the first two panels in Fig. 17 A are practically indistinguishable. Thus, the diagonal and off-diagonal components of the relaxed *in situ* CT FF indices act in phase, giving the same perfect matching between reactants and enhancing one another in the overall indices of panel c. Therefore, at least in this specific case, each type of the CT FF indices may serve as sensitive, adequate two-reactant criterion for diagnosing the CT-related charge reorganization in M.

The isolated reactant FF data give the composite FF diagram shown in panel d. There is only a minor differentiation between peripheral atoms and those in the interaction region, with most of the toluene (cluster) atoms loosing (gaining) electrons as a result of the toluene → cluster CT; the relaxed FF data give rise to a substantial differentiation

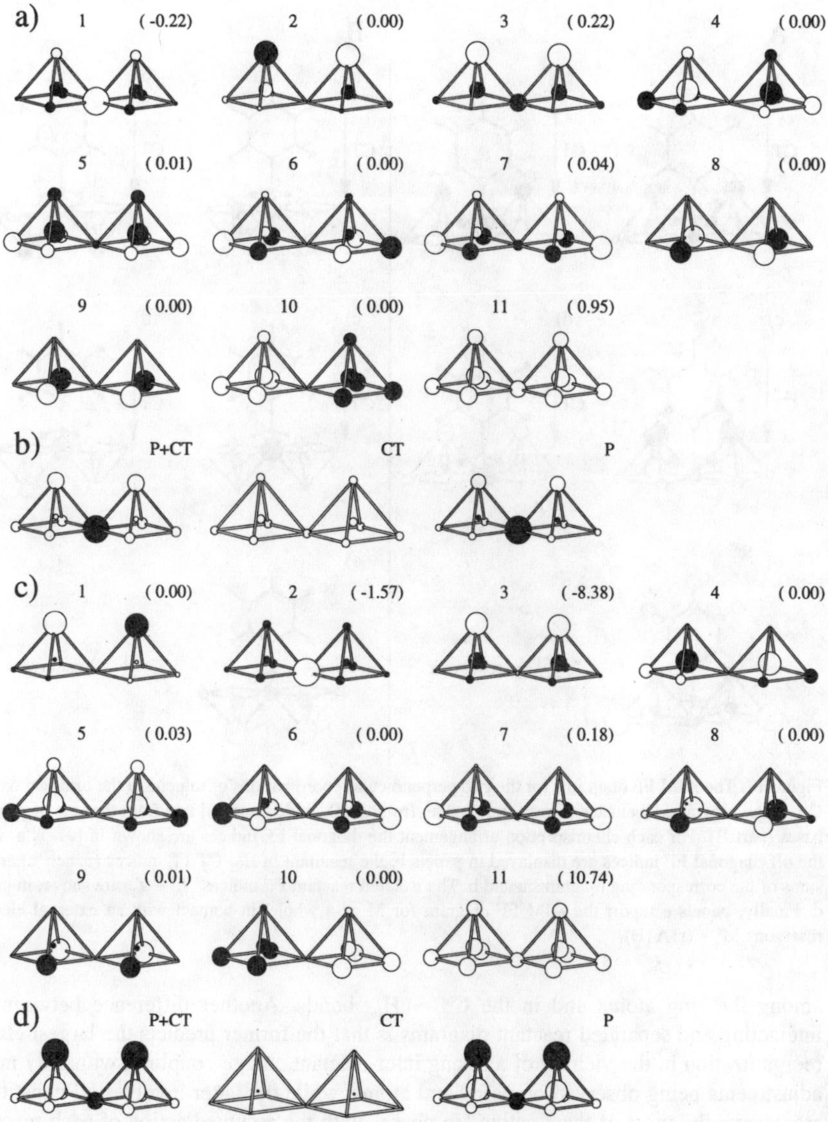

Figure 16. Diagrams of the PNM (part a) and the relaxed IDM (part c) of the bipyramidal cluster of the vanadium oxide surface in complex M of Fig. 15 a. The corresponding AIM FF plots are shown in parts b and d, together with their resolutions into P- and CT-components. Modes are ordered in accordance with increasing h_α (h_α^M) values. Numbers in parentheses are the corresponding w_α (w_α^M) values.

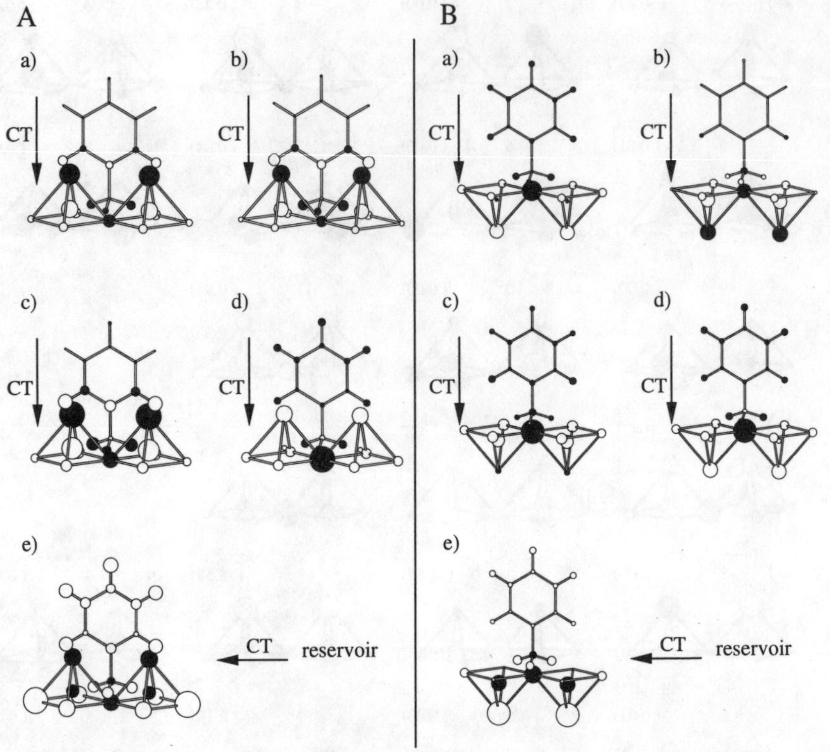

Figure 17. The AIM FF diagrams for the two perpendicular coordinations of toluene to the bridging oxygen $O_{(2)}$ of the bipyramidal cluster of vanadium oxide: from the $O_{(1)}$ side (part A)] and from the side of pyramid bases (part B). For each chemisorption arrangement the diagonal FF indices are shown in panels a, while the off-diagonal FF indices are displayed in panels b; the resultant *in situ* CT FF indices (panels c) are the sums of the corresponding diagrams a and b. The isolated reactant FF indices, $f_A - f_B$, are shown in panels d. Finally, panels e report the AIM FF diagram for M as a whole, in contact with an external electron reservoir: $M^r = (r|A|B)$.

among the ring atoms and in the $C_{(o)} - H_{(o)}$ bonds. Another difference between the interacting and separated reactant diagrams is that the former predicts the largest charge reorganization in the vicinity of a strong inter-reactant charge coupling, with only minor adjustments being observed on peripheral atoms, while the latter involves all constituent atoms equally, most of them acting "in phase" with the assumed action of each reactant. The interacting subsystem diagrams resemble more the minimum energy charge displacements, which exhibit the same general trend of a fast decay of charge shifts away from the primary populational perturbation, with an appreciable short range polarization marked by a large number of nodes. A comparison between the panels c and d of Fig.

17 A exhibits a reversal of the local CT in the $O_{(1)}$-$H_{(o)}$ region in the non-interacting approximation. This indicates that the inclusion of all charge couplings in a chemisorption system is vital for adequately predicting all adsorbate activation trends.

In panel e the AIM FF diagram for M in contact with an external reservoir ($N \neq$ const.), M^r = (r¦A¦B), is shown. As in the preceding diagrams the AIM charges before the B → A CT have been used in order to facilitate a comparison with the *in situ* (N = const.) FF contours. A reference to panels c and e of the figure indicates that the AIM FF pattern of M as a whole for $dN > 0$ resembles the *in situ* distribution in the region of a strong interaction between $H_{(o)}$ and $O_{(1)}$ atoms, while the methyl group charge displacements exhibit the opposite phases in both cases. Notice, that for $dN < 0$ (i.e., for the opposite phases in panel e the open system diagram closely resembles the FF plot of panel d. This indicates, that an outflow of electrons from M (its chemical oxidation) enhances the toluene → cluster coordination, i.e., strengthens the overall effect of the toluene – cluster chemisorption bond. Thus, the opposite influence, of this bond being weakened, should be expected for an inflow of electrons to M (its chemical reduction); this should facilitate a toluene desorption.

Consider now part B of Fig. 17, which corresponds to the perpendicular coordination of toluene to the $O_{(2)}$ site from the pyramid bases side of the vanadium oxide cluster. In this case the charge coupling between reactants is much weaker, so that the FF pattern is dominated by the AIM charge displacements on the cluster (softer reactant). An inspection of panels a and b reveals that some AIM displacements of the diagonal and off-diagonal contributions, e.g., on the methyl hydrogens and $O_{(1)}$ atoms, exhibit opposite phases, thus cancelling each other to some degree in the overall *in situ* diagram of panel c. This resultant FF plot exhibits mainly a transfer of electrons from the whole [toluene -$O_{(2)}$] unit to vanadium atoms and the remaining pyramid base oxygens. Since amount of charge removed from the bridging oxygen is large one detects favourable conditions for the overall toluene → cluster coordination. There is a marked similarity between the *in situ* FF pattern of panel c and that resulting from the isolated FF data of reactants (panel d); the only difference is observed in the $C(H_3)$ fragment, and in the signs of the FF indices of the vanadyl oxygens. Generally speaking, the composite FF diagram of the isolated reactants resembles the diagonal, relaxed FF distribution of panel a. Again, one detects that many local CT trends are reversed in the FF pattern of the open adsorption complex, shown in panel e. For $dN > 0$ (phases as shown in the figure) this charge reorganization reverses the CT effects of panel c, thus weakening the adsorption bond and facilitating the associative desorption of toluene. Accordingly, for $dN < 0$ (opposite phases to those shown in the figure) the original charge displacements and the adsorption bond are predicted to be strengthened.

Thus, the interaction does indeed facilitate the intuitively expected charge reorganization accompanying the toluene → cluster CT, when the hypothetical barrier preventing it is removed. A comparison between Figs. 12 a and 15 b, and between panels b and d of Fig. 16, shows that the very structure of the FF distribution has changed as a result of the other reactant presence. Namely, in the isolated reactants the CT and P components exhibit comparable magnitudes of atomic populational shifts, while the FF indices of the interacting reactants are strongly P-dominated.

A reference to Figs. 11 a and 15 a indicates that the IDM spectrum and FF indices exhibit dramatic changes in comparison with the corresponding PNM data of isolated toluene. The interaction with the V_2O_5 cluster has concentrated the CT-reactivity information in two modes only (compare the w_α values in parentheses of Fig. 15 a): $\alpha = 2$ IDM, which dominates the CT-induced polarization (see the third diagram of Fig. 15 b), and the nonselective, hardest mode $\alpha = 15$, which accounts for the external CT to the surface cluster (see the second diagram in Fig. 15 b). Thus, the toluene – cluster interaction preselects the $\alpha = 2$ IDM (resembling the $\alpha = 12$ PNM of the isolated toluene) as the most important selective reactivity channel, which differentiates between local sites in the adsorbed molecule. The topology of this crucial mode reflects a tendency of the physisorbed toluene to donate electrons to the cluster vanadium atoms through the methyl group hydrogens and to accept electrons (from the terminal $O_{(1)}$ oxygens of the cluster) through the ortho-hydrogens. A similar domination of the relaxed FF pattern of the vanadium oxide cluster is seen in Fig. 16 c, where two modes, $\alpha = 3$ and 11 explain the main features of the first diagram in part d of the same figure. Here, a relative role the mode $\alpha = 3$ has been strengthened as a result of the interaction, in comparison with the isolated cluster case.

These illustrative results have profound implications for the theory of chemical reactivity and for catalysis in particular. They indicate that for strongly interacting, large reactants the isolated reactant charge sensitivity data should be supplemented by the response criteria of interacting species. One has to resort to this truly two-reactant information, as the above IDM and their charge sensitivity characteristics, which *preserves the memory of the actual interaction in the reactive system*. The presence of the other reactant "filters out" the most important reaction channels, which are not "recognized" by the charge sensitivity criteria of the isolated species. This relaxational (interaction) promotion of reactants for their current separation and orientation has been shown to generate an almost perfect matching of their relaxed FF data.

As we have already shown in the first section of this chapter, both the f and f^{CT} AIM FF indices can be partitioned in terms of contributions from the collective charge displacement modes. The corresponding rigid IDM partitionings of the AIM FF indices for the illustrative case of the reactive system M = toluene (B) --- [V_2O_5 - cluster] (A) of Figs. 15 a and 17 A are shown in Figs. 18 and 19.

A comparison between Figs. 17 A-e and 18 shows that indeed the resolution of f in the rigid IDM framework leads to rather strong participation of a relatively few PNM of reactants. Namely, the cluster part of f is seen to be represented satisfactorily already by the $\{w_{a,5}\}$ distribution in Fig. 18 a, with the modes $\alpha = 1$ and 3 also strongly influencing the resultant pattern of Fig. 17 A-e in this region of a strong charge coupling with toluene. It should be noticed that the remaining part of the cluster is also affected by the modes $\alpha = 7$ and 11. The toluene part of the FF diagram of the whole chemisorption system generally exhibits generally a similar degree of concentration of the IDM components. The region of a strong interaction with the vanadium oxide is seen to originate mainly from the $\{w_{b,\beta}\}$ contributions for $\beta = 2$, 7, 14, and 15. However, the resultant pattern of Fig. 17 A-e cannot be related in this case to a single, crucial mode. As a rule, the dominating modes of the vanadium oxide cluster are relatively soft; the

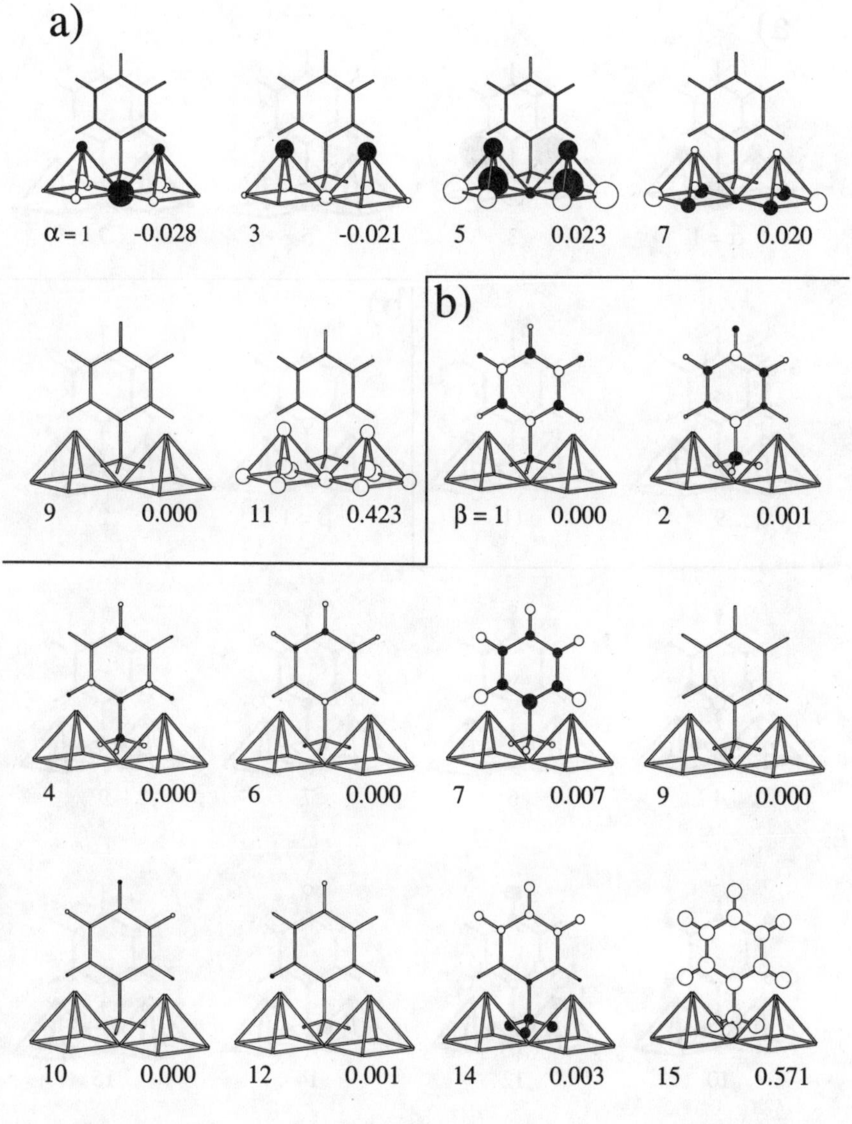

Figure 18. The rigid IDM (see Figs. 11 a and 16 a) resolution of the AIM FF diagram of Fig. 17 A-e: part a shows the $\{w_{a,\alpha},\ (a,\ \alpha)\in A\}$ distribution contours, while part b presents the corresponding $\{w_{b,\beta},\ (b,\ \beta)\in B\}$ data. Only the external CT-active modes are shown, for which w_α (w_β) (second number below the diagram) does not vanish identically (by symmetry).

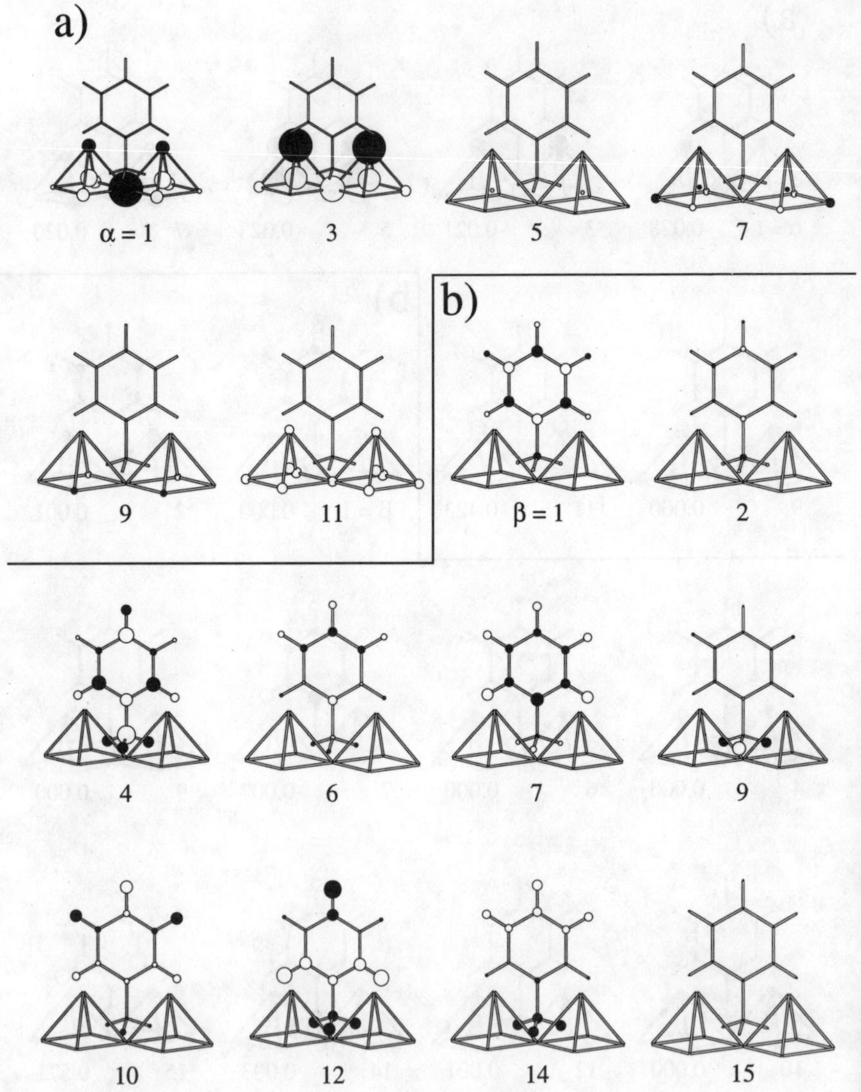

Figure 19. The rigid IDM resolution, $\{ \omega^{CT}_{a,\alpha} \}$ - part a and $\{ \omega^{CT}_{b,\beta} \}$ - part b, of the AIM FF diagram of Fig. 17 A-c.

opposite is true in the toluene case, in which relatively hard modes determine the overall FF distribution.

It follows from Fig. 19 that the rigid IDM resolution of $\boldsymbol{f}^{CT}$ (Fig. 17 A-c) is more mode concentrated in comparison to that shown in Fig. 18. The cluster part of the toluene → cluster isoelectronic CT FF indices have large components $\omega_{a,\alpha}^{CT}$ only for the softest CT-active modes α = 1 and 3. The toluene $\boldsymbol{f}^{CT}$ part in the strong interaction region is also relatively condensed, being dominated by the β = 4 and 12 modes, which alone explain the main qualitative features of Fig. 17 A-c.

One concludes from this analysis that the rigid IDM provide quite adequate reference frame for describing, in a reasonably compact form, both external and internal CT processes. However, as carrying no information whatever about the reactant interaction, they are by definition one-reactant concepts and therefore may not constitute the optimum collective charge displacement coordinates for reactive systems. In section 3 and 4 we extend this search into alternative modes for the charge reorganization in M = A --- B, shapes of which are influenced by the actual interaction between A and B, similarly to the relaxed IDM.

In Figs. 20 and 21 we have examined the effect of the relaxational contributions to the reactant hardness diagonal blocks, due to the interaction with the other reactant. They present the relaxed IDM (see Figs. 15 a and 16 c) resolution of the $\boldsymbol{f}$ and $\boldsymbol{f}^{CT}$ distributions of Figs. 17 A-e and 17 A-c, respectively, in the same toluene−[V_2O_5 - cluster] reactive system. A comparison with the previous, rigid IDM resolution of Figs. 18 and 19, respectively, indicates that the IDM components of $\boldsymbol{f}$ are only slightly modified by the mutual relaxational influence of reactants, with the $\boldsymbol{f}$ vector being dominated by the IDM exhibiting a similar topology and approximately the same $w_{\alpha(\beta)}$ values in both the rigid and relaxed cases. It is also seen that the inclusion of relaxational effects still gives rise to a strong dominance of both the toluene and cluster parts by the non-selective modes α = 11 and β = 15 (Fig. 20), which together practically account for the total electron transfer to/from the chemisorption system; similar selective modes are seen to determine the external CT-induced internal polarization within each reactant.

It follows from a comparison between Figs. 19 and 21 that the relaxational influence leads to a more compact resolution of the $\boldsymbol{f}^{CT}$ (toluene → cluster) distribution, particularly on toluene, where the accompanying charge redistribution is now well described by a single mode (β = 2) only. The two modes, α = 2, 3, dominate the cluster part, with the former accounting for the negative $f_{O_{(2)}}^{CT}$ and the latter reproducing the negative $f_{O_{(1)}}^{CT}$ and positive f_V^{CT} features of the $\boldsymbol{f}^{CT}$ diagram. This result demonstrates that the interaction-dependent reference frames are better suited to describe the internal CT between reactants in M, than the corresponding isolated reactant (PNM) coordinate systems.

3. Externally Decoupled Modes

Another, complementary set of partially decoupled populational modes is defined within the external hardness decoupling scheme [59, 88]. The *externally decoupled mode* (EDM) transformation, $\boldsymbol{\hat{U}}^{ext} = \boldsymbol{\hat{e}}_M \mathbf{U}_{ext}$, transforms the AIM hardness matrix $\boldsymbol{\eta}$ of the reactive system $M^+ = (A^+|B^+)$ into the block-diagonal form:

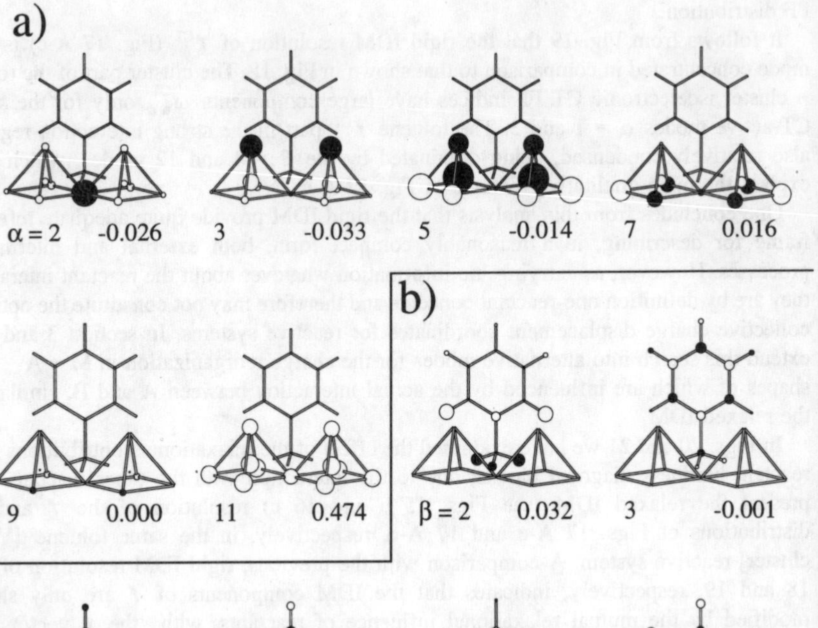

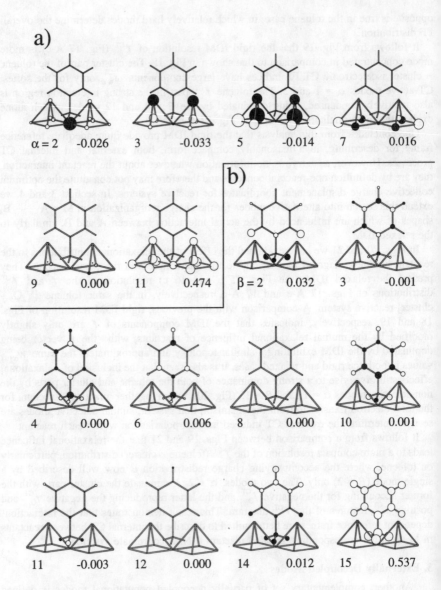

Figure 20. The relaxed IDM resolution of *f* (Fig. 17 A-e); see also caption to Fig. 18.

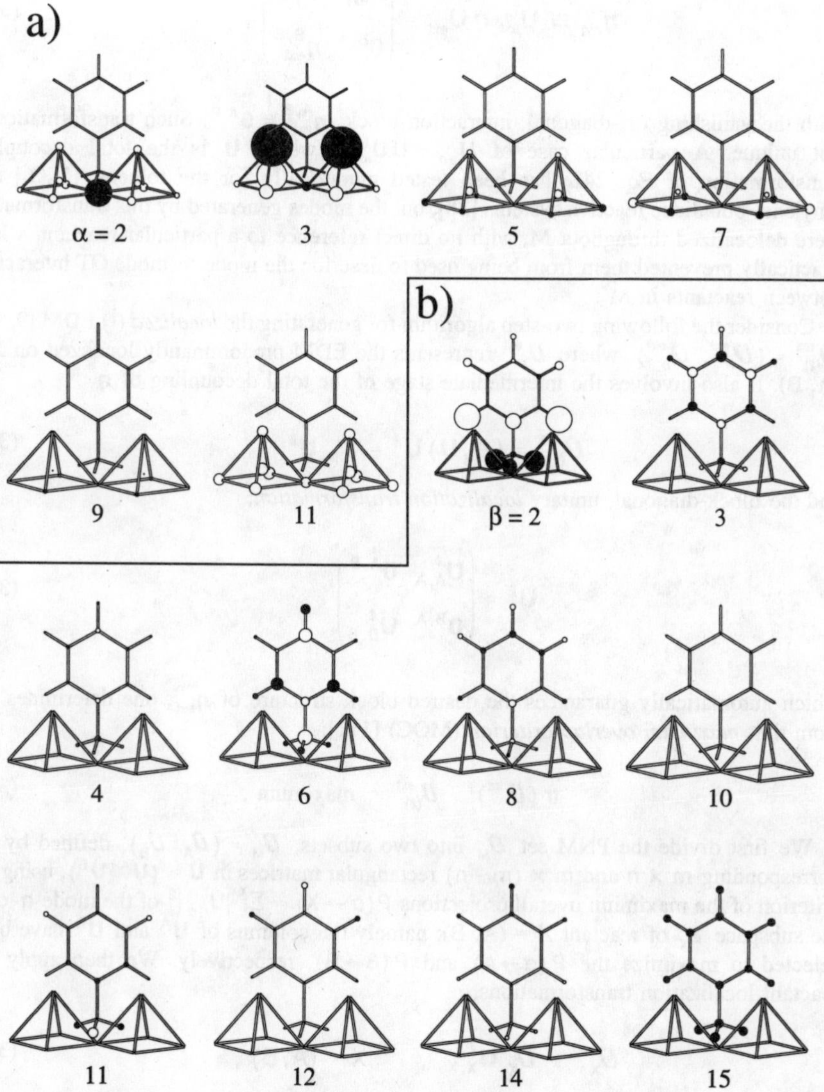

Figure 21. The relaxed IDM resolution of f^{CT} (Fig. 17 A-c); see also caption to Fig. 19.

$$\boldsymbol{\eta}_{ext} \equiv \mathbf{U}_{ext}^{\dagger}\, \boldsymbol{\eta}\, \mathbf{U}_{ext} = \begin{pmatrix} \boldsymbol{\eta}_{ext}^{A,A} & \mathbf{0}^{A,B} \\ \mathbf{0}^{B,A} & \boldsymbol{\eta}_{ext}^{B,B} \end{pmatrix}, \tag{325}$$

with the vanishing off-diagonal, interaction block: $\boldsymbol{\eta}_{ext}^{A,B} = \mathbf{0}^{A,B}$. Such transformation is not unique. A particular case of $\mathbf{U}_{ext} = \mathbf{U}\,\mathbf{U}_{int}^{\dagger}$, where $\mathbf{U}$ is the total decoupling transformation of Eq. 242, has been tested numerically for the toluene–[V_2O_5] and ethylene–butadiene reactive systems [88], but the modes generated by this transformation were delocalized throughout M, with no direct reference to a particular reactant, which practically prevented them from being used to describe the mode-to-mode CT interaction between reactants in M^+.

Consider the following two-step algorithm for generating the *localized* (ℓ) EDM [9, 92], $\vec{\boldsymbol{U}}_{(\ell)}^{ext} = (\vec{\boldsymbol{U}}_A^{ext}, \vec{\boldsymbol{U}}_B^{ext})$, where $\vec{\boldsymbol{U}}_X^{ext}$ represents the EDM predominantly localized on X = (A, B). It also involves the intermediate stage of the total decoupling of $\boldsymbol{\eta}$,

$$\vec{\boldsymbol{U}}_{(\ell)}^{ext} = (\vec{\boldsymbol{e}}_M \mathbf{U})\, \mathbf{U}^{\ell} = \vec{\boldsymbol{U}}_M\, \mathbf{U}^{\ell}\,, \tag{326}$$

and the block-diagonal, unitary *localization transformation*,

$$\mathbf{U}^{\ell} = \begin{pmatrix} \mathbf{U}_{A,A}^{\ell} & \mathbf{0}^{A,B} \\ \mathbf{0}^{B,A} & \mathbf{U}_{B,B}^{\ell} \end{pmatrix}, \tag{327}$$

which automatically guarantees the desired block structure of $\boldsymbol{\eta}_{ext}$; one determines $\mathbf{U}^{\ell}$ from the *maximum overlap criterion* (MOC) [111]:

$$\mathrm{tr}\,(\vec{\boldsymbol{U}}^{int})^{\dagger} \cdot \vec{\boldsymbol{U}}_{(\ell)}^{ext} = \text{maximum}\,, \tag{328}$$

We first divide the PNM set $\vec{\boldsymbol{U}}_M$ into two subsets, $\vec{\boldsymbol{U}}_M = (\vec{\boldsymbol{U}}_A, \vec{\boldsymbol{U}}_B)$, defined by the corresponding $m \times n$ and $m \times (m-n)$ rectangular matrices in $\mathbf{U} = (\mathbf{U}^A \,|\, \mathbf{U}^B)$, using the criterion of the maximum overall projections $P(\alpha \to X) = \Sigma_i^X |U_{i,\alpha}|^2$ of the mode α onto the subspace $\vec{\boldsymbol{e}}_X$ of reactant X = (A, B); namely the columns of $\mathbf{U}^A$ and $\mathbf{U}^B$ have been selected to maximize the $P(\alpha \to A)$ and $P(\beta \to B)$, respectively. We then apply the reactant localization transformations,

$$\vec{\boldsymbol{U}}_X^{ext} = \vec{\boldsymbol{U}}_X\, \mathbf{U}_{X,X}^{\ell}\,, \qquad X = (A, B)\,, \tag{329}$$

which simultaneously maximize

$$\mathrm{tr}\,(\vec{\boldsymbol{U}}_X^{int})^{\dagger} \cdot \vec{\boldsymbol{U}}_X^{ext} \equiv \mathrm{tr}\,\boldsymbol{\Lambda}_X\,, \qquad X = (A, B)\,. \tag{330}$$

where $\mathcal{U}_X^{int} = \boldsymbol{e}_X \mathbf{U}_X$ (see Eq. 323). As shown by Gołębiewski [111] this localization transformation is defined by the following expression:

$$\mathbf{U}^{\ell}_{X,X} = \boldsymbol{\Delta}_X^{\dagger} (\boldsymbol{\Delta}_X \boldsymbol{\Delta}_X^{\dagger})^{-1/2} , \qquad X = (A, B) , \tag{331}$$

where $\boldsymbol{\Delta}_X = (\mathcal{U}_X^{int})^{\dagger} \mathcal{U}_X = \mathbf{U}_X^{\dagger} \mathbf{U}^{X,X}$ and $\mathbf{U}^{X,X}$ is the corresponding diagonal block of $\mathbf{U}$.

This way of generating the EDM (→ IDM) collective modes guarantees their one-to-one correspondence to the respective IDM of reactants, and as such may provide an alternative, convenient framework for describing the CT processes. Similarly to the IDM, such localized EDM preserve the memory of the reactant interaction in M^+, and they should lead to a substantial hardness decoupling. This expectation is due to their resemblance to the PNM (IDM) of reactants, for which the diagonal (reactant) blocks of the relevant hardness matrix are exactly diagonal. Thus, with no external hardness interaction between the A and B subsets of EDM, it comes as no surprise that these collective, delocalized charge displacement modes also bear some resemblance to the PNM of M as a whole.

Besides a dominating component of one reactant, each EDM also exhibits the matching components of the other reactant. Therefore, the EDM provide a mode-to-mode matching between reactants; as practically independent, with only a small hardness coupling within each subset, they should be useful in an interpretation of the overall B→A CT, providing the mode-to-mode perspective on the accompanying charge reorganization in reactants.

Illustrative localized EDM (→ IDM) for M = (toluene --- $[V_2O_5]$) are shown in Fig. 22, with the sets (a) and (b) grouping the subsets $\{\alpha\}$ and $\{\beta\}$ associated with the $[V_2O_5]$ cluster (A) and toluene (B), respectively. A comparison between these CT reactive channels and the corresponding IDM of reactants (Figs. 11 a and 16 a) shows that the adopted MOC does indeed generate the inter-set decoupled displacements, which can be uniquely associated with the corresponding IDM of separated reactants.

Of interest is also a comparison between the localized EDM and the PNM (set $\{\gamma\}$) of M as a whole (Fig. 23), and between the corresponding diagrams representing the collective mode partitioning of the AIM FF indices (Figs. 24 and 25, respectively). It follows from the $w_\alpha^{(p)}$ values reported in Figs. 22, 24, and 25, that, when the system acts as a whole in an external CT process, the mode contributions are strongly dominated by the hardest, non-selective EDM β = 15 (Figs. 22 b and 24 b), which is practically indistinguishable from the γ = 26 PNM of Figs. 23 and 25. Next in this $w_\alpha^{(p)}$ hierarchy are the modes α = 5 (Figs, 22 a and 24 a), β = 12 (Figs. 22 b and 24 b), and γ = 20 (Figs. 23 and 25); important contributions to the FF pattern are also due to the modes α = 11, β = 4, and γ = 1. As explicitly shown in Figs. 24 and 25, the $w_\alpha^{(p)}$ parameters do not fully reflect the mode participation in shifting the AIM charges, since some polarizational modes ($w_\alpha^{(p)} \approx 0$) also exhibit large amplitudes in their AIM electron population displacements. When one is interested in specific channels for shifting the electrons between reactants, their relative importance is reflected by the $w_\alpha^{A(p)}$ and $w_\beta^{B(p)}$ quantities, shown in parentheses above each mode diagram.

It follows from a comparison between Figs. (22, 24) and Figs. (23, 25), respectively, that the localized EDM already strongly resemble the totally decoupled PNM. In most

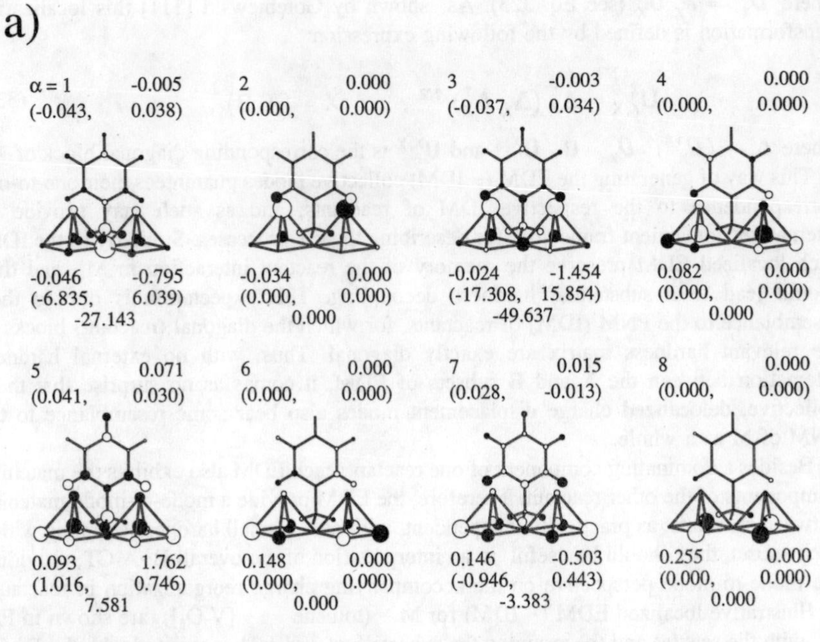

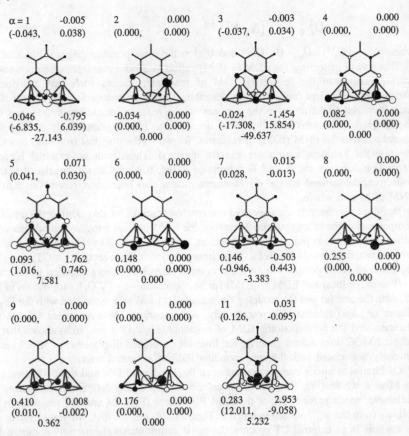

Figure 22. The localized EDM (→IDM) for the M = (toluene --- [V_2O_5]) system, numbered in accordance with the isolated reactant IDM they resemble the most (see Figs. 11 a and 16 a). The usual phase convention $\varphi_\alpha^{(EDM)} \equiv \partial N/\partial p_\alpha > 0$ has been adopted. Part a groups the EDM associated with the (V_2O_5) cluster, X = A, and part b corresponds to toluene, X = B. The numerical data above each diagram report: the number of the IDM giving the maximum projection on the EDM in question, $w_\alpha^{(EDM)}$, followed by its resolution into reactant contributions $w_\alpha^{A(EDM)}$ and $w_\alpha^{B(EDM)}$, in parentheses. The bottom numbers include: $(\eta_{ext}^{X,X})_{\alpha,\alpha}$, $\omega_\alpha^{CT(EDM)}$ (first row), $\Omega_{A,\alpha}^{CT(EDM)}$, $\Omega_{B,\alpha}^{CT(EDM)}$ (second row, in parentheses), and $F_\alpha^{CT(EDM)}$.

cases a clear one-to-one correspondence can be established by inspection. Compare, e.g., contributions from the CT-active modes in Figs. 24 and 25. The patterns α = 1, 3, 5, 7, 9, and 11 EDM of Fig. 24 a may be related to those corresponding to those from the PNM γ = 1, 3, 9, 13, 21, and 20 of Fig. 25, respectively. Similarly, the most important

b)

β = 1 0.000
(0.000, 0.000)

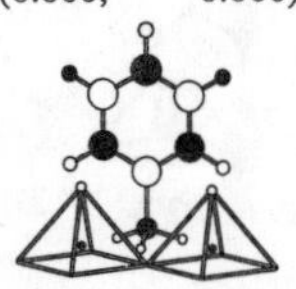

0.025 0.013
(0.015, -0.002)
1.317

2 -0.002
(-0.005, 0.003)

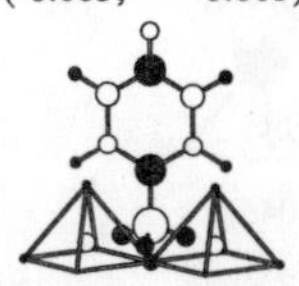

0.037 0.406
(0.774, -0.368)
8.413

3 0.000
(0.000, 0.000)

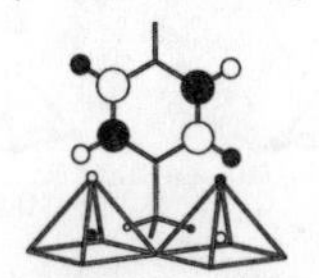

0.041 0.000
(0.000, 0.000)
0.000

4 -0.020
(-0.035, 0.015)

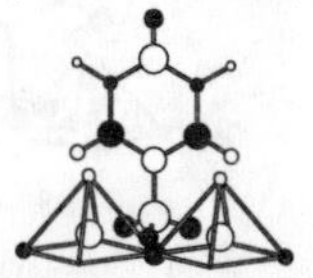

0.064 4.572
(8.078, -3.506)
21.802

5 0.000
(0.000, 0.000)

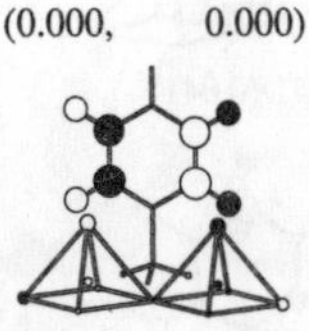

0.094 0.000
(0.000, 0.000)
0.000

6 -0.001
(-0.001, 0.000)

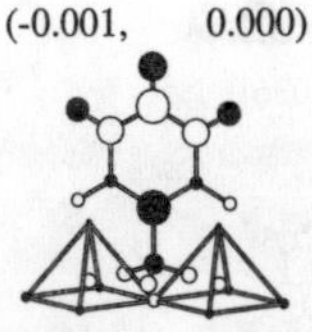

0.107 -0.004
(-0.006, 0.002)
-0.330

7 0.000
(0.000, 0.000)

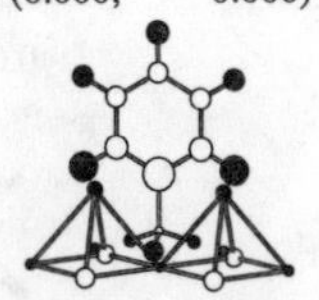

0.188 0.244
(0.775, -0.503)
3.988

8 0.000
(0.000, 0.000)

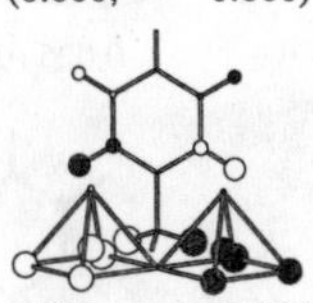

0.513 0.000
(0.000, 0.000)
0.000

9 0.001
(0.002, -0.001)

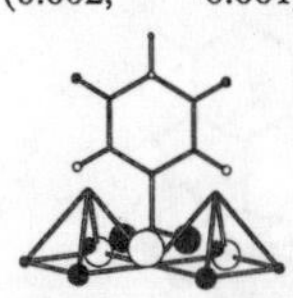

0.200 0.127
(0.178, -0.051)
2.325

10 0.000
(0.000, 0.000)

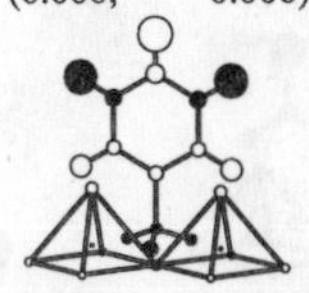

0.319 0.106
(0.268, -0.162)
1.660

11 0.000
(0.000, 0.000)

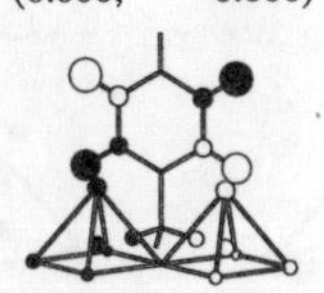

0.455 0.000
(0.000, 0.000)
0.000

12 0.051
(0.059, -0.008)

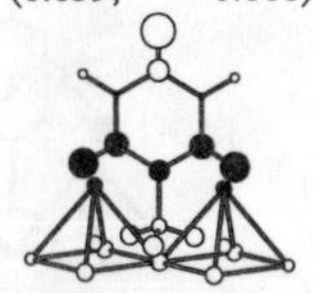

0.535 2.183
(2.540, -0.357)
3.422

13 0.000
(0.000, 0.000)

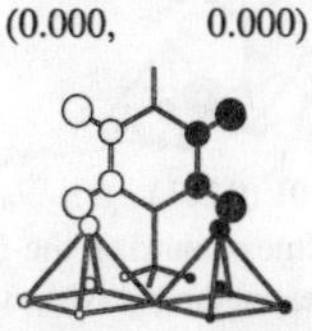

0.790 0.000
(0.000, 0.000)
0.000

14 0.010
(0.016, -0.006)

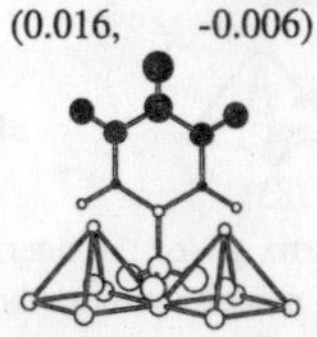

1.332 2.749
(4.288, -1.539)
2.809

15 0.851
(0.265, 0.586)

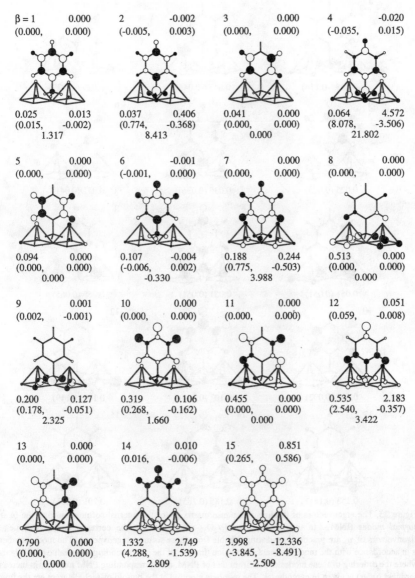

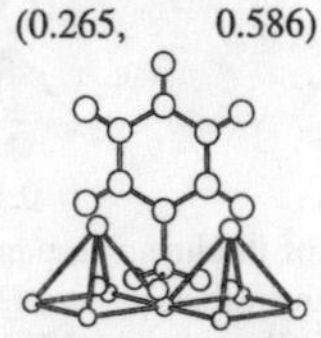

3.998 -12.336
(-3.845, -8.491)
-2.509

Figure 22. Continued

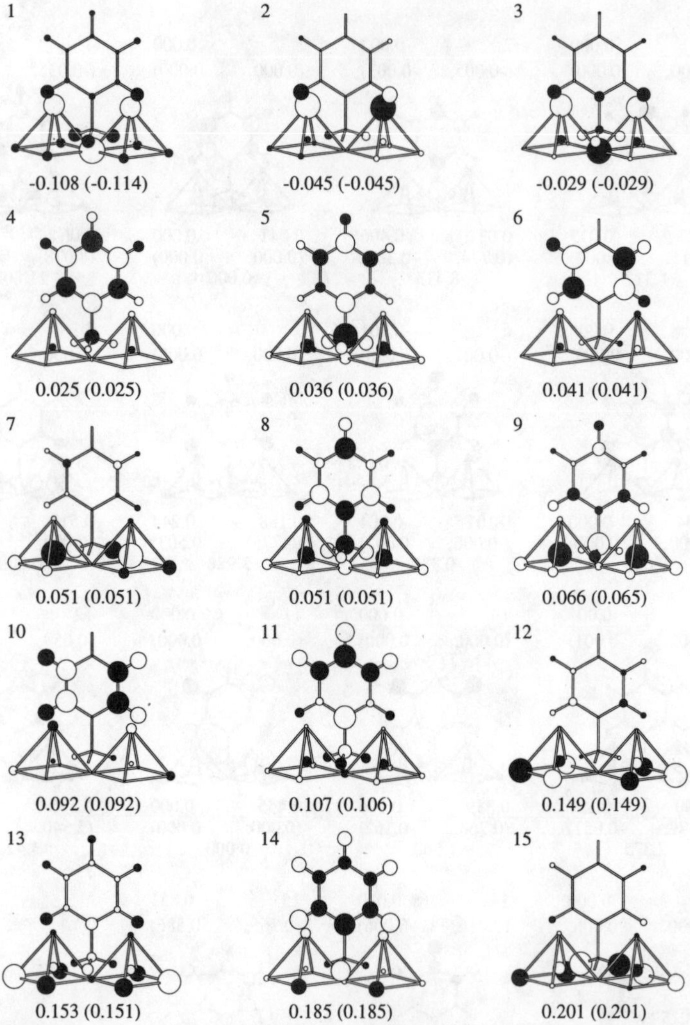

Figure 23. The eigenvectors of the linear response matrix $\boldsymbol{\beta}$ (or the internal softness matrix), the *internal normal modes* (INM) in the toluene --- $[V_2O_5]$ model system; the corresponding PNM, i.e., the eigenvectors of $\boldsymbol{\eta}$, are practically indistinguishable from these softness eigenvectors. The mode numbering is in accordance with the mode hardness eigenvalue (inverse of the corresponding eigenvalue of $\boldsymbol{\beta}$). In cases where the ordering of these modes differs from that of PNM, the corresponding PNM number (in increasing hardness order) is given in parentheses. The numbers reported at the bottom of each diagram are the inverse of the $\boldsymbol{\beta}$-eigenvalue and the principal hardness of $\boldsymbol{\eta}$. The nonselective, hardest (pure CT) mode exhibits infinite $\boldsymbol{\beta}$-hardness eigenvalue, thus being excluded from the remaining internal polarizational modes.

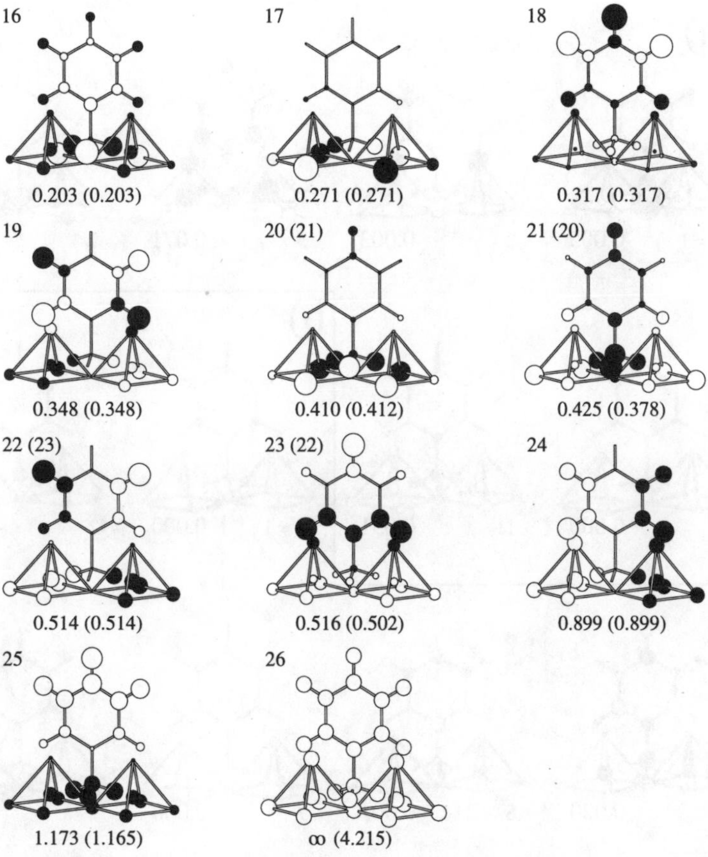

Figure 23. Continued

EDM patterns β = 1, 2, 4, 6, 9, 12, 14, and 15 of Fig. 24 b are seen to closely resemble those observed for γ = 4, 5, 8, 11, 16, 22, 25, and 26, in Fig. 25.

We therefore conclude that our algorithm for the *externally decoupled modes*, IDM-*localized*, EDM (→ IDM), does indeed generate reactive charge displacement modes, which can be easily correlated with both the IDM of reactants and the PNM of M as a whole. It follows from our illustrative results that the EDM (→ IDM) set gives rise to a strong concentration of the AIM-resolved FF information in a few crucial modes; this mode-resolution closely follows that corresponding to the PNM description. The remarkable similarity between the PNM and these localized EDM also indicates that there is little of the intra-set hardness coupling left within the $\{\alpha\}$ and $\{\beta\}$ subsets. This is explicitly demonstrated in the relevant contour diagrams of Fig. 27 a.

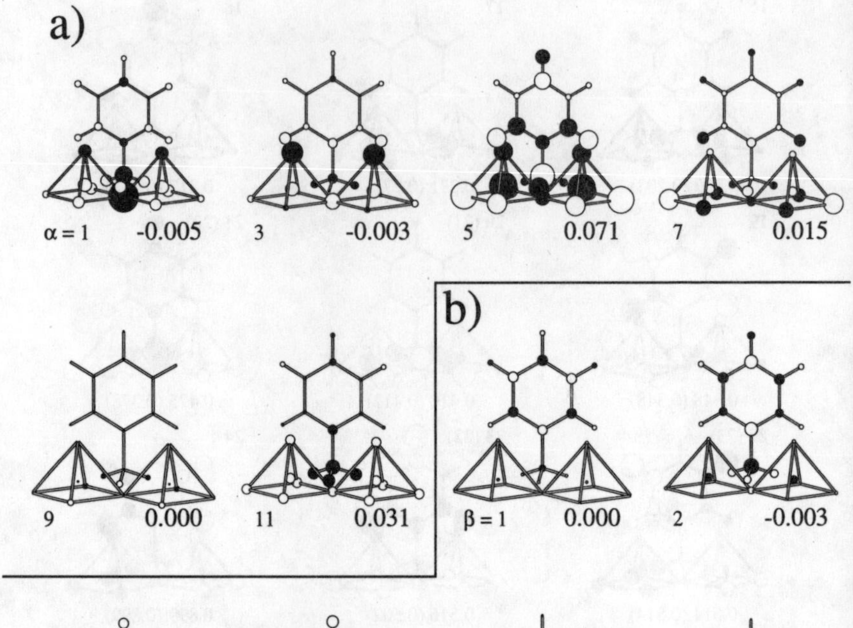

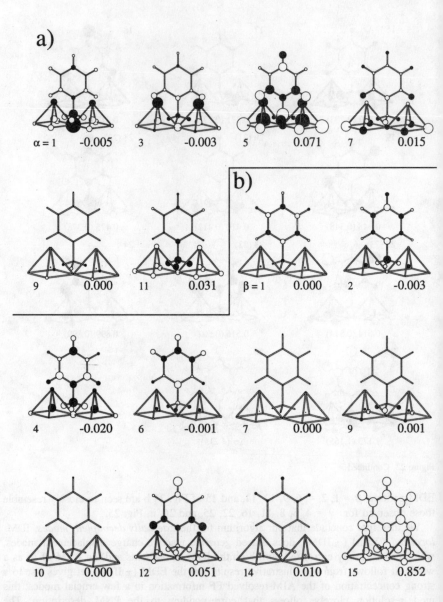

Figure 24. The EDM (→IDM) resolution ,$\{ w_{i,\alpha}^{(EDM)} \}$, of the FF diagram of Fig. 17 A-e; only the CT-active modes ($w_\alpha^{EDM} \neq 0$, shown below the mode diagram) are reported. The mode numbers are the same as in the corresponding parts of Fig. 16.

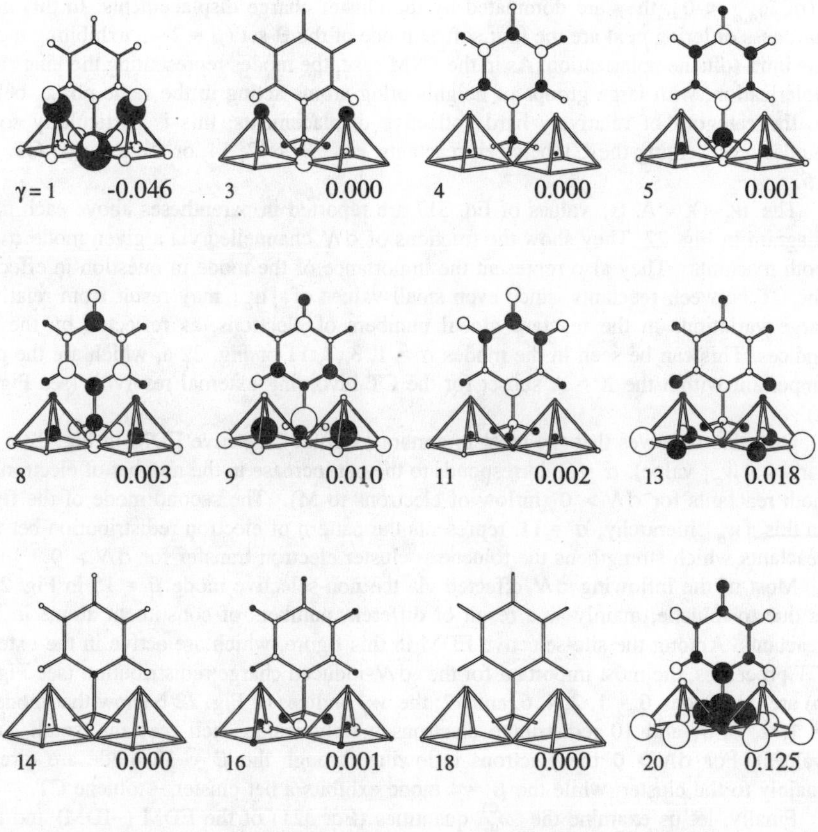

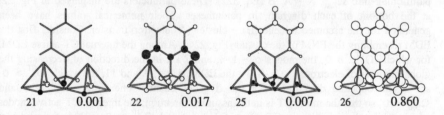

Figure 25. The PNM resolution, $\{w_{i,\alpha}\}$, of the FF diagram of Fig. 17 A-e; only the CT-active modes ($w_\alpha \neq 0$) are shown. The mode numbers and the overall w_α index are reported at the bottom of each contour diagram.

A reference to the EDM hardnesses reported in Fig. 22 (the first number at the bottom of each diagram) shows that the three softest EDM, α = (1-3) are unstable $[(\eta_{ext}^{A,A})_{\alpha,\alpha} < 0]$; they are dominated by the cluster charge displacements. In this mode hardness ordering next are the four softest mode of the B-set (β = 1-4), exhibiting mostly the intra-toluene polarization. As in the PNM case, the modes representing the long-range polarization, with large groups of neighbouring atoms acting in the same phase, belong to the category of relatively hard collective displacements; this is particularly so for modes representing the CT between reactants, e.g., β = 12, 14, or the external CT, β = 15.

The w_α^X (X = A, B) values of Eq. 317 are reported in parentheses above each mode diagram in Fig. 22. They show the fractions of dN channelled via a given mode α into both reactants. They also represent the importance of the mode in question in effecting the CT between reactants, since even small values of $|w_\alpha|$ may result from relatively large variations in the reactant global numbers of electrons, as reflected by the w_α^X indices. This can be seen in the modes α = 1, 3, 7, 11 of Fig. 22 a, which are the most important within the X = A subset for the CT involving external reservoir (see Fig. 24 a).

One also observes that the most important external CT-active EDM in Fig. 22 a (the largest $|w_\alpha|$ value), α = 5, corresponds to the net increase in the number of electrons on both reactants for $dN > 0$ (inflow of electrons to M). The second mode of the figure in this $|w_\alpha|$ hierarchy, α = 11, represents the pattern of electron redistribution between reactants which strengthens the toluene → cluster electron transfer for $dN > 0$.

Most of the inflowing dN effected via the non-selective mode β = 15 in Fig. 22 b, is due to toluene, mainly as a result of different numbers of constituent atoms in both reactants. Among the site-selective EDM in this figure, which are active in the external CT processes, the most important for the dN-induced charge redistribution (see Fig. 24 b) are the modes β = 1, 2, 4, 6, and 12; the w_α^X indices of Fig. 22 b show that modes β = 1, 2, 6, 7, and 10 redistribute electrons mainly inside each reactant (small $|w_\alpha^X|$ values). For $dN > 0$ the electrons inflowing through the β = 12 mode are directed mainly to the cluster, while the β = 4 mode exhibits a net cluster → toluene CT.

Finally, let us examine the ω_α^{CT} quantities (Eq. 321) of the EDM (→IDM) and their partitioning into $\Omega_{X,\alpha}^{CT}$, X = A, B (Eq. 322). These parameters are displayed in Fig. 22, at the bottom of each diagram, in parentheses. Their numerical values have been generated for the assumed toluene (B) → cluster (A) electron transfer. Consider first the EDM resembling the PNM of the cluster (Fig. 22 a). Among the internal CT-active EDM, for which $\Omega_{X,\alpha}^{CT} \neq 0$, the modes α = 1, 3, and 7 act in the direction of decreasing the global number of electrons ($\omega_\alpha^{CT} < 0$); the EDM α = 5, 9, and 11, for which $\omega_\alpha^{CT} > 0$, act in the opposite direction. For the α = 11 channel one finds $\Omega_{A,\alpha}^{CT} > 0$ and $\Omega_{B,\alpha}^{CT} < 0$, so that the mode CT is in the assumed direction; the internal CT-active modes which exhibit the opposite net CT effect, $\Omega_{A,\alpha}^{CT} < 0$ and $\Omega_{B,\alpha}^{CT} > 0$, include the modes α = 1, 3, 7; for α = 5 one finds positive values of both these reactant contributions.

In the second part of Fig. 22 one also detects that a majority of modes redistribute global electron numbers of reactants in the direction of the assumed CT (α = 2, 4, 7, 9, 10, 12, 14). It should be noticed that the fully symmetrical, background channel α = 15 represents an external CT, with both reactants loosing electrons for the assumed CT direction.

Clearly, one could also define the EDM, which resemble the MEC/REC associated with the constituent atoms of both reactants, $\boldsymbol{r}_M = (\boldsymbol{r}_A, \boldsymbol{r}_B)$ (Section 6.2 of Chapter 3). Then, the corresponding localization transformation blocks, which maximize tr $\boldsymbol{r}_X^{\dagger} \cdot \boldsymbol{U}_X^{ext}$, immediately follow from the last equation, when one appropriately redefines $\boldsymbol{\Delta}_X$.

In Fig. 26 we have compared the external MEC in the model toluene---[V_2O_5] system with the EDM localized to resemble the corresponding MEC: EDM (→ MEC). One should recall that the *non-orthogonal* MEC are all dominated by the softest, short-range polarization modes. In the *orthogonal* EDM set only the softest modes, localized mainly on the vanadium oxide cluster, do exhibit such polarization, thus generally resembling the corresponding MEC. One also detects a degree of such short-range polarization in the EDM of part c. However, as seen in part b, the EDM localized on the carbon MEC, are quite different from their MEC analogs since, due to the EDM orthogonality, the PNM exhibiting a short-range polarization do not contribute to these relatively hard modes.

These differences are well reflected by the mode hardness parameters. When a pair of the (EDM, MEC) diagrams exhibits a similar topology, e.g., selected modes in parts a and c, the hardnesses are of comparable magnitude. The opposite conclusion follows from part b, where hardnesses of the EDM are much higher than those of their MEC analogs.

Another obvious conclusion following from Fig. 26 is that the EDM (→ MEC) modes are localized, as are the MEC modes they resemble. This is particularly so in part b of the figure, where modes α = 14 and 15 represent the respective CH fragments of toluene. A reference to the w_α values, which reflect a partitioning of the external CT, dN, among the most important CT-active EDM (→ IDM) (Fig. 24), PNM (Fig. 25) and the EDM (→ MEC) (Fig. 26), indicates that the EDM (→ IDM) give almost as compact description of the external CT phenomena as do the PNM; in this respect the charge inflow (or outflow) from the system as a whole is seen to be scattered into a relatively large subset of the EDM (→ MEC): α = 18, 12, 15, 14, 25, 26, 6, 21, and 13, in order of decreasing $|w_\alpha|$. Therefore, the EDM (→IDM) set appears as much more suitable collective populational coordinate system for describing the external CT, in comparison to the EDM (→ MEC) reference frame.

A similar conclusion follows from an examination of the non-vanishing blocks $\eta_{ext}^{X,X}$ (X = A, B) of the EDM hardness matrices. Their structures for the two EDM sets in the toluene---[V_2O_5] system of Figs. 22 and 26, are represented in the contour maps of Fig. 27. This comparison shows quite convincingly, that the EDM (→ IDM) set indeed generates almost decoupled EDM, with only a few modes exhibiting appreciable non-diagonal hardness matrix elements. The other EDM set gives rise to strongly coupled modes in each subset, and particularly so in toluene. Therefore, the EDM (→ IDM) set indeed appears as the superior reference frame for describing charge redistribution phenomena in molecular systems.

Finally, let us examine the EDM (→ IDM) and PNM resolutions of the AIM *in situ* FF indices (see Fig. 17 A-c), in accordance with Eq. 319; they are reported graphically in Figs. 28 and 29, respectively. One concludes from these results that the EDM (→ IDM) set gives the most condensed collective mode interpretation of the inter-reactant (N-

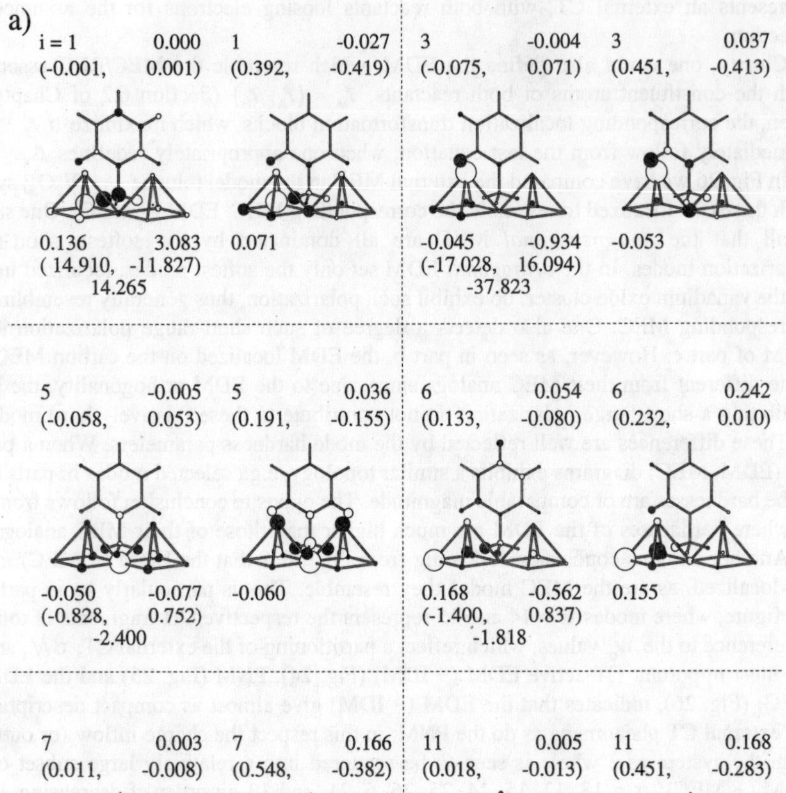

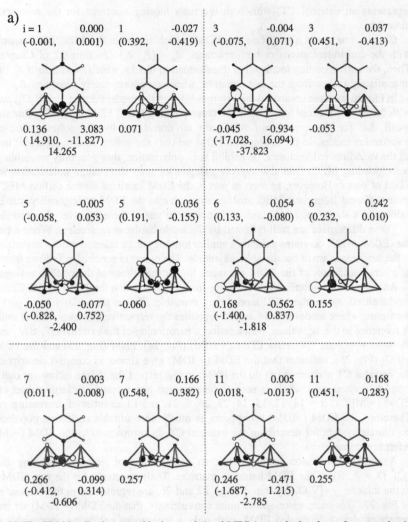

Figure 26. The EDM localized to resemble the populational MEC (external, also shown for comparison): EDM (→MEC). The modes are divided into subsets associated with constituent atoms in the $[V_2O_5]$–cluster (reactant A, part a), ring carbons (part b) and hydrogen atoms (part c) of toluene (reactant B), in the model toluene---$[V_2O_5]$ system. The isolated reactant charges (MNDO–toluene, scaled-INDO–cluster) have been used to generate the hardness tensor. The primarily displaced atom is identified by the largest open circle in the specified group of atoms, or it is marked by an asterisk (the $H_{(o)}$ case). In each two-diagram area the first diagram represents the EDM localized on the MEC shown in the second diagram. The numerical data reported in the EDM plots are arranged in the same way as in Fig. 22. The numbers above each MEC diagram are $\varphi_i \equiv (dN/dN_i)_{E=\min} = \varphi_i^A + \varphi_i^B$, $\varphi_i^A = (dN_A/dN_i)_{E=\min}$, and $\varphi_i^B = (dN_B/dN_i)_{E=\min}$. Below the MEC diagram the Γ_i value is also reported.

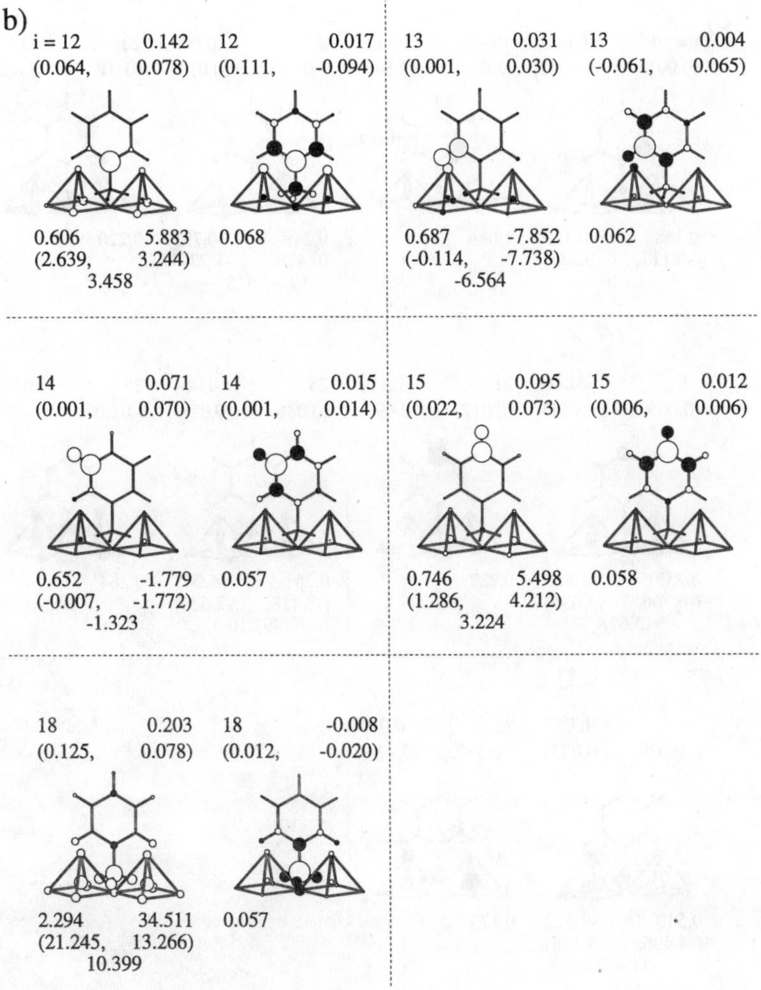

Figure 26. Continued.

constrained) CT, with only two unstable modes (negative mode hardness) α = 1 and 3 strongly participating in the toluene → cluster electron transfer; the two modes β = 2 and 4, provide only a slight correction needed to reproduce the positive feature of $f^{CT}_{C(H_3)}$, seen in Fig. 17 A-c. The PNM set gives a resolution of a comparable compactness, with the modes γ = 1 and 3, also unstable, generating the gross features of f^{CT}, and the modes γ = 5, 8, 9, and 20 generating a slight correction, mainly to produce negative values at the methyl hydrogens. It follows from comparisons of Figs. (24, 25) and (28, 29) that the

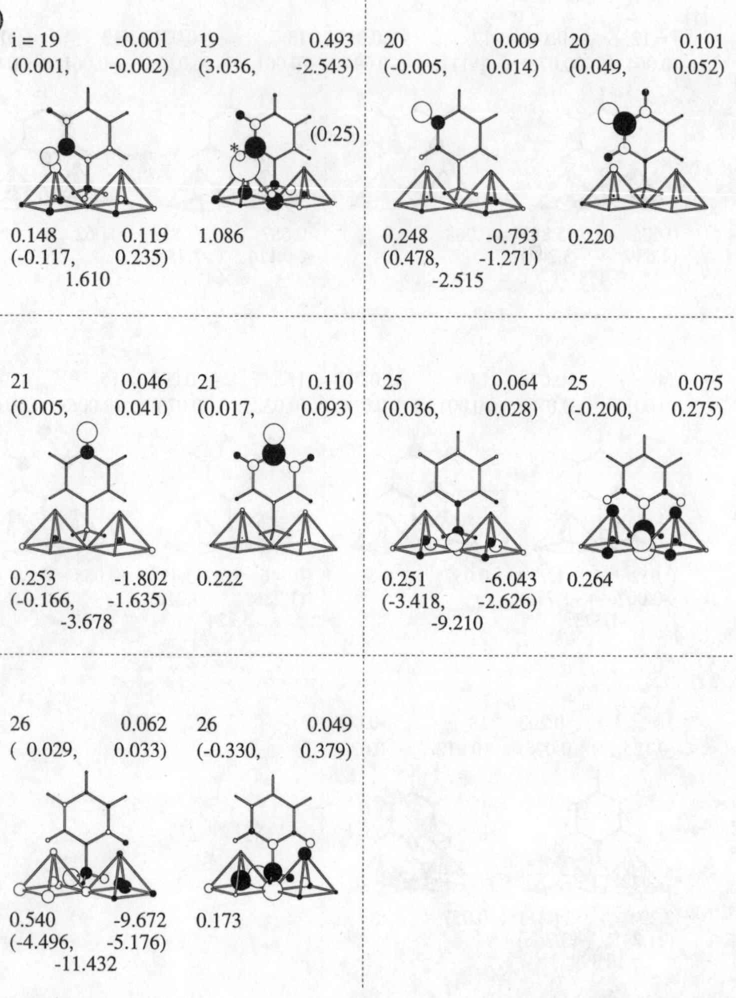

Figure 26. Conclusion.

most compact collective mode resolution of f, representing the open M as a whole, is obtained in the PNM representation (total decoupling), while the N-constrained CT process of the electron transfer between interacting species, a resolution of f^{CT}, is best described in the EDM($\rightarrow$ IDM) framework. In this regard the PNM represent "natural" collective displacements for an external CT involving hypothetical electron reservoir, while the EDM ($\rightarrow$ IDM) are the "natural" reaction channels for the B $\rightarrow$ A isoelectronic CT; they both preserve a memory of the actual charge couplings between reactants.

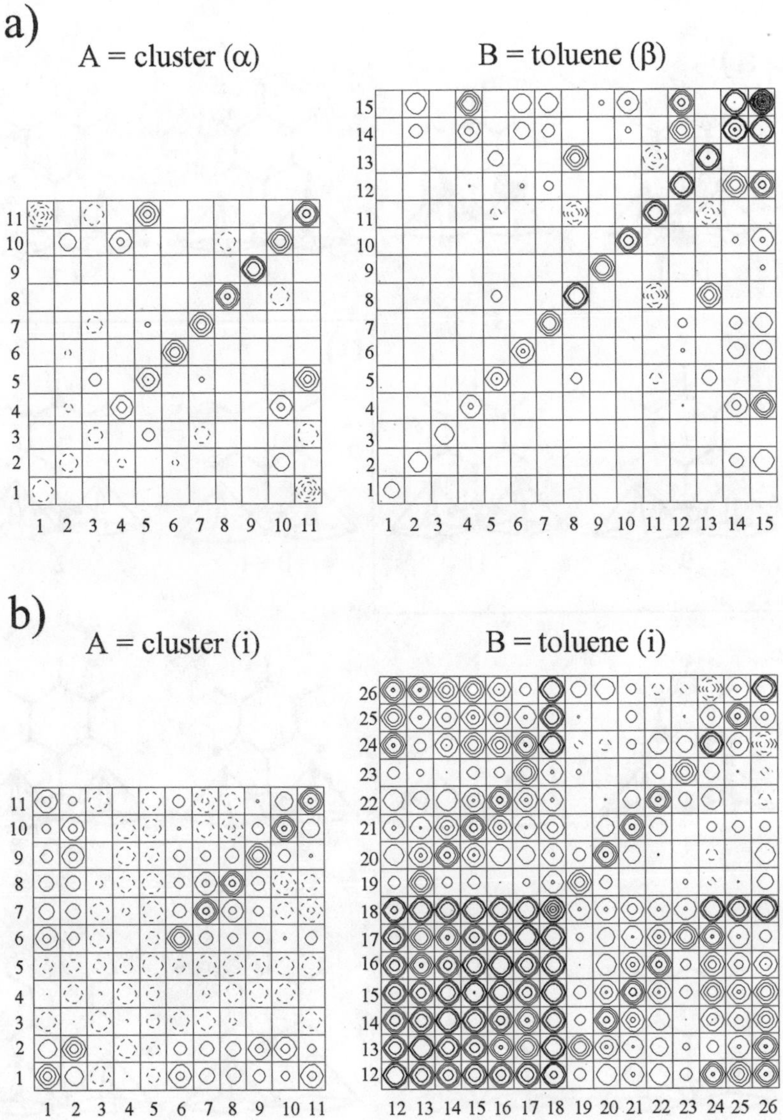

Figure 27. The contour map representations of the $\eta_{ext}^{X,X}$ (X = A, B) blocks in the EDM (→ IDM) (part a) and EDM (→ MEC) (part b) representations; for the mode numbering and diagrams see Figs. 22 and 26, respectively. For reasons of clarity the small hardness values are removed from the maps. The contour spacings $\Delta = 0.04$ for $0.01 \leq |\eta| \leq 0.1$ and $\Delta = 0.5$ for $0.2 \leq |\eta|$ have been used; the negative values are represented by broken lines.

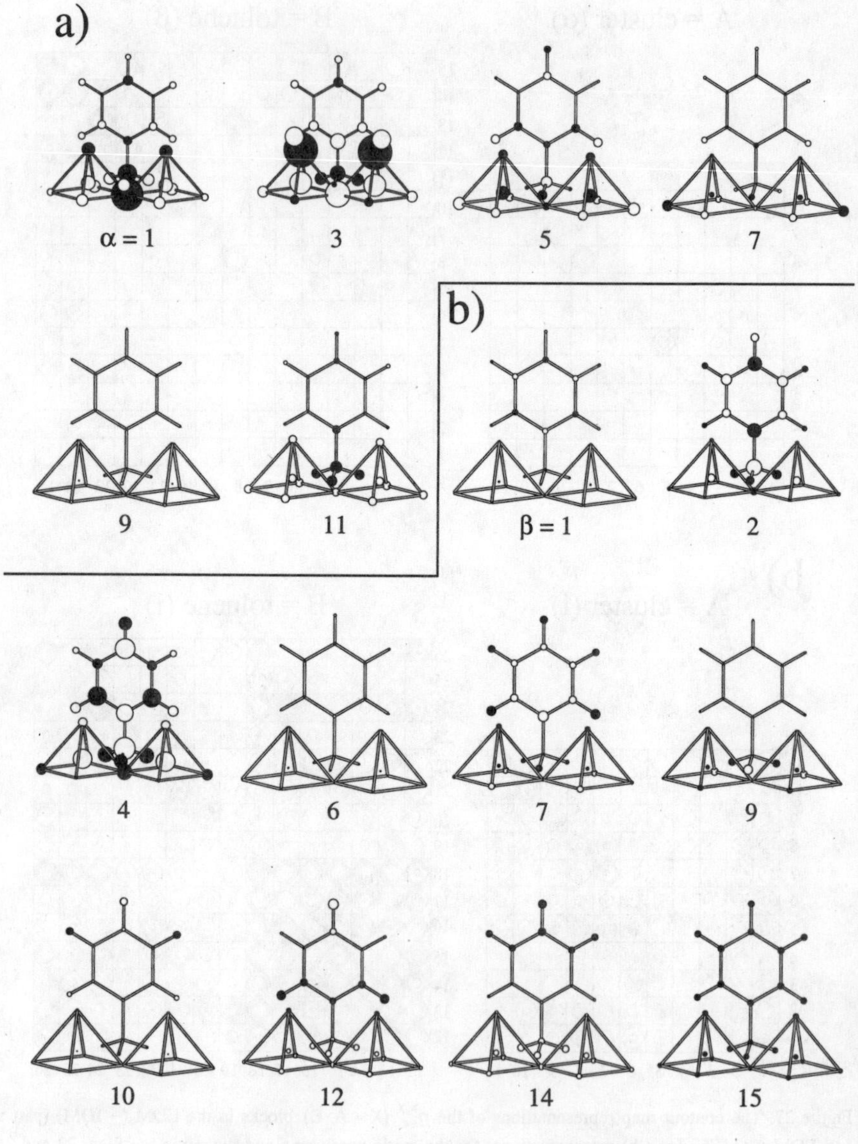

Figure 28. The EDM (→ IDM) resolution of the f^{CT} (see Figs. 17 A-c and 22). The diagrams represent the $\omega^{CT(EDM)}_{i,\alpha(\beta)}$ quantities, i= 1, ... , m, and $\alpha \in A$, $\beta \in B$; only the $F_{\alpha(\beta)} \neq 0$ modes are shown.

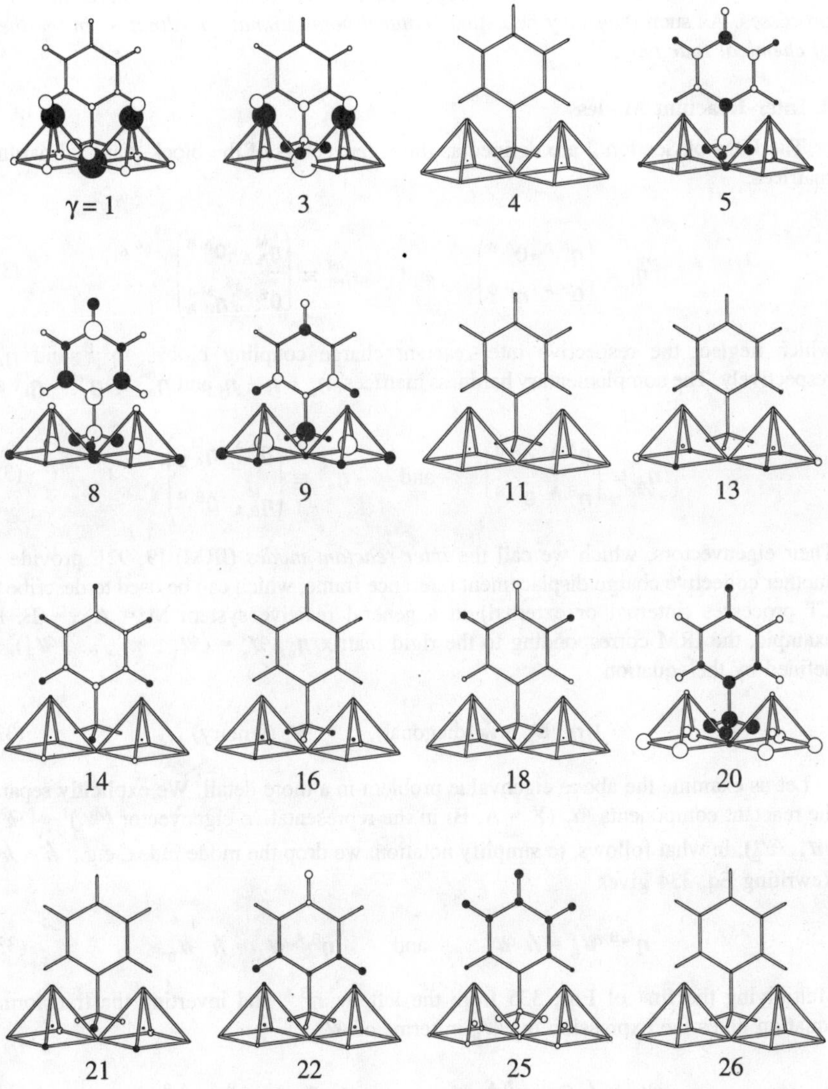

Figure 29. The PNM resolution of f^{CT} (Eq. 308) (see Figs. 17 A-c, 24, and caption to Fig. 28).

Hence, these two sets constitute valuable interpretative and diagnostic tools, designed to provide the most condensed description of the charge rearrangement accompanying external (between M and the reservoir) and internal (between A and B in M) CT processes. As such they may be called *"natural populational coordinates" in the theory of chemical reactivity.*

4. Inter-Reactant Modes

The IDM of Section 2 are defined as the eigenvectors of the block-diagonal hardness matrices,

$$\boldsymbol{\eta}_i \equiv \begin{pmatrix} \boldsymbol{\eta}^{A,A} & \mathbf{0}^{A,B} \\ \mathbf{0}^{B,A} & \boldsymbol{\eta}^{B,B} \end{pmatrix} \qquad \text{and} \qquad \boldsymbol{\eta}_i^{rel} \equiv \begin{pmatrix} \boldsymbol{\eta}_{A,A}^{M} & \mathbf{0}^{A,B} \\ \mathbf{0}^{B,A} & \boldsymbol{\eta}_{B,B}^{M} \end{pmatrix}, \tag{332}$$

which neglect the respective inter-reactant charge coupling blocks, $\boldsymbol{\eta}^{A,B}$ and $\boldsymbol{\eta}_{A,B}^{M}$, respectively. The complementary hardness matrices, $\boldsymbol{\eta}_e = \boldsymbol{\eta} - \boldsymbol{\eta}_i$ and $\boldsymbol{\eta}_e^{rel} \equiv \boldsymbol{\eta}^{rel} - \boldsymbol{\eta}_i^{rel}$ are:

$$\boldsymbol{\eta}_e \equiv \begin{pmatrix} \mathbf{0}^{A,A} & \boldsymbol{\eta}^{A,B} \\ \boldsymbol{\eta}^{B,A} & \mathbf{0}^{B,B} \end{pmatrix} \qquad \text{and} \qquad \boldsymbol{\eta}_e^{rel} \equiv \begin{pmatrix} \mathbf{0}^{A,A} & \boldsymbol{\eta}_{A,B}^{M} \\ \boldsymbol{\eta}_{B,A}^{M} & \mathbf{0}^{B,B} \end{pmatrix}. \tag{333}$$

Their eigenvectors, which we call the *inter-reactant modes* (IRM) [9, 92], provide yet another collective charge displacement reference frame, which can be used to describe the CT processes (internal or external) in a general reactive system M = A ---B. For example, the IRM corresponding to the rigid matrix $\boldsymbol{\eta}_e$, $\mathcal{U}_e = (\mathcal{U}_1, \mathcal{U}_2, \ldots, \mathcal{U}_m)$, are defined by the equation

$$\mathcal{U}_e^{\dagger}\, \boldsymbol{\eta}_e\, \mathcal{U}_e = \boldsymbol{h} \text{ (diagonal)}, \qquad \mathcal{U}_e \text{ (unitary)}. \tag{334}$$

Let us examine the above eigenvalue problem in a more detail. We explicitly separate the reactant components $\mathcal{U}_X$ (X = A, B) in the representative eigenvector $(\mathcal{U}_\gamma)^{\dagger} \equiv \mathcal{U}^{\dagger} = (\mathcal{U}_A^{\dagger}, \mathcal{U}_B^{\dagger})$; in what follows, to simplify notation, we drop the mode index, e.g., $h \equiv h_{\gamma,\gamma}$. Rewriting Eq. 334 gives

$$\boldsymbol{\eta}^{A,B}\, \mathcal{U}_B = h\, \mathcal{U}_A \qquad \text{and} \qquad \boldsymbol{\eta}^{B,A}\, \mathcal{U}_A = h\, \mathcal{U}_B. \tag{335}$$

Multiplying the first of Eqs. 335 from the left by $\boldsymbol{\eta}^{B,A}$ and inverting the transformed equation gives the expression for $\mathcal{U}_B$ in terms of $\mathcal{U}_A$,

$$\mathcal{U}_B = h\, \mathbf{G}_B^{-1}\, \boldsymbol{\eta}^{B,A}\, \mathcal{U}_A, \qquad \mathbf{G}_B = \boldsymbol{\eta}^{B,A}\, \boldsymbol{\eta}^{A,B}, \tag{336}$$

which can be used to obtain from the second Eq. 335 the following eigenvalue equation for the $\mathcal{U}_A$ component:

$$\mathbf{A}\,\mathcal{U}_A = \hbar^2\,\mathcal{U}_A \quad , \quad \mathbf{A} \equiv (\boldsymbol{\eta}^{A,B}\mathbf{G}_B^{-1}\boldsymbol{\eta}^{B,A})^{-1}\mathbf{G}_A \ , \tag{337}$$

where $\mathbf{G}_A = \boldsymbol{\eta}^{A,B}\boldsymbol{\eta}^{B,A}$ (see Eq. 336). A similar elimination process for the $\mathcal{U}_B$ component gives the corresponding eigenvalue problem,

$$\mathbf{B}\,\mathcal{U}_B = \hbar^2\,\mathcal{U}_B \quad , \quad \mathbf{B} \equiv (\boldsymbol{\eta}^{B,A}\mathbf{G}_A^{-1}\boldsymbol{\eta}^{A,B})^{-1}\mathbf{G}_B \ , \tag{338}$$

with the same eigenvalue as in Eq. 337.

Therefore, the $\mathcal{U}_A$ and $\mathcal{U}_B$ components of $\mathcal{U}$ are the eigenvectors of the $(n \times n)$ - $\mathbf{A}$ and $[(m-n) \times (m-n)]$ - $\mathbf{B}$ matrices, respectively, which exhibit the same eigenvalue $\hbar^2$. Let us assume that, e.g., $m > 2n$. Hence, the n eigenvectors $\mathcal{U}_A$ of $\mathbf{A}$ will combine with their n conjugates $\mathcal{U}_B$ among the eigenvectors of $\mathbf{B}$, which exhibit the same (positive) eigenvalue, and the remaining $m-2n$ eigenvectors of $\mathbf{B}$, $\mathcal{U}_B{}^{\ell}$, will combine with the zero component on A, giving rise to the *B - localized* (ℓ) IRM,

$$(\mathcal{U}_{\gamma}{}^{\ell})^{\dagger} = [\ \boldsymbol{o}_A^{\dagger}, (\mathcal{U}_B{}^{\ell})^{\dagger}\ ] \quad , \qquad \gamma = 2n+1, \dots, m; \tag{339}$$

it can be quickly verified using Eq. 334 that the eigenvalues of such localized modes vanish identically, $\hbar^{\ell} = 0$. Thus, the B-localized modes can only polarize B and they play no part in the CT between reactants. However, they can lead to a net change in N_B, thus exhibiting a nonvanishing external CT component of reactant B, i.e., generally nonvanishing w_{γ} and w_{γ}^{B} values, and $w_{\gamma}^{A} \equiv 0$. When the CT between reactants is considered, these modes will participate only in polarizing B, thus exhibiting generally nonvanishing F_{γ}^{CT}, ω_{γ}^{CT}, $\Omega_{B,\gamma}^{CT}$, and identically vanishing $\Omega_{A,\gamma}^{CT}$. The remaining (reactive) $2n$ eigenvectors will be delocalized throughout the whole M. They can be grouped in complementary pairs of IRM exhibiting the same magnitude of the opposite eigenvalues, $\hbar_+ = |\hbar|$ and $\hbar_- = -|\hbar|$:

$$\mathcal{U}(\pm|\hbar|) = [\mathcal{U}_A(\hbar^2) \pm \mathcal{U}_B(\hbar^2)\,]\,2^{-1/2}\ ; \tag{340}$$

again, one can verify these eigenvalues using Eq. 334.

There is a close analogy between this IRM representation and the localized molecular orbitals of the SCF MO theory. Namely, the qualitative localized MO, bonding and antibonding, can also be constructed by combining a pair of the directed hybrid orbitals of the bonded atoms, with the non-bonding hybrids describing lone electron pairs on constituent atoms. Following this analogy one could consider the above inter-reactant decoupling scheme as constructing the $\mathcal{U}_A$ and $\mathcal{U}_B$ charge displacement "hybrids" of reactants, prepared to reflect the actual reactant interaction in M.

In Fig. 30 we present the illustrative IRM diagrams for the toluene---$[V_2O_5]$ system, together with the relevant characteristics: their eigenvalues, hardness parameters, and quantities reflecting their participation in the external and internal (toluene → $[V_2O_5]$) CT. The corresponding $\{w_{i,\gamma}\}$ and $\{\omega_{i,\gamma}^{CT}\}$ diagrams, providing the IRM resolution of the f

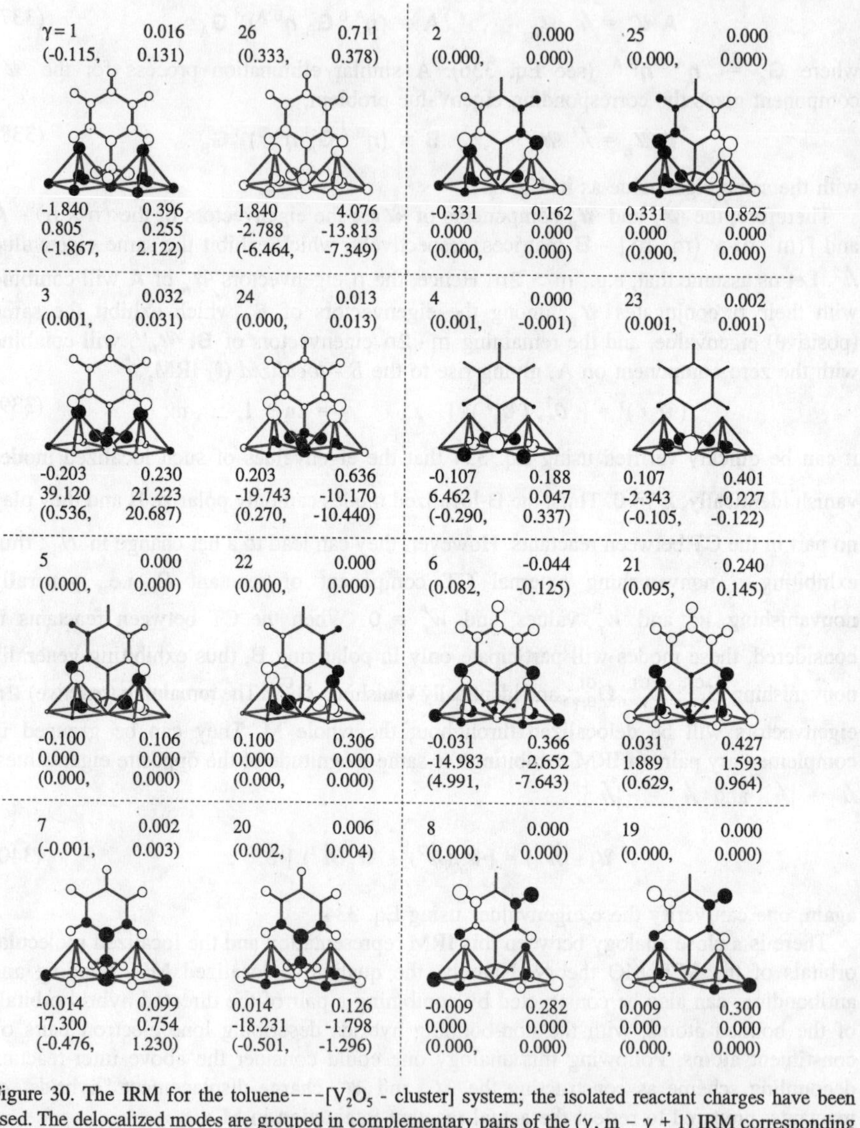

Figure 30. The IRM for the toluene---[V_2O_5 - cluster] system; the isolated reactant charges have been used. The delocalized modes are grouped in complementary pairs of the (γ, m − γ + 1) IRM corresponding to the same magnitude of the eigenvalue, and the B-localized modes, γ = 12-15, are shown in the last row of the figure. The numerical data above the mode diagram include: the mode number (ordering in accordance with increasing $h_{\gamma,\gamma}$), w_γ (first row), and its reactant resolution into (w_γ^A, w_γ^B) (second row). Below each diagram the following quantities are listed: $h_{\gamma,\gamma}$, $\tilde{\eta}_{\gamma,\gamma} \equiv [(\mathcal{U}_e)^\dagger\ \eta\ \mathcal{U}_e]_{\gamma,\gamma}$ (first row), F_γ^{CT}, ω_γ^{CT}, and ($\Omega_{A,\gamma}^{CT}$, $\Omega_{B,\gamma}^{CT}$) (third row). The CT parameters are calculated for the assumed toluene → [V_2O_5] electron transfer.

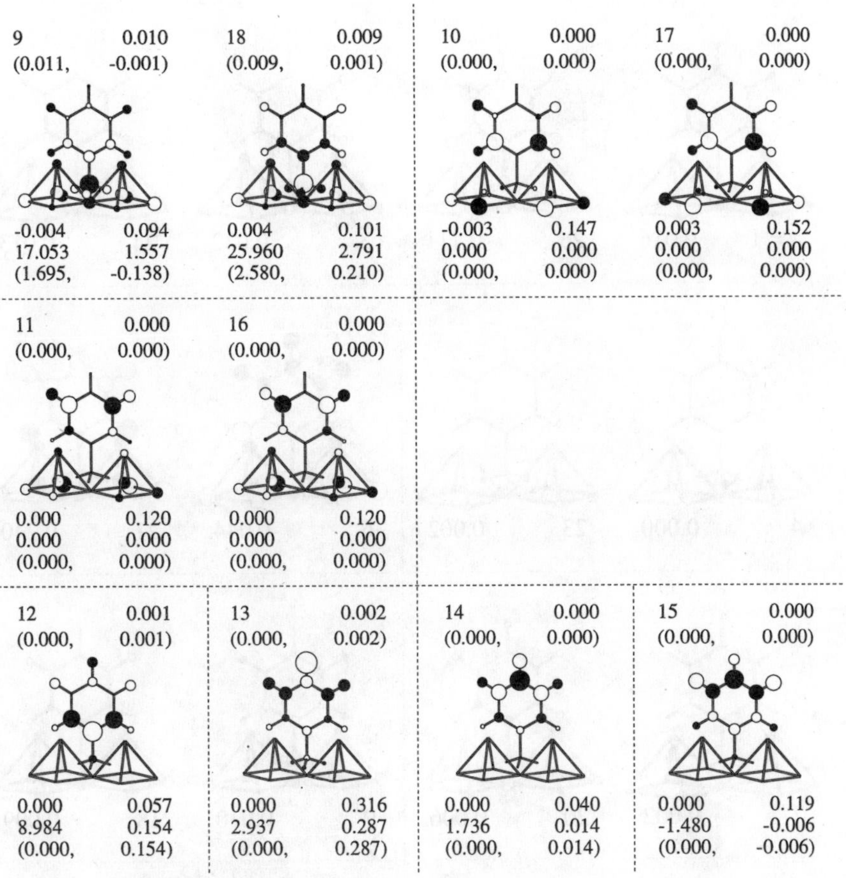

Figure 30. Continued.

and $\boldsymbol{f}^{CT}$ vectors, respectively, are shown in Figs. 31 and 32. Finally, a general structure of the hardness matrix in the IRM representation is characterized by the contour map diagram of Fig. 33.

It follows from Fig. 31 that the external FF indices, $\boldsymbol{f}$, are determined by the two complementary IRM, γ = (6, 21), which originate from the single $\mathcal{U}_A$ and $\mathcal{U}_B$ components. A reference to Fig. 17 A-e shows that indeed their combined effect correctly reproduces the cluster part of $\boldsymbol{f}$, with only minor charge shifts on the toluene constituent atoms. The toluene part of the FF diagram is seen to be mainly due to the background, nonselective mode γ = 26. The four toluene-localized IRM, γ =12-15, are seen to be only slightly involved in the external CT.

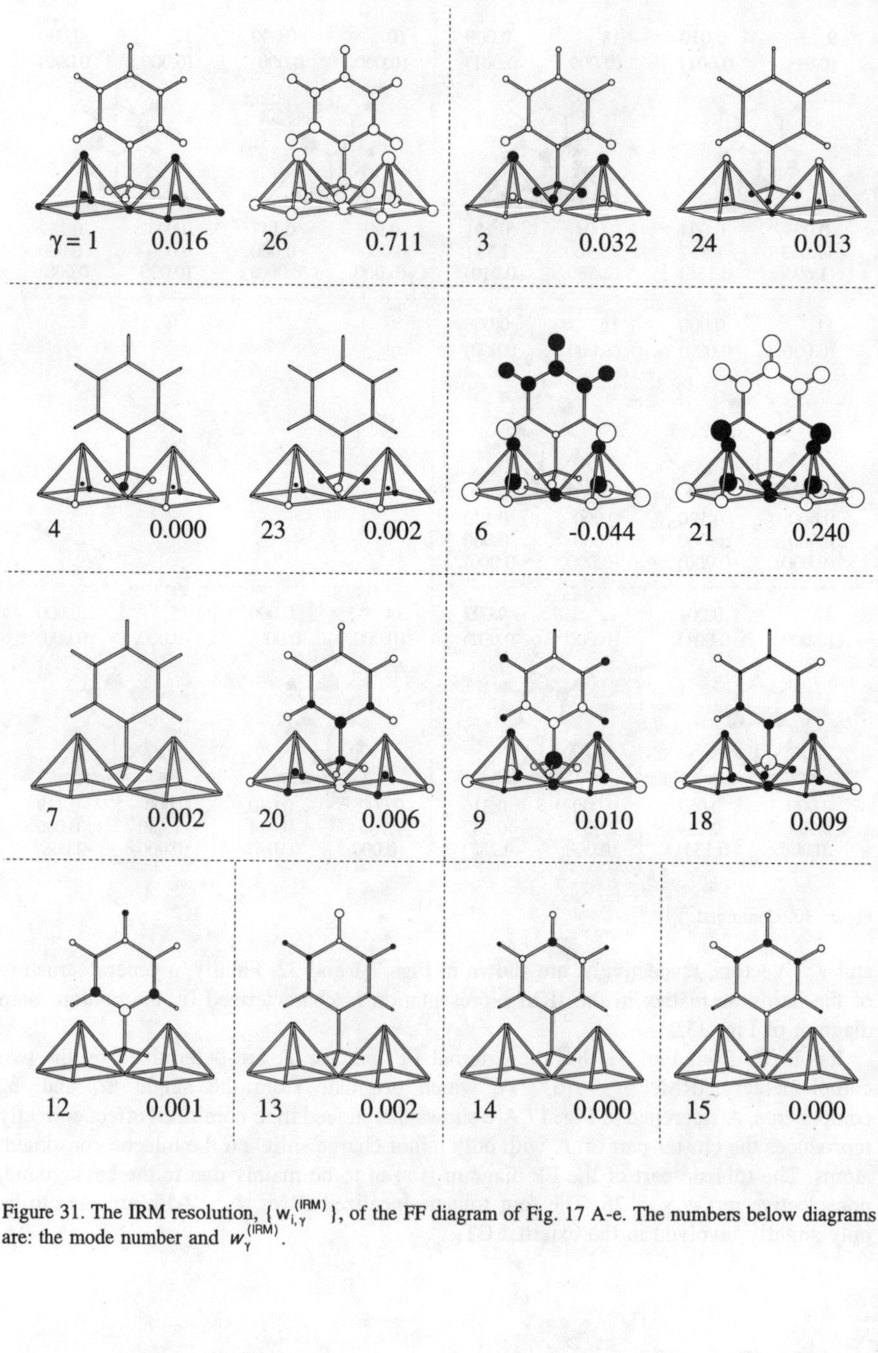

Figure 31. The IRM resolution, $\{ w_{i,\gamma}^{(IRM)} \}$, of the FF diagram of Fig. 17 A-e. The numbers below diagrams are: the mode number and $w_{\gamma}^{(IRM)}$.

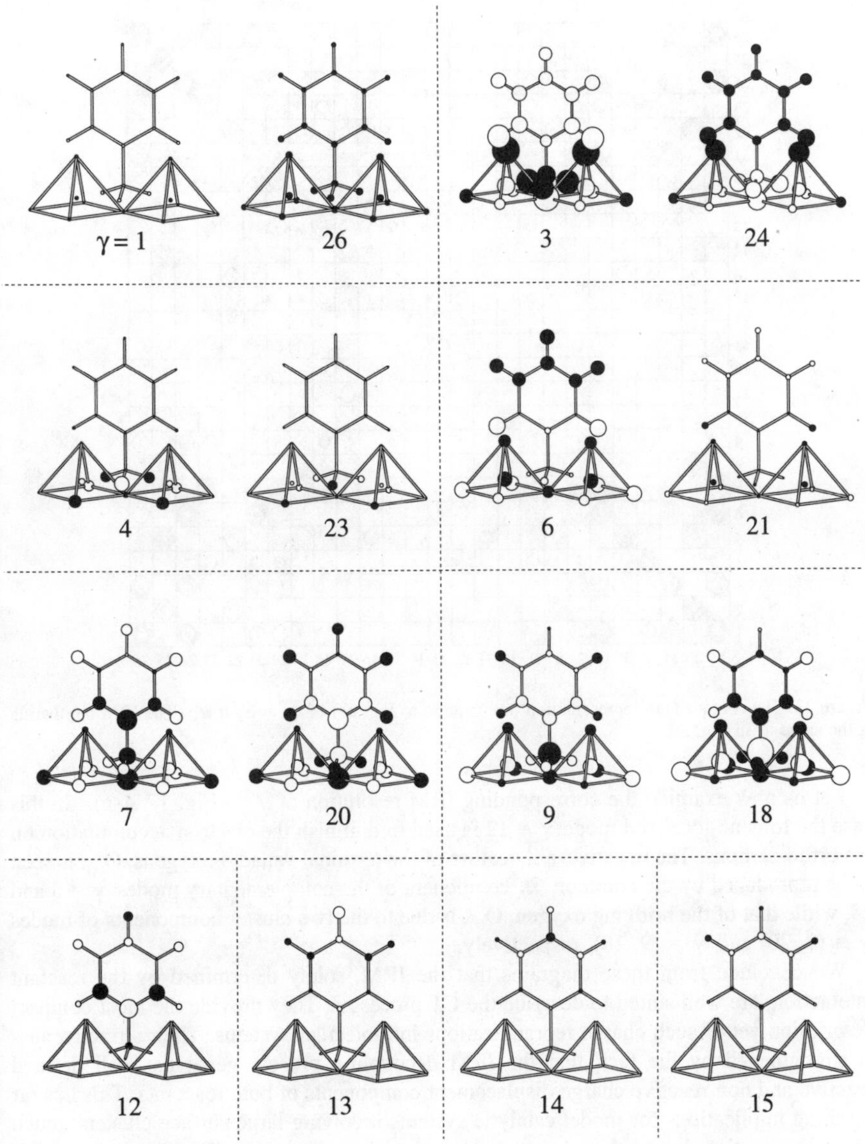

Figure 32. The IRM resolution, $\{\omega_{i,\gamma}^{CT(IRM)}\}$, of the f^{CT} diagram of Fig. 17 A-c.

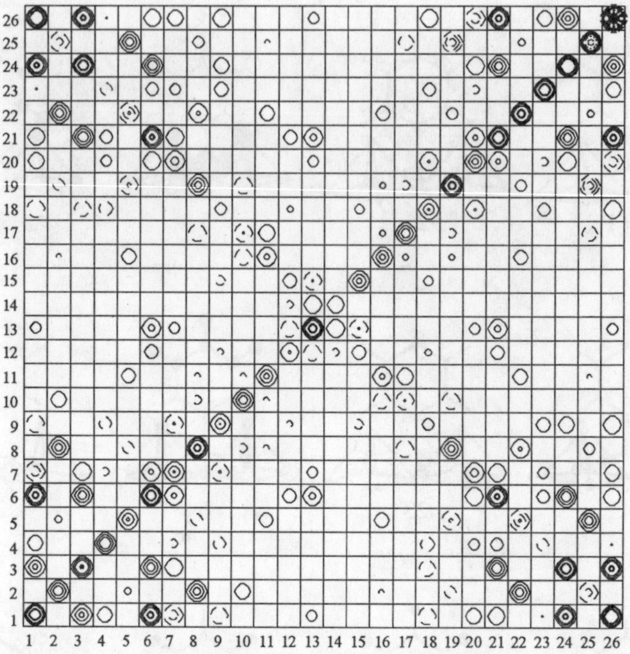

Figure 33. The contour map representation (see caption to Fig. 27) of $\tilde{\eta} \equiv \mathcal{U}_e \eta \mathcal{U}_e$. The IRM numbering is the same as in Fig. 30.

Let us now examine the corresponding IRM resolution of $\boldsymbol{f}^{CT}$ (Fig. 17 A-c). In this case the toluene-localized mode $\gamma = 12$ is used to diminish the electron accumulation on the ortho-carbons. The negative f_i^{CT} feature of the terminal vanadyl oxygens, $O_{(1)}$, is seen to be reproduced by the common $\mathcal{U}_A$ component of the complementary modes $\gamma = 3$ and 24, while that of the bridging oxygen, $O_{(2)}$, is due to the two cluster components of modes $\gamma = (7, 20)$ and $\gamma = (9, 18)$, respectively.

We conclude from these diagrams that the IRM, solely determined by the reactant interaction, are well suited to describe the CT processes. They provide the most compact description yet of such charge reorganizations in molecular systems. Their attractiveness is strengthened by the fact, that the IRM decoupling process generates well defined reactive and non-reactive charge displacement components of both reactants. This has far reaching implications for model catalytic systems involving large surface clusters, much exceeding in size that of the adsorbate. In such structures the localized (environmental) modes will solely involve the substrate and delocalized (reactive) modes will involve both the cluster and the substrate. The substrate-localized modes, which substantially contribute to $\boldsymbol{f}$ or $\boldsymbol{f}^{CT}$ could then be used to diagnose the influence of the cluster environment upon the active site of the surface or the adsorbate.

5. Mapping Relations

The ground-state relation between the geometric and electronic structure quantities provides a map between the nuclear position and electronic population degrees of freedom of molecular systems. An understanding of a subtle interplay between these parameters has been a major goal of theoretical chemistry [95, 112-114]. It has recently been demonstrated [32, 86, 92, 115], that CSA provides a theoretical framework for constructing explicit *mapping relations* between the AIM (or collective) charge (electron population) coordinates and the corresponding nuclear (or collective) displacement modes. The crucial element of this development is the linear response (internal softness) matrix $\boldsymbol{\beta} = -\mathbf{S}$ (Eq. 215), which links these two (electronic and geometrical) structural aspects of molecular systems in the AIM resolution.

Let $\boldsymbol{Q}$ denotes the minimum set of bonds and angles that uniquely specify the system structure. The AIM population displacements for constant N, $(\mathrm{d}\boldsymbol{N})_N$, are linked with the corresponding geometrical displacements via the chain rule transformation [86]:

$$(\mathrm{d}\boldsymbol{N})_N = \mathrm{d}\boldsymbol{Q}(\partial \boldsymbol{v}/\partial \boldsymbol{Q})\boldsymbol{\beta} \equiv \mathrm{d}\boldsymbol{Q}\mathbf{G}\boldsymbol{\beta} \equiv \mathrm{d}\boldsymbol{Q}\,\mathrm{T}(\boldsymbol{Q}\to\boldsymbol{N})_N, \tag{341}$$

where $\boldsymbol{\beta} = (\partial \boldsymbol{N}/\partial \boldsymbol{v})_N$ and the *geometrical transformation* $\mathbf{G}$ represents the external potential response per unit changes in geometrical parameters. The latter is analytically available for the internuclear distances, $\boldsymbol{Q} = \boldsymbol{R} \equiv (\boldsymbol{R}^{\mathrm{b}}, \boldsymbol{R}^{\mathrm{nb}})$, including the bonding (b) and nonbonding (nb) internuclear separations, respectively, in the "frozen" AIM charges approximation, $\boldsymbol{q} = \boldsymbol{q}^0 = \boldsymbol{Z} - \boldsymbol{N}^0$, where $\boldsymbol{Z}$ groups the nuclear atomic numbers and $\boldsymbol{N}^0$ denotes the equilibrium AIM populations at the starting geometric structure $\boldsymbol{Q}^0$.

The charges from EEM calculations [10] where shown to provide an adequate basis for the model force field, which determines realistic bond-stretch Hessian matrix, $\mathcal{H}^{\mathrm{b}} = \partial^2 E/\partial \boldsymbol{R}^{\mathrm{b}} \partial \boldsymbol{R}^{\mathrm{b}}|_{\boldsymbol{Q}^0}$, calculated for the fixed bond angles at the starting geometry [86]. This force field uses for each atomic pair the effective point-charge electrostatic interactions as the potential component of this pair contribution to the total Born-Oppenheimer energy, and supplements it with a matching electronic kinetic energy function, obtained from the virial theorem, in the spirit of the familiar simple-bond-charge model [116]. Thus, one attributes in this approximation the Fues potential-like contribution to each atomic pair (bonding or nonbonding) to the overall potential energy surface for nuclear motions. The full geometric Hessian, $\mathcal{H} = \partial^2 E/\partial \boldsymbol{Q}\, \partial \boldsymbol{Q}|_{\boldsymbol{Q}^0}$, or $\mathcal{H}^{\mathrm{b}}$ can also be generated from independent SCF MO/CI or DFT calculations.

Clearly, the $\mathrm{d}\boldsymbol{N}$ and $\mathrm{d}\boldsymbol{Q}$ vectors have different dimensions. Should one be interested in relating the $(m-1)$ populational degrees of freedom for constant N, $(\mathrm{d}\boldsymbol{N})_N$, to the equidimensional subset, $\boldsymbol{Q}^{\mathrm{s}}$ of $\boldsymbol{Q}$, e.g., $\boldsymbol{Q}^{\mathrm{b}} \equiv \boldsymbol{R}^{\mathrm{b}}$ in noncyclic systems, for the fixed bond angles, one requires an additional *constraint transformation*, $\boldsymbol{\tau}^{\mathrm{s}} = \partial \boldsymbol{Q}/\partial \boldsymbol{Q}^{\mathrm{s}}$, which is available from geometrical considerations [86]. For example, the $\mathrm{d}\boldsymbol{R}^{\mathrm{b}} \to (\mathrm{d}\boldsymbol{N})_N$ mapping is given by the product (chain rule) transformation:

$$(\mathrm{d}\boldsymbol{N})_N = \mathrm{d}\boldsymbol{R}^{\mathrm{b}}\boldsymbol{\tau}^{\mathrm{b}}\mathrm{T}(\boldsymbol{Q}\to\boldsymbol{N})_N \equiv \mathrm{d}\boldsymbol{R}^{\mathrm{b}}\,\mathrm{T}(\boldsymbol{R}^{\mathrm{b}}\to\boldsymbol{N})_N, \tag{342}$$

These *local mapping transformations*, relating AIM electron populations (charges) to bond lengths, can be easily generalized into relations involving collective electron population and/or nuclear position displacements, e.g., the PNM and nuclear normal modes. The bond-stretch normal vibrations, $\mathcal{Q}^b$, defined by the components of the $\mathcal{H}^b$ principal directions, $\mathbf{O} = \partial \mathcal{Q}^b / \partial \boldsymbol{R}^b$,

$$\mathbf{O}^\dagger\, \mathcal{H}^b\, \mathbf{O} = \mathbf{k}^b = \{k_m^b \delta_{mn}\}, \qquad \mathbf{O}^\dagger\, \mathbf{O} = \mathbf{I}, \tag{343}$$

are related to the AIM charges through the $d\mathcal{Q}^b \to (d\boldsymbol{N})_N$ mapping:

$$(d\boldsymbol{N})_N = d\mathcal{Q}^b\, \mathbf{O}^\dagger\, \mathbf{T}(\boldsymbol{R}^b \to \boldsymbol{N})_N \equiv d\mathcal{Q}^b\, \mathbf{T}(\mathcal{Q}^b \to \boldsymbol{N})_N\,. \tag{344}$$

Consider now the set of $(m-1)$ collective internal PNM (INM), represented by all the eigenvectors of $\boldsymbol{\beta}$ but the zero-eigenvalue ($d\boldsymbol{N}$, external CT) mode $\alpha = m$ (see Fig. 23) for which all components are identical:

$$\mathbf{c}^\dagger \boldsymbol{\beta}\, \mathbf{c} = \mathbf{s} = \{s_\alpha \delta_{\alpha\gamma}\}, \qquad \mathbf{c}^\dagger\, \mathbf{c} = \mathbf{I}\,. \tag{345}$$

The transformation $\mathbf{c}$ may be partitioned into the matrix $\mathbf{c}'$ grouping the first $(m-1)$ columns, representing independent internal charge redistribution modes, and the last column representing the pure CT displacement: $\mathbf{c} = (\mathbf{c}', \boldsymbol{c}_m)$. The $(m-1)$ collective populational variables,

$$d\boldsymbol{p}' = (d\boldsymbol{N})_N \mathbf{c}', \tag{346}$$

can now be related to geometrical degrees of freedom. For example, multiplying Eq. 342 by $\mathbf{c}'$ immediately gives the $\mathbf{T}(\boldsymbol{R}^b \to \boldsymbol{p}')_N$ map:

$$d\boldsymbol{p}' = d\boldsymbol{R}^b \mathbf{T}(\boldsymbol{R}^b \to \boldsymbol{N})_N \mathbf{c}' \equiv d\boldsymbol{R}^b \mathbf{T}(\boldsymbol{R}^b \to \boldsymbol{p}')_N\,. \tag{347}$$

Finally, by expressing $d\boldsymbol{R}^b$ in terms of $d\mathcal{Q}^b$, $d\boldsymbol{R}^b = d\mathcal{Q}^b\, \mathbf{O}^\dagger$, gives the $d\mathcal{Q}^b \to d\boldsymbol{p}'$ mapping transformation:

$$d\boldsymbol{p}' = d\mathcal{Q}^b\, \mathbf{O}^\dagger \mathbf{T}(\boldsymbol{R}^b \to \boldsymbol{p}')_N \equiv d\mathcal{Q}^b\, \mathbf{T}(\mathcal{Q}^b \to \boldsymbol{p}')_N\,. \tag{348}$$

Obviously, one can derive similar mapping relations involving the populational and/or nuclear MEC or REC [32]. However, one has to recognize their non-orthogonal character when generating the relevant transformations.

It should be observed at this point that the whole MEC/REC development of Sections 6.2 and 6.3 in Chapter 3 can also be carried out within the nuclear position space $\boldsymbol{Q}$. Consider, e.g., the bond-stretch MEC defined in the subspace spanned by the orthogonal basis vectors $\boldsymbol{\epsilon}^b = (\epsilon_1^b, \ldots, \epsilon_{m-1}^b)$, associated with the independent bond lengths: $(\boldsymbol{\epsilon}^b)^\dagger \cdot \boldsymbol{\epsilon}^b = \mathbf{I}^b$. They can be expressed in terms of the basis vectors $\boldsymbol{\epsilon}^b$: $\boldsymbol{r}_b \equiv \boldsymbol{\epsilon}^b\, \mathbf{m}^b$;

the transformation matrix can be expressed in terms of the nuclear compliant matrix $\mathscr{C}^b \equiv (\mathscr{H}^b)^{-1}$ (see Eqs. 272, 273, 276, and 277):

$$\vec{r}_b^{\,k} = \sum_{\ell} \left(\partial R_\ell^b / \partial R_k^b\right)_{E=\min} \vec{\epsilon}_\ell^{\,b} = \sum_{\ell} (\mathscr{C}^b_{\ell,k} / \mathscr{C}^b_{k,k})\, \vec{\epsilon}_\ell^{\,b}$$
$$\equiv \sum_{\ell} \{\ell\}_k^b\, \vec{\epsilon}_\ell^{\,b} \equiv \sum_{\ell} m_{\ell,k}^b\, \vec{\epsilon}_\ell^{\,b}, \qquad k = 1, \ldots, m-1 \ . \tag{349}$$

The radius vector of an arbitrary point, $\vec{R}^{(b)} = \vec{\epsilon}^{\,b}\,(\boldsymbol{R}^b)^\dagger = \vec{r}_b\,(\boldsymbol{r}_b)^\dagger$, depends upon the bond length MEC $\boldsymbol{r}_b$, related to $\boldsymbol{R}^b$ through the transformation $\boldsymbol{R}^b = \boldsymbol{r}_b\,(\mathbf{m}^b)^\dagger$. Thus, the metric tensor of the bond-stretch MEC space is given by $\boldsymbol{m}^b = (\mathbf{m}^b)^\dagger\,\mathbf{m}^b$, since

$$(ds)^2 \equiv d\vec{R}^{(b)} \cdot d\vec{R}^{(b)} = \sum_i \sum_j \left(\frac{\partial \vec{R}^{(b)}}{\partial r_b^i}\right) \cdot \left(\frac{\partial \vec{R}^{(b)}}{\partial r_b^j}\right) dr_b^i\, dr_b^j$$
$$= \sum_i \sum_j (\vec{r}_b^{\,i} \cdot \vec{r}_b^{\,j})\, dr_b^i\, dr_b^j = \sum_i \sum_j \left(\sum_\ell m_{\ell,i}^b\, m_{\ell,j}^b\right) dr_b^i\, dr_b^j$$
$$\equiv \sum_i \sum_j \boldsymbol{m}_{i,j}^b\, dr_b^i\, dr_b^j \ . \tag{350}$$

Therefore,

$$d\boldsymbol{R}^b = d\boldsymbol{r}^b\,(\mathbf{m}^b)^\dagger \equiv d\boldsymbol{r}^b\,\mathscr{B} \ . \tag{351}$$

Consider now the $(m-1)$ independent, internal MEC, associated with the primary populational displacements on independent atoms: $i = 1, \ldots, m-1$ (see Section 6 of Chapter 3). Thus, we regard the population of the atom m as dependent, through the closure constraint N = const., on populations of the remaining atoms, $\boldsymbol{N}' = (N_1, \ldots, N_{m-1})$. These populations are the projections of the general population vector, $\vec{\mathscr{N}}$, onto the populational directions of independent AIM, defined by the corresponding (orthogonal) basis vectors $\vec{\boldsymbol{e}}' = (\vec{e}_1, \ldots, \vec{e}_{m-1})$: $\vec{\mathscr{N}} = \vec{\boldsymbol{e}}'(\boldsymbol{N}')^\dagger$. Let the $\vec{\boldsymbol{e}}'$ components of the corresponding (non-orthogonal) internal MEC, $\vec{\boldsymbol{n}}' = (\vec{n}_1, \ldots, \vec{n}_{m-1})$, be defined by the transformation $\mathbf{M}'$: $\vec{\boldsymbol{n}}' = \vec{\boldsymbol{e}}'\,\mathbf{M}'$. The population vector has the following MEC coordinates: $\vec{\mathscr{N}} = \vec{\boldsymbol{n}}'\,(\mathscr{N}')^\dagger$, where $\boldsymbol{N}' = \mathscr{N}'\,(\mathbf{M}')^\dagger$. Therefore, the metric tensor of the vector space spanned by these populational MEC is given by:

$$\mathscr{M}' = (\vec{\boldsymbol{n}}')^\dagger\,\vec{\boldsymbol{n}}' = (\mathbf{M}')^\dagger\,\mathbf{M}' \ , \tag{352}$$

and

$$d\boldsymbol{N}' = d\mathscr{N}'\,(\partial \boldsymbol{N}' / \partial \mathscr{N}') = d\mathscr{N}'\,(\mathbf{M}')^\dagger \equiv d\mathscr{N}'\mathscr{A} \ . \tag{353}$$

Equations 351 and 353 when inserted into the previously derived mapping transformations involving $\mathrm{d}\boldsymbol{R}^{b}$ and/or $(\mathrm{d}\boldsymbol{N})_{N}$ give their respective MEC analogs. For example, Eqs. 342 and 351 give the mapping between nuclear MEC and atomic electron populations in the N-constrained process:

$$(\mathrm{d}\boldsymbol{N})_{N} = \mathrm{d}\boldsymbol{r}^{b}\,\mathscr{B}\,\mathbf{T}(\boldsymbol{R}^{b}\rightarrow\boldsymbol{N})_{N} \equiv \mathrm{d}\boldsymbol{r}^{b}\,\mathbf{T}(\boldsymbol{r}^{b}\rightarrow\boldsymbol{N})_{N}\,. \tag{354}$$

Moreover, by the closure condition,

$$(\mathrm{d}\boldsymbol{N})_{N} = [\mathrm{d}\boldsymbol{N}', -\mathrm{d}\boldsymbol{N}'(\boldsymbol{1}')^{\dagger}] \equiv \mathrm{d}\boldsymbol{N}'\,\mathscr{Z}\,, \tag{355}$$

where the $(m-1)\times m$ rectangular matrix $\mathscr{Z}$ consists of the $(m-1)\times(m-1)$ identity matrix $\mathbf{I}'$ and the negative unit column vector, $-(\boldsymbol{1}')^{\dagger}$, as the additional m-th column generating the dependent atom populational displacement.

Combining Eqs. 353-355 gives the implicit relation linking the populational and nuclear (bond-stretch) MEC:

$$\mathrm{d}\boldsymbol{N}'\,\mathscr{Z} = \mathrm{d}\mathscr{N}'\,\mathscr{A}\,\mathscr{Z} = \mathrm{d}\boldsymbol{r}^{b}\,\mathbf{T}(\boldsymbol{r}^{b}\rightarrow\boldsymbol{N})_{N}\,. \tag{356}$$

To obtain an explicit mapping relation between these MEC displacements, one has to define the $\mathbf{T}(\boldsymbol{r}^{b}\rightarrow\boldsymbol{N}')_{N}$ transformation, which uses the independent atom matrix $\boldsymbol{\beta}' = (\partial\boldsymbol{N}'/\partial\boldsymbol{v})_{N}$, consisting of all but the last columns of $\boldsymbol{\beta}$ in the transformation of Eq. 341. Such a change modifies the left hand sides of equations involving $(\mathrm{d}\boldsymbol{N})_{N}$ into $\mathrm{d}\boldsymbol{N}'$, which can be explicitly expressed in terms of $\mathrm{d}\mathscr{N}'$, (see Eq. 353). The explicit mapping between the nuclear and populational MEC reads

$$\mathrm{d}\mathscr{N}' = \mathrm{d}\boldsymbol{r}^{b}\,\mathbf{T}(\boldsymbol{r}^{b}\rightarrow\boldsymbol{N}')_{N}\,\mathscr{A}^{-1} \equiv \mathrm{d}\boldsymbol{r}^{b}\,\mathbf{T}(\boldsymbol{r}^{b}\rightarrow\mathscr{N}')_{N}\,. \tag{357}$$

In a similar way one can derive the relevant mapping transformations linking the MEC of one space with the normal modes of the other space. We would like to stress that the mapping relations between equidimensional nuclear and populational space displacements should be invertible, so that one can easily write down the nuclear displacements in terms of the relevant populational shifts.

In Fig. 34 we report illustrative mapping "translators", shown as contour maps, between the geometrical (bond-stretch) and populational degrees of freedom in CH_3OH [86]. They have been obtained from the MNDO bond-stretch Hessian matrix and the relevant charge sensitivity data from the CSA-EEM calculations in the AIM resolution.

The first panel (a1) shows the bond length responses per unit populational displacements (chemical reductions, $\mathrm{d}N_i = 0$) of constituent AIM. It can be seen in the diagram that such test displacements on methyl hydrogens (reversing original bond polarization) are predicted to elongate both the CO (secondary response) and CH (primary response) bonds. A similar direct geometrical relaxation of stretching the CO bond is predicted, when electron population on C is increased. The opposite, bond shortening geometry responses are predicted, when the hydroxyl atoms are reduced, with the major

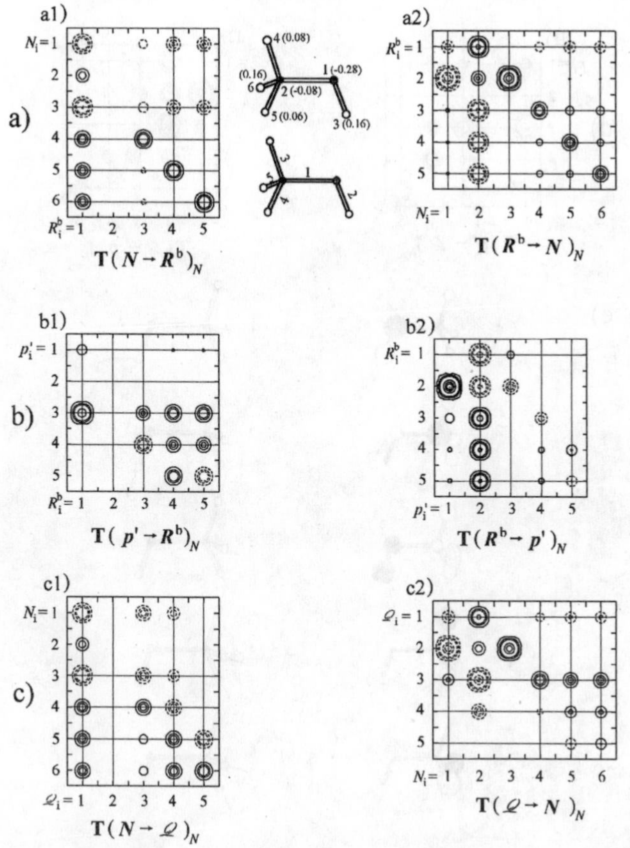

Figure 34. Illustrative mapping relations [from Ref. 86] between the populational and geometrical (bond-stretch) degrees of freedom in methanol: local-local (part a), collective-local (part b), local-collective (part c), and collective-collective (part d). The corresponding local and collective structure variables as well as the initial EEM charges are defined in additional diagrams of parts a and e. The matrix "translators" $\mathsf{T}(\boldsymbol{x} \rightarrow \boldsymbol{y})$ are interpreted as "surfaces", represented graphically as the corresponding contour maps, with the row (independent) and column (dependent) variables $\boldsymbol{x}$ and $\boldsymbol{y}$ being fixed along the horizontal and vertical lines, respectively. In part e the same convention as that previously used to illustrate collective charge displacements (first column) has been adopted, with the circle areas being proportional to the $|c_{i,\alpha}|$ coefficients (Eq. 345); the lengths of arrows in the second column of part e are proportional to the dominating $|O_{r,n}|$ components (Eq. 343) in the n-th *vibrational normal mode* (VNM). There are two sets of contour values shown in the figure, common for panels in each column, respectively: in the first column (a1-d1) the starting value of $|T_{k,l}| = 2$, with equal spacing between 3 consecutive contours $\Delta = 2$; above $|T_{k,l}| = 6$ the $\Delta = 10$ increment has been used; the contour values in the second column (a2-d2) are divided by 100, relative to those used in the first column. Negative values are represented by broken lines. In part e the INM contours have been arranged to exhibit the maximum visual resemblance to the corresponding VNM.

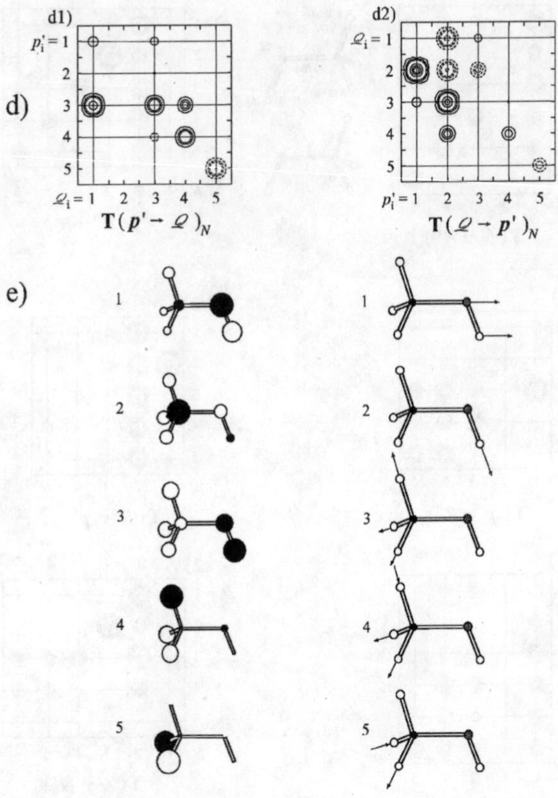

Figure 34. Continued.

relaxational responses involving a dominating increase in R^b_{CO} for both cases (i = 1, 3); there is also a secondary structural change, $dR^b_{CH} < 0$, accompanying a reduction of the OH group. The assumed chemical reduction on oxygen atom implies an increase in the initial polarity of the CO bond, so that the diagram a1 associates the CO bond shortening with the increased bond polarity. A similar physical implication follows from the $dN_C > 0$ displacement, which decreases the initial CO bond polarity and as such is related to the CO bond lengthening response. One could interpret the CH bond length responses to populational (charge) shifts on methyl hydrogens in the same spirit: test reductions on hydrogens decrease bond polarity and, therefore, are predicted to stretch the CH bonds. Finally, one also concludes from the map a1 that the OH bond remains relatively resistant to all AIM charge displacements. Clearly, all above responses are reversed when the test chemical oxidations ($dN_i < 0$) are considered.

Let us now examine the inverse mapping transformation (panel a2), which has direct connection to the Gutmann *bond-length variation rules* [112]. This map exhibits the dominant and intuitively expected *direct charge response effect* of a decreasing polarity of the elongated bond. For example, a test stretch of the CO bond mainly increases the electron population on C and decreases it on O, thus resulting in a lower bond polarity. Similarly, $\mathrm{d}R^{b}_{OH} > 0$ generates $\mathrm{d}N_{O} < 0$ and $\mathrm{d}N_{H(O)} > 0$ thus diminishing the initial bond polarization. The same, though weaker effect is detected in all CH bonds, with the carbon atom decreasing its electron population and the corresponding hydrogen increasing it when the bond is elongated. All these primary charge relaxation effects are reversed when the $\mathrm{d}R^{b}_{i} < 0$ primary structural displacements are examined.

There are several weak indirect bond polarization responses seen in the map a2 of Fig. 34. Namely, the CO stretch is seen to diminish the original polarity of the OH bond, since the magnitude of the net (negative) charge on O decreases and that on the hydroxyl hydrogen remains approximately constant. Similarly, the OH stretch is predicted to result in diminished initial CO bond polarity, since it implies the secondary responses: $\delta N_{C} > 0$ and $\delta N_{O} < 0$.

In non-cyclic systems each bond, say i, divides the molecule into two complementary subsystems. Let b^{D}_{i} and a^{A}_{i} denote the donor (basic) and acceptor (acidic) atoms of the heteronuclear, coordination bond $b^{D}_{i} \rightarrow a^{A}_{i}$ with D_i and A_i standing for the donor (consisting b^{D}_{i}) and acceptor (including a^{A}_{i}) subsystems, respectively:

$$\cdots \overbrace{\underset{\delta-}{a^{D}_{i}} \;——\; \underset{\delta+}{b^{D}_{i}}}^{D_i} \;------\; \overbrace{\underset{\delta-}{a^{A}_{i}} \;——\; \underset{\delta+}{b^{A}_{i}}}^{A_i} \cdots \tag{358}$$

here a^{D}_{i} and b^{A}_{i} represent the acidic and basic atoms in D and A, respectively, bonded to the two coordinating atoms of the perturbed bond, generally exhibiting fractional negative and positive charges, respectively, as a result of the intra-D_i and intra-A_i bond formations. The intuitively expected electron distribution response to a lengthening of bond i can be summarized by the following Gutmann-like diagram:

$$\cdots \underbrace{a^{D}_{i} \overset{\curvearrowleft}{\longleftarrow} b^{D}_{i}}_{D_i} \quad \underset{\mathrm{d}R^{b}_{i} > 0}{\overset{\curvearrowleft}{\longleftarrow\!\!\longleftarrow}} \quad \underbrace{a^{A}_{i} \overset{\curvearrowleft}{\longrightarrow} b^{A}_{i}}_{A_i} \cdots \tag{359}$$

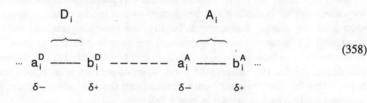

where the arrows connecting the bonded nuclei point toward the induced electron shift (bent arrows) or an increase in bond lengths (straight arrows). This diagram takes into account a decreased CT component of the bond being stretched (the directly coupled response) and the polarizations inside D_i and A_i (indirectly coupled, induced responses), which moderate the effects of the primary charge shift, in the spirit of the Le Châtelier-Braun principle.

Paraphrasing the first bond-length variation rule of Gutmann, which summarizes changes in the molecule D and A when the intermolecular distance D --- A is increased, suggests the following general behaviour: the larger the distance R_i^b, the lower polarization of the stretched bond and the greater coordination (shorter bond lengths) of bonds involving b_i^D and a_i^A. As a result of this charge redistribution the a_i^D atom becomes less acidic and the b_i^A atom becomes less basic. However, since the direct charge responses are expected to be larger than the secondary ones, the outflow (inflow) of electrons from a_i^A (to b_i^D) should be larger than the moderating (induced) inflow (outflow) due to remaining atoms, when R_i^b is increased. This also suggests a decrease in the initial bond polarity of the other bonds involving constituent atoms of the primarily stretched bond.

Some of these qualitative predictions are indeed observed in Fig. 34 (panel a2). For example, as we have already observed, the $dR_2^b > 0$ perturbation of the OH bond does indeed results in a decreased CO bond polarity; a similar effect is observed for the OH bond when R_{CO}^b is increased.

We now turn to the mixed translators of part b. The first of them (b1) maps the collective charge mode displacements (along INM, part e) into the bond length variations. The modes α = 1, 2 are seen to have a relatively weak effect upon all bonds. The α = 3 mode, describing the [OH] → [CH_3] internal CT, has the strongest geometry relaxation response, with all bonds but OH being predicted to increase their lengths as a result of such a forced charge displacement. Finally, the remaining internal modes, α = 4, 5, describing CT between methyl hydrogens, are seen to give rise to the CH bond length differentiation, as intuitively expected. It should be noticed, that the previously observed rule of lengthening (shortening) the bond, when its polarity is decreased (increased), also holds when the charge redistribution results from the INM displacement, in accordance with the mode contours shown in part e.

The inverse mapping from the forced bond stretches to the INM displacements (panel b2) shows that the CO stretch in methanol translates almost exclusively into the conjugate displacement in the negative direction of the second INM, if one neglects a small positive component from the α = 3 mode. The α = 2 INM is seen to strongly participate in charge responses to all bond elongations. This is not surprising, since this softest mode (exhibiting largest number of nodes) represents the very (short-range) polarization which is responsible for the minimum energy charge reorganization following a given bond length increase. One also concludes from this diagram that the localized OH stretch requires a combination of displacements along the α = 1 - 3 INM, to describe a charge response to such a localized geometrical perturbation. This is consistent with the local mapping relations. Namely, to account for both direct and indirect charge responses to an increase in R_{OH}^b one needs components from the OH (mode α = 1) and CO (CH) bond polarizations (modes α = 2, 3). This is also the reason why the softest mode α = 2, including alternating polarizations of the neighbouring atoms, is so manifestly present in all charge responses to localized bond stretches.

Now, let us examine the mixed translators of part c, involving bond-stretch *vibrational normal modes* VNM (part e) and the AIM electron populations (part a). A comparison between diagrams a1 and c1 shows that the local charge perturbations translate in general into a combination of many VNM. We observe, that the localized nature of the first two VNM (practically the localized CO and OH stretches) implies that the first two columns of the c1 translator should be almost identical with those in panel a1, as indeed seen in the figure. It should be emphasized at this point that such local (test) reduction of a given atom i generates the charge response along the associated i -th internal populational MEC, which triggers the structural change represented by diagrams a1 and c1. Thus, the structural responses reflected by these mapping relations do in fact represent geometry relaxation effects associated with the populational (internal) MEC in question.

Turning now to the transformation c2 we again observe that the first two rows are almost identical with those in the map a2, due to the localized character of $\mathcal{Q}_1^b$ and $\mathcal{Q}_2^b$. The symmetrical stretch of the CH bonds ($\mathcal{Q}_3^b$) is again seen to result in their diminished polarity; there is also a slight increase in the polarity of the CO bond observed for a shift along this VNM. In the asymmetric stretches $\mathcal{Q}_4^b$ and $\mathcal{Q}_5^b$ one detects that the stretched (shortened) CH bonds are mapped into their decreased (increased) bond polarity changes.

The most compact mapping is obtained in panel d1, in which displacements along the populational INM are translated into the VNM bond-stretches. It follows from this diagram that the softest INM (α = 2) does not generate any appreciable movement of nuclei along five VNM. This prediction is consistent with our earlier observation relating the bond polarity to its length. Namely, since this INM changes polarity of all bonds it should not be expected to translate into the VNM representing localized bond stretches. The α = 1 INM slightly increases the $CH_{(4)}$ and OH bond lengths, polarities of which become decreased as a result of the positive INM displacement. The third INM, which shifts electrons from the OH to CH_3 groups, is seen to be mapped into a combination of positive displacements along $\mathcal{Q}_1^b$, $\mathcal{Q}_3^b$, and $\mathcal{Q}_4^b$. Again, the dominating $\mathcal{Q}_1^b$ (CO stretch) component is in accordance with the most important decrease in the CO polarity along this INM. The remaining components can also be explained in terms of this approximate bond polarity - bond length relationship. Namely, the α = 3 INM is now seen to generate a decreased polarity of the $OH_{(4,\ 5)}$ non-bonded interactions. Thus, this electron redistribution should also translate into an appreciable increase in length of bonds 4 and 5, and a smaller increase in R_3^b. This is indeed the net displacement of the methyl hydrogens represented by the ($\mathcal{Q}_3^b$, $\mathcal{Q}_4^b$)-components in the α = 3 row of panel d1. Finally, the α = 4 and 5 INM translate, almost on the one-to-one basis, into the $\mathcal{Q}_4^b$ and $\mathcal{Q}_5^b$, respectively. These mappings are partly due to the symmetry constraints and they fully conform to the general bond polarity-bond length relationship.

Finally, let us interpret the physical implications following from the inverse collective-collective transformation of panel d2. It relates the independent (unit) displacements along the VNM into the respective charge responses in the INM representation. As expected, the $\mathcal{Q}_1^b$ (CO stretch) mode is mapped into the minimum energy response represented by the dominating (negative) shift along the α = 2 INM; this sign is necessary to decrease the CO bond polarity. A more complicated charge shift accompanies a unit displacement along the $\mathcal{Q}_2^b$ (OH stretch). Clearly, the dominating INM in this nuclear motion, α = 1, accounts for a decreased OH polarity. The remaining, negative shifts along the α = 2 and

3 INM are required to represent the short- and long-range polarizations, describing charge responses on remaining atoms, which act in the direction to diminish the charge reorganization due to the $\alpha = 1$ component, conjugated directly to the $d\mathcal{Q}_1^b > 0$ displacement. This Le Châtelier-Braun-type interpretation can also rationalize the small component in the next $\mathcal{Q}_3^b$ (symmetric CH_3 stretch) row of panel d2. In this case the directly coupled normal charge response, $\alpha = 2$, dominates the charge relaxation, as describing mainly a collective charge shift diminishing the initial polarity of bonds in the CH_3 group; a small positive component from the $\alpha = 1$ INM is to account for the induced polarizational shift inside the OH bond. An intuitively expected charge reorganization is obtained for the geometry change along the $\mathcal{Q}_4^b$, since a combination of positive shifts along the $\alpha = 2$ and 4 INM generate the required CH bond polarity changes: increase for the $CH_{(4)}$ (squeezing the bond) and decrease for $CH_{(5,6)}$ (stretching the bonds). Finally, due to the symmetry, the $\mathcal{Q}_5^b$ VNM is seen to be translated into the single conjugated shift (negative) along the $\alpha = 5$ INM.

All chemical reactions involve both the nuclear displacements and the concomitant electron redistributions. By the Hohenberg-Kohn theorems the ground-state electron density is in the one-to-one correspondence with the underlying external potential due to the system nuclei. In a given change of the system geometry, depending on what is considered the perturbation and what a response to it, the following two perspectives can be adopted [95]: (1) the *electron-cloud following*, when the electron density shift is viewed as the ground-state *response* to the primary nuclear displacement (*perturbation*), in the spirit of the Born-Oppenheimer approximation; (2) the *electron-cloud preceding*, in which a hypothetical electron density displacement (*perturbation*) is regarded as preceding (accelerating) changes in nuclear positions (*response*). Under some simplifying assumptions it is possible to rationalize main electron redistributions processes in terms of the MO transition densities, although many of them are expected to contribute to even the simplest responses of the system [113, 114]. The symmetry rules of chemical reactions have also been formulated [113, 114] by considering the symmetry restrictions on the excited state contributions to the perturbed ground-state wave-function, due to the perturbation created by a given normal mode displacement of nuclei. The mapping relations of the CSA in the AIM resolution provide a more explicit and transparent framework for discussing general ground-state relations between the *Electron Population Space* (EPS) and the *Nuclear Position Space* (NPS) characteristics. First of all there is an urgent need to have means to directly *"translate"* the reaction coordinate representations in both spaces one into another. As we have demonstrated above, the CSA development generates such very "translators" between attributes of alternative reference frames in both spaces.

The explicit mapping relations have profound implications for catalysis and the theory of chemical reactivity in general, since a given, desired geometrical (bond-stretch) NPS displacement of a system from its initial structure can now be translated into the associated charge EPS shifts, and *vice versa*. One can also transform selected reactive EPS channels, e.g., the softest PNM or INM of reactants, into their NPS conjugates. The numerical data reported for methanol also provide an additional validation of the physical significance of the PNM reference frame, which is practically indistinguishable from the

INM coordinate system, having only an additional external CT channel. The VNM are known to have physical meaning at low energies (amplitudes) (*cf. "natural vibrations"*); by the mapping relations they are now almost uniquely related to their EPS conjugates (*cf. "natural polarizations"*). The system geometrical/electronic structure relaxations are seen to take place along a very few normal modes, in both the NPS and EPS representations. The one-to-one character of some normal mapping relations is also important for interpretative (predictive) purposes. The successful applications of the PNM reactivity criteria can thus be given a more conventional NPS rationalization when the collective-collective translators are dominated by the one-to-one relations: electrons in such molecular systems do indeed tend to be redistributed along a single INM/PNM in response to a geometrical distortion along a single conjugated VNM.

Both the electron-cloud preceding (accelerating nuclear movement) and electron-cloud incomplete following (resisting nuclear movement) mechanisms of Nakatsuji [95] can now be viewed as particular manifestations of the present mapping relations.

A possibility to map the local (or collective) displacements in one space onto the local (or collective) shifts in the other space can be used to quickly asses the feasibility of various populational displacements in a molecule, by testing whether they translate into one of the normal vibrational modes. One also achieves a direct connection between the charge sensitivities and spectroscopic data, which eventually could help to establish additional ways of verifying the reactivity trends predicted within CSA. These mapping relations also provide a sound phenomenological basis for a systematic manipulation of the surface structure, to eventually facilitate the desired charge redistribution in the adsorbate. This opens new possibilities to affect reactivities and selectivities along a given reaction channel.

One could use the NPS-EPS translators to identify natural vibrational modes associated with a given normal or local charge polarization channels. This is potentially important for catalysis, in which polarizational promotions of an adsorbate molecule by an active site of the surface can be translated into bond-stretch vibrations, possibly those leading to bond cleavages. One could also attempt a search for the optimum matching of two reactants towards achieving the desired charge polarization, ultimately leading to the required bond-forming-bond-weakening concerted mechanism.

As emphasized by Gutmann [112] the structure of a molecular system is not rigid but rather adaptable to the environment. The CSA/EEM mapping relation development allows one to view this electronic-geometrical structure relationship from many chemically interesting perspectives. Changes in the charge density pattern are reflected by changes in structural pattern, which in turn may lead to a new reactivity pattern in the system. By applying an appropriate molecular environment, e.g., a catalyst, one can induce desired electronic structure conditions for "activating" bonds of interest, through changing their polarity.

6. Stability Considerations

In this section we shall summarize charge stability criteria and their physical implications [32, 57, 61]. For reasons of clarity let us consider a general partitioning of the system under consideration, $\mathcal{S}$, into two complementary subsystems, $\mathcal{S}_1$ and $\mathcal{S}_2$. Following the previous MEC development of Section 6.2 in Chapter 3, we shall separately examine the external and internal cases. The former refers to the combined system consisting of $\mathcal{S}$ and a hypothetical electron reservoir, $\mathcal{S}^{\mathrm{r}} = (\mathcal{S} \mid \mathrm{r})$, while the latter corresponds to the closed $\mathcal{S}^{\mathrm{e}} = (\mathcal{S}_1 \mid \mathcal{S}_2)$. In the N-constrained case the complementary subsystems denote, e.g., the A and B reactants in a general acceptor-donor reactive system, the adsorbate and substrate in the heterogenous catalysis, etc. Let $N_1^{\mathcal{S}}$ and $N_2^{\mathcal{S}}$ denote the respective numbers of electrons, grouped in the vector $\boldsymbol{N}_{\mathcal{S}} = (N_1^{\mathcal{S}}, N_2^{\mathcal{S}})$, with $N = N_1^{\mathcal{S}} + N_2^{\mathcal{S}}$ denoting the global number of electrons in $\mathcal{S}$. The relevant condensed hardness tensor in the subsystem resolution, $\boldsymbol{\eta}^{\mathcal{S}}_{(1|2)} = \{\eta_{i,j}^{\mathcal{S}}\}$, is defined by Eqs. 111 and 181 in the relaxed and rigid approximations, respectively. Its elements define the subsystem effective hardnesses, $\eta_1^{\mathcal{S}} = \eta_{1,1}^{\mathcal{S}} - \eta_{1,2}^{\mathcal{S}}$, etc., $\eta_{CT}^{\mathcal{S}} = \eta_1^{\mathcal{S}} + \eta_2^{\mathcal{S}}$, and the related response properties, e.g., the subsystem FF indices of Eq. 114: $f_1^{\mathcal{S}} = \eta_2^{\mathcal{S}}/\eta_{CT}^{\mathcal{S}}$ and $f_2^{\mathcal{S}} = \eta_1^{\mathcal{S}}/\eta_{CT}^{\mathcal{S}}$.

Let us assume the internal equilibrium in $\mathcal{S}$, which corresponds to the mutually open subsystems, $\mathcal{S}^{\mathrm{e}} = (\mathcal{S}_1 | \mathcal{S}_2)$, with equalized chemical potentials, $\mu_1^{\mathcal{S}} = \mu_2^{\mathcal{S}} = \mu \equiv \partial E/\partial N$, at the global chemical potential level. The internal stability refers to the intra-$\mathcal{S}$ (hypothetical) charge displacements, $\mathrm{d}\boldsymbol{N}_{\mathcal{S}}(\Delta) = (\Delta, -\Delta)$, preserving N. The corresponding quadratic energy change due to this polarizational displacement from the initial internal equilibrium state:

$$\mathrm{d}E(\Delta) = (\mu_1^{\mathcal{S}} - \mu_2^{\mathcal{S}})\,\Delta + \frac{1}{2}\,\eta_{CT}^{\mathcal{S}}\,\Delta^2 = \frac{1}{2}\,\eta_{CT}^{\mathcal{S}}\,\Delta^2 \equiv \mathrm{d}^2E(\Delta)\,, \tag{360}$$

implies the following internal stability criterion, $\mathrm{d}^2E(\Delta) > 0$:

$$\eta_{CT}^{\mathcal{S}} > 0 \quad \text{or} \quad a \equiv (\eta_{1,1}^{\mathcal{S}} + \eta_{2,2}^{\mathcal{S}})/2 > \eta_{1,2}^{\mathcal{S}}\,. \tag{361}$$

One similarly considers $\mathcal{S}$ in equilibrium with an external electron reservoir in $\mathcal{S}^{\mathrm{r}}$, characterized by its chemical potential μ_{r}, $\mathcal{S}^{\mathrm{r}} = (\mathcal{S}_1|\mathcal{S}_2|\mathrm{r})$, which implies the chemical potential equalization: $\mu_1^{\mathcal{S}} = \mu_2^{\mathcal{S}} = \mu_{\mathcal{S}} = \mu_{\mathrm{r}}$. The virtual flow of electrons between $\mathcal{S}$ and r, $\mathrm{d}N = -\mathrm{d}N_{\mathrm{r}}$, gives rise to the associated quadratic energy change,

$$\mathrm{d}E(\mathrm{d}N) = (\mu_{\mathcal{S}} - \mu_{\mathrm{r}})\,\mathrm{d}N + \frac{1}{2}\,\eta_{\mathcal{S}}\,(\mathrm{d}N)^2 = \frac{1}{2}\,\eta_{\mathcal{S}}\,(\mathrm{d}N)^2 \equiv \mathrm{d}^2E(\mathrm{d}N)\,, \tag{362}$$

where the second-order energy change of the infinitely soft (macroscopic) reservoir is neglected, and the global hardness $\eta_{\mathcal{S}}$ (Eq. 115 and Fig. 8) is:

$$\eta_{\mathcal{S}} = \partial\mu/\partial N = [\eta_{1,1}^{\mathcal{S}}\,\eta_{2,2}^{\mathcal{S}} - (\eta_{1,2}^{\mathcal{S}})^2]/\,\eta_{CT}^{\mathcal{S}}\,. \tag{363}$$

Thus, the external stability criterion $\mathrm{d}E(\mathrm{d}N) > 0$ for the internally stable charge distribution implies (Eq. 361)

$$g^2 \equiv \eta_{1,1}^{\mathscr{S}} \eta_{2,2}^{\mathscr{S}} > (\eta_{1,2}^{\mathscr{S}})^2 , \tag{364}$$

and this inequality is reversed for the internally unstable ($\eta_{CT}^{\mathscr{S}} < 0$) charge distribution in $\mathscr{S}$. Moreover, since

$$|g| = [4a^2 - (\eta_{2,2}^{\mathscr{S}} - \eta_{1,1}^{\mathscr{S}})^2]^{1/2} / 2 \leq |a| , \tag{365}$$

one obtains the following stability regimes for the typical region of the coupling hardness between subsystems, $0 < \eta_{1,2}^{\mathscr{S}} < \eta_{1,1}^{\mathscr{S}}$ (we denote $\mathscr{S}_1$ as the harder subsystem, $\eta_{1,1}^{\mathscr{S}} \geq \eta_{2,2}^{\mathscr{S}}$):

$$\begin{aligned} &0 \leq \eta_{1,2}^{\mathscr{S}} < g , && \text{internally (I) and externally (E) stable } \mathscr{S}: (+,+); \\ &g \leq \eta_{1,2}^{\mathscr{S}} < a , && \text{I - stable and E - unstable } \mathscr{S}: (+,-); \\ &a \leq \eta_{1,2}^{\mathscr{S}} < \eta_{1,1}^{\mathscr{S}} , && \text{I - unstable and E - stable } \mathscr{S}: (-,+) . \end{aligned} \tag{366}$$

They are shown in a qualitative diagram of Fig. 35 a.

Clearly, since the subsystem FF indices are closely related to the structure of the condensed hardness tensor $\boldsymbol{\eta}^{\mathscr{S}}_{(1|2)}$, these stability/instability regions can also be identified in terms of the $f_1^{\mathscr{S}}$ and $f_2^{\mathscr{S}}$ subsystem FF indices, defining orientation of the associated FF vector in the subsystem populational space (see Fig. 10 b) [61].

These qualitative conclusions have important consequences for catalytic systems [57, 117]. Let us assume, e.g., $\mathscr{S}$ = (adsorbate|substrate) with a large substrate representing the soft reactant ($\mathscr{S}_2$) and a small adsorbate corresponding to the hard reactant ($\mathscr{S}_1$). The substrate exerts a strong softening influence on the adsorbate (large, negative relaxational correction to $\eta_{1,1}^{\mathscr{S}}$ due to $\mathscr{S}_2$), $\delta\eta_{1,1}^{\mathscr{S}}(\mathscr{S})$ (see Fig. 35 b); the opposite effect $\delta\eta_{2,2}^{\mathscr{S}}(\mathscr{S})$, of the adsorbate on $\eta_{2,2}^{\mathscr{S}}$, is negligible and exactly vanishing. Moreover, the corresponding relaxational contributions to the coupling hardness $\delta\eta_{1,2}^{\mathscr{S}}(\mathscr{S})$ vanishes identically, since there is no third subsystem to relax. This implies a lowering of both a and g levels relative to the fixed coupling hardness, eventually generating the system instabilities. This mechanism is illustrated in Fig. 35 b. Thus, in this CSA, phenomenological perspective the catalyst acts as a generator of the I- and/or E-instabilities, through its softening influence upon the adsorbate. The instability conditions correspond to reactive charge distributions, which spontaneously readjust themselves through both the intra-adsorbate and the adsorbate-substrate charge flows, eventually leading to an adsorbate reconstruction and/or desorption (dissociative or molecular).

Consider now the illustrative example of alternative toluene chemisorption systems shown in Fig. 36 [57, 91], involving bipyramidal cluster of the V_2O_5 "active site" of the (010)-surface shown in Fig. 37 a. The molecularly adsorbed structures (panels a, c, e, g, i) correspond to the scaled INDO-optimized distances between the "frozen" geometries of the adsorbate and the active site [118]. They include the perpendicular molecular adsorptions of toluene, through the methyl group, to the vanadyl oxygen $O_{(1)}$ (terminal,

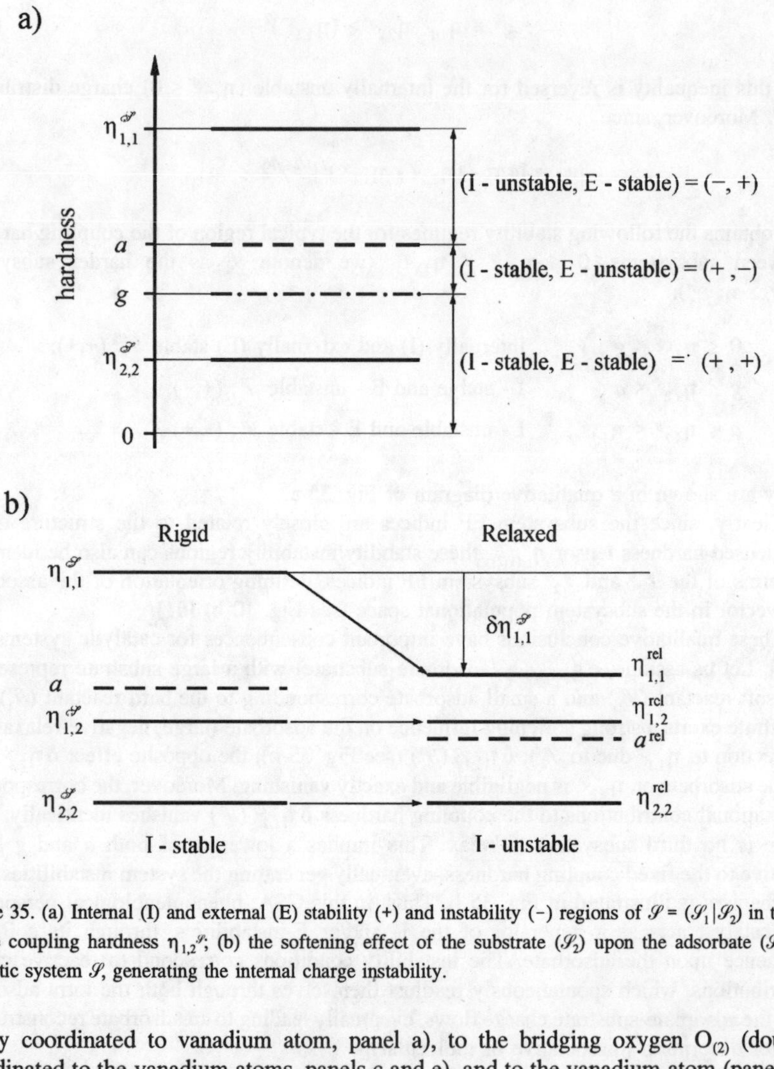

Figure 35. (a) Internal (I) and external (E) stability (+) and instability (−) regions of $\mathcal{S} = (\mathcal{S}_1|\mathcal{S}_2)$ in terms of the coupling hardness $\eta_{1,2}^{\mathcal{S}}$; (b) the softening effect of the substrate ($\mathcal{S}_2$) upon the adsorbate ($\mathcal{S}_1$) in catalytic system $\mathcal{S}$, generating the internal charge instability.

singly coordinated to vanadium atom, panel a), to the bridging oxygen $O_{(2)}$ (doubly coordinated to the vanadium atoms, panels c and e), and to the vanadium atom (panel g), as well as the parallel adsorption of the benzene ring to the pyramid bases (panel i). The remaining structures represent hypothetical dissociated forms (panels b, d, f, and h), with the two hydrogens of the methyl group being already bonded to the pyramid base oxygens (the rigid structure of the remaining toluene fragment has been assumed) [118]. We have used the scaled INDO charges [118, 119] to generate the condensed hardnesses $\boldsymbol{\eta}^{\mathcal{S}}_{(\text{toluene}|\text{cluster})}$, to be used to diagnose the (I, E) - stabilities of these model

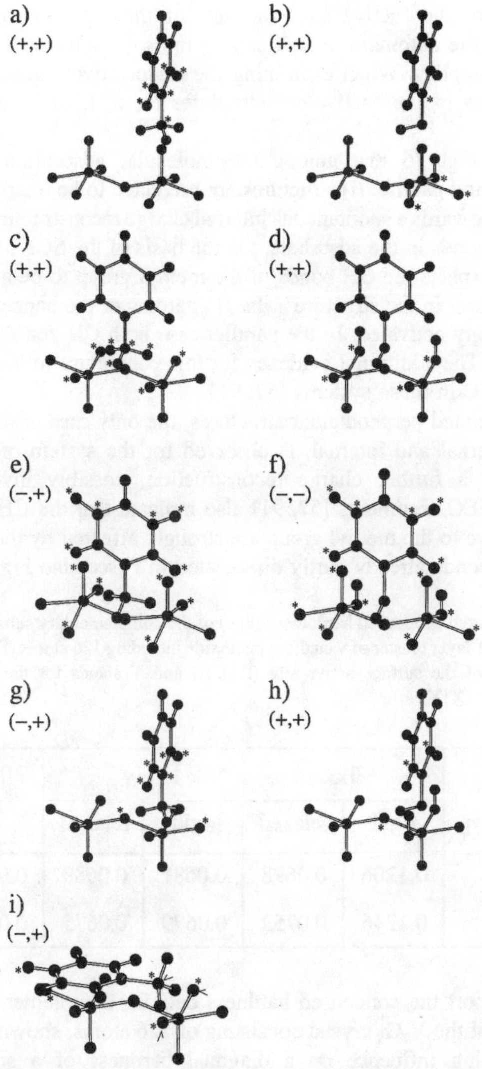

Figure 36. Alternative toluene chemisorption systems involving the bipyramidal surface cluster of vanadium pentoxide. The structures (a, c, e, g, i) represent molecularly adsorbed toluene, while the remaining systems (b, d, f, h) model the dissociative adsorption with two methyl hydrogens being chemically bonded to the surface oxygens at the pyramid bases. Asterisks indicate positions which give rise to the unstable MEC. The three atoms of the parallel complex (i), marked with an arrow, denote the extra instabilities appearing when the closed-system constraint ($dN = 0$) is imposed. The results are taken from Ref. 57. At each diagram the (I, E)-stability diagnosis is also indicated (see Fig. 35).

chemisorption systems. The predictions are reported in Fig. 36. We would like to stress, that in this model of the "active site" the size of the V_2O_5 cluster and toluene are comparable, so that the automatic classification: the soft (substrate) and hard (toluene) reactants no longer applies. When examining the dissociatively adsorbed structures the dissociated hydrogens of the "activated" structures (b, d, f, h) have been included in toluene.

It follows from Fig. 36 that among the molecular adsorption arrangements the perpendicular (e, g) and parallel (i) structures are predicted to be internally unstable. This suggests a tendency towards a spontaneous internal charge reconstruction, possibly leading to a dissociation of bonds in the adsorbate. On the basis of the SCF MO studies [118] of these systems, one expects the CH bonds of the methyl group to be affected the most in the perpendicular cases; in the structure e the $H_{(o)}$ atoms of the benzene ring can also be expected to be strongly activated. In the parallel case both CH and CC bonds should be strongly influenced. The additional evidence for this conjucture follows from examining the populational MEC in these systems [57, 91].

Among the dissociated perpendicular structures the only case of the predicted charge instability, both external and internal, is observed for the system of panel f; this may suggest a need for a further charge reconstruction, possibly involving the cluster environment. The MEC arguments [57, 91] also indicate that the CH ring bonds in the *ortho*-position relative to the methyl group are strongly affected by the vanadyl oxygens, besides the methyl bonds already partly dissociated in f (see also Fig. 17).

TABLE 5. Rigid and relaxed condensed hardnesses (a.u.) of the complementary subsystems X and Y in the stoichiometric, two-(010)-layer cluster of vanadium pentoxide including 126 atoms (Fig. 37); here X denotes the bipyramidal model of the surface active site (I or II) and Y stands for the corresponding cluster remainder: $\mathscr{S} = [V_2O_5] = (X|Y)$.

X	$\eta_{X,X}$		$\eta_{Y,Y}$		$\eta_{X,Y}$
Fig. 37b	rigid	relaxed	rigid	relaxed	
I	0.1206	0.0698	0.0681	0.0680	0.0661
II	0.1246	0.0753	0.0679	0.0675	0.0664

In Table 5 we report the condensed hardness data for complementary subsystems in much larger cluster of the V_2O_5 crystal consisting of 126 atoms, shown in Fig. 37 b. They illustrate the softening influence on a diagonal hardness of a small subsystem X (bipyramidal units shown in Fig. 37 b), due to its interaction with its large complementary subsystem Y. Notice that the relaxational softening changes the structure of the rigid condensed hardness matrix $\boldsymbol{\eta}^{\mathscr{S}}_{(X|Y)}$, producing rather soft (close to instability) charge distribution, with the off-diagonal, coupling hardness $\eta_{X,Y}{}^{\mathscr{S}}$ having values comparable with those of the two relaxed diagonal hardnesses.

The AIM FF indices of the large $[V_2O_5]$- cluster are shown in Fig. 37 c. It is seen that

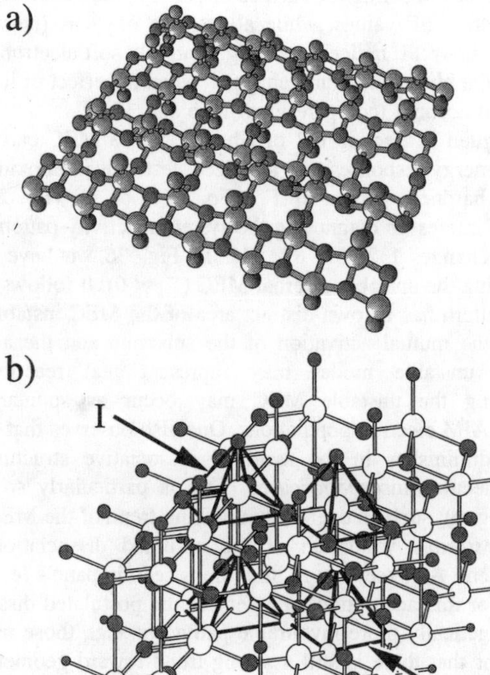

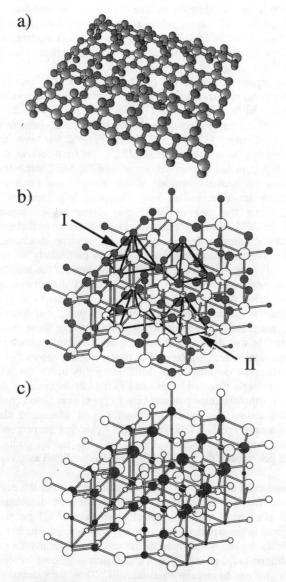

Figure 37. Perspective view of the idealized (010)-surface of the vanadium pentoxide, with different mutual arrangements of neighbouring pyramidal [VO_5] units (panel a). Panels b and c corresponds to the neutral, stoichiometric surface cluster including two layers of the (010)-surface pyramids (126 atoms); in panel b, which illustrates the SINDO AIM net-charge distribution, the bipyramidal subsystems I and II are shown (see Table 5); panel c represents the AIM FF distribution diagram.

in the first (010)-layer, modeling the surface of panel a, the positively charged vanadium atoms exhibit negative FF values, while all surface oxygens (negatively charged) are characterized by positive FF indices, marking relatively soft electron distributions. Thus, the FF diagram of the bipyramidal unit changes when the effect of its environment in the crystal is taken into account (compare Figs. 16 b and 37 c).

As we have argued in Section 6.2 of Chapter 3, the MEC can be used to test the system minimum energy responses to a localized, test reduction/oxidation. Both the sign of the coordinate hardness parameter Γ_k (Eq. 274) or γ_k (Eq. 278), and the mode topology can be of interest in diagnosing the system reactivity patterns and its sensitivity to environmental changes [32, 57, 87, 91]. In Fig. 36 we have indicated the AIM positions, k, defining the unstable external MEC ($\Gamma_k < 0$). It follows from the figure that each adsorption pattern has its own distinct area of the MEC instabilities, which can be used to diagnose the mutual activation of the substrate and the adsorbate due to the adsorption. Such unstable modes may represent real reaction pathways since displacements along the unstable MEC may occur as spontaneous responses to fluctuations in the AIM electron populations. One also observes that such an "activation" region generally diminishes in the assumed dissociative structures, relative to the corresponding molecular adsorption cases. This is particularly so in pairs of related panels: (c, d) and (g, h); such an overall diminishing trend of the MEC instability regions provides an *a posteriori* validation of the assumed dissociation pathways of the chemisorption system. An exception to this rule is seen in panels (e, f) of the figure; this may suggest that for this adsorption arrangement the postulated dissociation pattern has missed other, energetically more favourable pathways, e.g., those involving the toluene *ortho*-hydrogens, or that there is still a strong trend toward geometry relaxation in the assumed activated structure. A reference to panels a and b shows that after dissociating two methyl hydrogens the area of the MEC instability inside the toluene fragment is limited only to a single ring carbon (bonded to the CH fragment of the methyl group). One also observes that the three pyramid base oxygens have been activated as a result of the toluene dissociation; this indicates a possibility of subsequent dissociations of the lattice hydroxyls or oxygens. It follows from the figure that the perpendicular adsorption on the bridging oxygen selectively "activates" only the methyl group; in all remaining cases different positions in the benzene ring are also predicted as giving rise to unstable MEC.

In all perpendicular arrangements, the instability areas from the external and internal MEC have been found to be identical. Therefore, they are dominated by the internal polarization displacements with little effect of an external CT on predicted instability (activation) areas. In the parallel arrangement i, however, the internal MEC give rise to a wider instability region, which includes the methyl carbon and the neighboring lattice oxygens, in addition to the *para*-hydrogen and its near-neighbour lattice oxygen; the latter is predicted to give rise to unstable external MEC. Thus, the external CT may moderate the internal, purely polarizational MEC instabilities, by an intervention of the compensating charge flow involving the system environment.

7. Partitioning of the Interaction Energy

For interpretative purposes it is important to partition the overall electronic interaction energy $E^{e}_{AB}(d\boldsymbol{v}, N_{CT})$ into contributions attributed to reactants A (acidic) and B (basic) in the closed M = A---B system, or to their constituent fragments. Following Eq. 161 we separately consider the (electrostatic+polarization)-energy, $E^{+}_{AB}(d\boldsymbol{v}, N_{CT}=0) = E_{ES} + E_P = E(M^+) - E(M^0)$, and the CT energy (Eq. 93),

$$E_{CT} = E^{*}_{AB}(d\boldsymbol{v}, N_{CT}) = E(M^*) - E(M^+) = \frac{1}{2}\mu^{+}_{CT} N_{CT}, \tag{367}$$

$$N_{CT} = -\mu^{+}_{CT}/\eta_{CT}. \tag{368}$$

The first E^{+}_{AB} part can be naturally partitioned into the reactant X = (A, B) contributions:

$$E^{+}_{AB} = \sum_X^{A,B} \int d v_X(\mathbf{r}) \left[\rho^0_X(\mathbf{r}) + \frac{1}{2} d\rho^{+}_X(\mathbf{r})\right] d\mathbf{r} \equiv E^{+}_A + E^{+}_B, \tag{369}$$

with the first and second terms in parentheses giving rise to the E_{ES}- and E_P- energies, respectively. The CT interaction is also naturally partitioned into the reactant components:

$$E_{CT} = \sum_X^{A,B} \left\{ dN_X \left[\mu^0_X + \frac{1}{2}(d\mu^M_X + d\mu^*_X)\right] + \frac{1}{2}\int d v_X(\mathbf{r})\, d\rho^*_X(\mathbf{r})\, d\mathbf{r} \right\}$$
$$\equiv E^*_A + E^*_B. \tag{370}$$

Consider now the phenomenological AIM discretization of the above energy expressions, in which $d\boldsymbol{N}_X = \{dN_x\}$ and $d\boldsymbol{v}_X = \{dv_x\}$ replace $d\rho_X(\mathbf{r})$ and $dv_X(\mathbf{r})$, respectively. In this resolution

$$d\boldsymbol{N}^{+}_X = \sum_Y^{A,B} d\boldsymbol{v}_Y \boldsymbol{\beta}^{Y,X}, \qquad X = A, B, \quad \text{or} \quad d\boldsymbol{N}^{+} \equiv d\boldsymbol{v}\boldsymbol{\beta}, \tag{371}$$

$$d\boldsymbol{\mu}^{+}_X = \sum_Y^{A,B} d\boldsymbol{v}_Y \boldsymbol{f}^{X,Y}, \qquad X = A, B, \quad \text{or} \quad d\boldsymbol{\mu}^{+} \equiv d\boldsymbol{v}\,\mathbf{f}^{\dagger}, \tag{372}$$

where, in short notation $d\boldsymbol{N}^+ = (d\boldsymbol{N}^+_A, d\boldsymbol{N}^+_B)$, $d\boldsymbol{v} = (d\boldsymbol{v}_A, d\boldsymbol{v}_B)$, $d\boldsymbol{\mu}^+ = (d\boldsymbol{\mu}^+_A, d\boldsymbol{\mu}^+_B)$, $\boldsymbol{\beta} = \partial\boldsymbol{N}/\partial\boldsymbol{v}$, and $\mathbf{f}^{\dagger} = \partial\boldsymbol{\mu}^+/\partial\boldsymbol{v}$. Similarly, the CT displacements are:

$$d\boldsymbol{N}^*_X = \boldsymbol{f}^{CT}_X N_{CT}, \qquad X = A, B, \quad \text{or} \quad d\boldsymbol{N}^* = \boldsymbol{f}^{CT} N_{CT}, \tag{373}$$

$$d\boldsymbol{\mu}^*_X = \boldsymbol{\eta}^M_X N_{CT}, \qquad X = A, B, \quad \text{or} \quad d\boldsymbol{\mu}^* = \boldsymbol{\eta}^M N_{CT}, \tag{374}$$

where $\eta_X^M = \eta_{X,X}^M - \eta_{Y,X}^M$ $(X \neq Y)$, $\boldsymbol{\eta}^M = (\eta_A^M, \eta_B^M) = \partial\boldsymbol{\mu}/\partial N_{CT}$, $f_X^{CT} = f^{X,X} - f^{Y,X}$ $(X \neq Y) = \partial \boldsymbol{N}_X/\partial N_{CT}$, $\boldsymbol{f}^{CT} = (f_A^{CT}, f_B^{CT})$.

Thus, in the above AIM discretization,

$$E_X^+ = \left(\boldsymbol{N}_X^0 + \frac{1}{2}\mathrm{d}\boldsymbol{N}_X^+\right)\mathrm{d}\boldsymbol{v}_X^\dagger , \tag{375}$$

$$E_X^* = \left[\mu_X^0 + \frac{1}{2}\left(\mathrm{d}\mu_X^M + \mathrm{d}\mu_X^*\right)\right]\mathrm{d}N_X + \frac{1}{2}\mathrm{d}\boldsymbol{N}_X^*\,\mathrm{d}\boldsymbol{v}_X^\dagger , \qquad X = A,\ B \ , \tag{376}$$

where $\mathrm{d}N_A = N_{CT}$ and $\mathrm{d}N_B = -N_{CT}$. The above expressions provide the unique partitioning of E_X^+ and E_X^* into the constituent atom contributions.

One could also partition the external CT energy, when electrons are exchanged between a given molecular system M and the reservoir r. Let

$$\mu_{CT}^r \equiv \mu_M - \mu_r \quad , \qquad \eta_{CT}^r \equiv \eta_M - \eta_{M,r} > 0 , \tag{377}$$

where μ_r stands for the reservoir chemical potential and $\eta_{M,r}$ represents a weak coupling hardness between the system and the reservoir. We have neglected the reservoir diagonal hardness which, by assumption, vanishes (infinitely soft, macroscopic reservoir). Then, from Eqs. 100 and 101, $N_{CT}^r > 0$ for $\mu_{CT}^r < 0$ (inflow of electrons to M) and $N_{CT}^r < 0$ for $\mu_{CT}^r > 0$ (outflow of electrons from M), and the CT stabilization energy can be partitioned into the relevant AIM contributions by partitioning N_{CT}^r into atomic population changes,

$$E_{CT}^r = \frac{1}{2}\mu_{CT}^r N_{CT}^r = \left(\textstyle\sum_i^M f_i\right) E_{CT}^r \equiv \textstyle\sum_i^M \varepsilon_{CT,i}^r . \tag{378}$$

For the molecular fragment partitioning in the global equilibrium, M = (X¦Y¦...), one calculates the associated group contributions to E_{CT}^r by summation over all constituent atoms:

$$E_{CT}^r = \textstyle\sum_X^M \sum_x^X \varepsilon_{CT,x}^r \equiv \sum_X^M E_{CT,X}^r , \tag{379}$$

with $E_{CT,X}^r = f_X^M E_{CT}^r$, due to the additivity of the AIM FF indices.

Such a FF partitioning of N_{CT} can also be used to partition E_{CT} of Eq. 101. Besides the resulting *B*- and *A-resolved* divisions of E_{CT} (see Eq. 236),

$$E_{CT} = \left(\textstyle\sum_a^A f_{A,a}^{CT}\right) E_{CT} \equiv \sum_a^A \epsilon_{CT,a} = \left(\sum_b^B f_{B,b}^{CT}\right) E_{CT} \equiv \sum_b^B \epsilon_{CT,b} , \tag{380}$$

one could adopt a deeper division [59], which involves also the chemical potential difference (Eq. 238): $\overset{+}{\mu}_{CT} = \mu_A^M - \mu_B^M = f_B^{CT}\,\mu_B^M$. Namely, since $N_{CT} = -dN_B = -\Sigma_b^B dN_b$, $dN_b = f_{B,b}^{CT}\,dN_B$, the following *B-resolved* expression is obtained for E_{CT},

$$E_{CT} = -\frac{1}{2}\sum_b^B \mu_{B,b}^M\,dN_b \equiv \sum_b^B \varepsilon_b^{CT}\ ; \tag{381}$$

here ε_b^{CT} provides the CT hierarchy of constituent atoms in the basic reactant. The corresponding *A-resolved* partitioning, when one interprets the CT energy in terms of contributions attributed to constituent atoms in the acidic reactant, follows from analogous relations: $N_{CT} = dN_A = \Sigma_a^A dN_a$, $dN_a = f_{A,a}^{CT}\,dN_A$, $\overset{+}{\mu}_{CT} = f_A^{CT}\,\mu_A^M$:

$$E_{CT} = \frac{1}{2}\sum_a^A \mu_{A,a}^M\,dN_a \equiv \sum_a^A \varepsilon_a^{CT}\ . \tag{382}$$

The deepest partitioning of the AIM resolution is the *(A, B)-resolved* division, in which we use the combination formulas for $\mu_A^M = \mathbf{f}^{A,A}\,(\boldsymbol{\mu}_A^M)^\dagger$ and $\mu_B^M = \mathbf{f}^{B,B}\,(\boldsymbol{\mu}_B^M)^\dagger$:

$$\begin{aligned} E_{CT} &= \frac{1}{2}(\mu_B^M\,dN_B + \mu_A^M\,dN_A) = \frac{1}{2}\Big(\sum_b^B \mu_{B,b}^M\,dN_b^d + \sum_a^A \mu_{A,a}^M\,dN_a^d\Big) \\ &\equiv \sum_b^B \varepsilon_b + \sum_a^A \varepsilon_a \equiv E_B + E_A\ , \end{aligned} \tag{383}$$

where the direct (d) shifts in atomic electron populations, $dN_a^d = f_a^{A,A}\,dN_A$ and $dN_b^d = f_b^{B,B}\,dN_B$; here the first sum represents the *base charge withdrawal energy* (E_B, positive) and the second sum gives the *acid charge inflow energy* (E_A, negative).

Finally, we would like to comment upon the corresponding collective-mode partitionings of the CT energies, E_{CT}^r and E_{CT}. Clearly, the CT-projected w_α indices (Eq. 258) replace in the PNM-representation the FF indices of the AIM-representation, e.g.,

$$E_{CT}^r = \Big(\sum_\alpha^M w_\alpha\Big) E_{CT}^r \equiv \sum_\alpha^M \varepsilon_{CT,\alpha}^r\ , \qquad \text{etc.} \tag{384}$$

The E_{CT} energy in the closed A–B system can be similarly partitioned into the reactant PNM contributions. In this reference frame $N_{CT} = dN_A = \Sigma_\alpha^A d\bar{n}_\alpha = -dN_B = -\Sigma_\beta^B d\bar{n}_\beta$ (Eqs. 254-257). Again, the intra-reactant equilibrium state implies $E(A|B) = E[\bar{n}_A, \bar{n}_B(\bar{n}_A)] = E[\bar{n}_A(\bar{n}_B), \bar{n}_B]$; this defines the relaxed (CT-projected) mode chemical potentials $(\partial E/\partial \bar{n}_A)_{N_B} = \bar{\xi}_A^M$ and $(\partial E/\partial \bar{n}_B)_{N_A} = \bar{\xi}_B^M$, in terms of which $\overset{+}{\mu}_{CT} = \boldsymbol{\omega}_A^{CT}\,(\bar{\xi}_A^M)^\dagger = \boldsymbol{\omega}_B^{CT}\,(\bar{\xi}_B^M)^\dagger$,

$$\mu_A^M = \left(\frac{\partial \bar{n}_A}{\partial N_A}\right)_{N_B} (\bar{\xi}_A^M)^\dagger \equiv \mathbf{w}^{A,A}\,(\bar{\xi}_A^M)^\dagger\ , \qquad \mu_B^M = \left(\frac{\partial \bar{n}_B}{\partial N_B}\right)_{N_A} (\bar{\xi}_B^M)^\dagger \equiv \mathbf{w}^{B,B}\,(\bar{\xi}_B^M)^\dagger\ .$$

Thus, in this representation of the PNM of reactants E_{CT} can be partitioned as follows:

$$E_{CT} = -\frac{1}{2}\sum_{\beta}^{B} \overline{\xi}_{B,\beta}^{M}\, d\overline{n}_{\beta} \equiv \sum_{\beta}^{B} \overline{\varepsilon}_{\beta}^{CT} \tag{385}$$

$$= \frac{1}{2}\sum_{\alpha}^{A} \overline{\xi}_{A,\alpha}^{M}\, d\overline{n}_{\alpha} \equiv \sum_{\alpha}^{A} \overline{\varepsilon}_{\alpha}^{CT} \tag{386}$$

$$= \frac{1}{2}\left(\sum_{\beta}^{B} \overline{\xi}_{B,\beta}^{M}\, d\overline{n}_{\beta}^{d} + \sum_{\alpha}^{A} \overline{\xi}_{A,\alpha}^{M}\, d\overline{n}_{\alpha}^{d}\right) \equiv \sum_{\beta}^{B} \overline{\varepsilon}_{\beta} + \sum_{\alpha}^{A} \overline{\varepsilon}_{\alpha} \equiv E_{B} + E_{A}\,, \tag{387}$$

where $d\overline{n}_{\alpha}^{d} = w_{\alpha}^{A,A}\, dN_{A}$ and $d\overline{n}_{\beta}^{d} = w_{\beta}^{B,B}\, dN_{B}$. These equations closely follow the corresponding expressions (Eqs. 381-383) of the AIM representation.

8. Mutual Polarization of Reactants

When considering the *in situ* FF indices we have included the polarization induced by a transfer of a single electron, from the basic to acidic reactant in M: B → A. However, when the CT-component is relatively unimportant, i.e., when $N_{CT} \cong 0$, e.g., in the physisorption process, the *direct P-component*, P^{+}, i.e., the polarization of the mutually closed but interacting reactants (Eqs. 153 and 371), will dominate the charge redistribution pattern due to the presence of the reaction partner [stage (*ii*) discussed in Section 8 of Chapter 2]. This stage also determines the shifted equilibrium chemical potentials of interacting reactants (Eqs. 155, 227), which determine μ_{CT}^{+} (Eq. 231), the driving force for the final stage (*iii*) of the charge redistribution via the inter-reactant CT, which generates the associated, *CT-induced polarization*, P^{*} (Fig. 10), measured by $d\rho_{X}^{*} \equiv d\rho_{X}^{*}(N_{CT})$ (Eq. 159) or $d\boldsymbol{N}_{X}^{*} \equiv d\boldsymbol{N}_{X}^{*}(N_{CT})$ (Eq. 373). The sign of this *in situ* chemical potential difference, $\mu_{CT}^{+}(d\boldsymbol{v})$, determines the actual direction of the charge flow $N_{CT} = N_{CT}(d\boldsymbol{v})$, i.e., it ascribes to each polarized reactant the actual basic or acidic character in M^{+}. This is by no means a trivial matter in the system of strongly interacting large reactants, where the asymptotic approximation of Eq. 98 may often suggest a qualitatively incorrect direction of CT. As we have already demonstrated this direction is crucial for determining the CT-induced charge redistribution trends (Fig. 17), i.e., the reactivity information complementary to the P^{+} charge redistributions.

Thus, the polarization stage (*ii*) plays a vital role in the CSA approach to the problem of chemical reactivity, determining both the direct polarization displacements and conditions for the final CT between reactants. One needs both the P and CT populational changes on constituent atoms of both reactants to quickly diagnose the main reactivity trends, for a selection of their mutual separations and orientations.

In the classical theory of chemical reactivity this purely polarizational stage has also been very important for elucidating the reaction mechanism. Since the AIM-resolved CSA generates the linear response matrix $\boldsymbol{\beta} = \{\boldsymbol{\beta}^{X,Y}, (X,Y) = (A,B)\}$, one can predict the direct polarizational changes of reactants, $d\rho_{X}^{+}(\mathbf{r})$ (Eq. 153) or $d\boldsymbol{N}_{X}^{+} \equiv d\boldsymbol{N}_{X}^{+}(d\boldsymbol{v})$ (Eq. 371), provided the realistic local or AIM representations of the displacements in external potential, $d\boldsymbol{v} = (d\boldsymbol{v}_{A}, d\boldsymbol{v}_{B})$ (polarizational stimuli) are readily available.

As we have mentioned before, the CSA/EEM approach can both determine the net

AIM charges $\boldsymbol{q}^0 = \boldsymbol{Z}^0 - \boldsymbol{N}^0 = (\boldsymbol{q}_A^0, \boldsymbol{q}_B^0)$ of isolated reactants (EEM), or use the charges from another, e.g., SCF MO source (CSA), to predict relevant charge sensitivities and interaction energies. Therefore, one can always realistically approximate the perturbing potential values at positions of constituent atoms of one reactant, via the resultant electrostatic potential per electron, due to the net charges of all constituent atoms of the other reactant:

$$dv_a \cong -\sum_b^B q_b / |\vec{R}_a - \vec{R}_b|, \qquad dv_b \cong -\sum_a^A q_a / |\vec{R}_a - \vec{R}_b|. \tag{388}$$

In such a point charge approximation the electrostatic and polarizational contributions to the total quadratic energy change are:

$$E_{ES} = \sum_a^A \sum_b^B q_a\, q_B / |\vec{R}_a - \vec{R}_b|, \qquad E_P = \frac{1}{2} d\boldsymbol{v}\, \boldsymbol{\beta}\, d\boldsymbol{v}^\dagger, \tag{389}$$

while the CT energy, E_{CT}, is defined by Eq. 367. Together this three contributions define the CSA interaction energy between reactants in the AIM resolution.

The direct polarization pattern $d\boldsymbol{N}^+(d\boldsymbol{v})$ includes both the reactant diagonal (due to $\boldsymbol{\beta}^{X,X}$ or $\boldsymbol{\beta}^{Y,Y}$) and off-diagonal (due to $\boldsymbol{\beta}^{X,Y}$ or $\boldsymbol{\beta}^{Y,X}$) contributions. In Figs. 38 and 39, reporting the $d\boldsymbol{N}^+(d\boldsymbol{v})$ data obtained from the point-charge perturbing potentials of Eq. 388 for the two models of the toluene-[V_2O_5] chemisorption systems of Fig. 17, we have separately displayed these two types of contributions, together with the respective resultant (total) polarizational diagrams. In Fig. 38 the first column corresponds to the same inter-reactant distance as that in Fig. 17 A; this distance has been increased by 1.5 and 2.5 Å, respectively, in the two remaining columns of the figure. The last panels in Figs. 38 and 39 also include the illustrative contour maps of the perturbing potentials in the ring plane, to be compared with the $N_{X(Y)}$-restricted charge redistributions they generate.

It follows from Fig. 38 that at larger inter-reactant separations (columns 2 and 3) the total polarization patterns (third row) agree with intuitive expectations. Namely, the two terminal [$O_{(1)}$] atoms of the surface, the most perturbed by the electron attractive potential due to the methyl hydrogens, are seen to accumulate electrons, mainly at the expense of the bridging [$O_{(2)}$] oxygen. The same is true in the toluene case, where the methyl hydrogens, the most exposed to the perturbation by the surface, loose electrons due to the repulsive potential from the vanadyl oxygens of the cluster. The accompanying responses of the remaining, less perturbed atoms, are then basically determined by the closed-reactants constraints. This intuitively expected pattern changes at the close approach model of the first column, where both the terminal and bridging oxygens of the surface are predicted to loose electrons, which are being shifted to the remaining oxygens and vanadium atoms. This diagram is practically identical with that implied by the toluene → cluster CT (Fig. 17 A-c). Finally, as also intuitively expected, the largest amplitudes of the reactant polarizations are seen to gradually diminish with increasing inter-reactant separation.

Another general conclusion from Fig. 38 is that the total polarization diagram (third row) is roughly a combination of the diagonal contribution of toluene (adsorbate) and the off-diagonal contribution of the cluster (substrate). This is in contrast with the similar

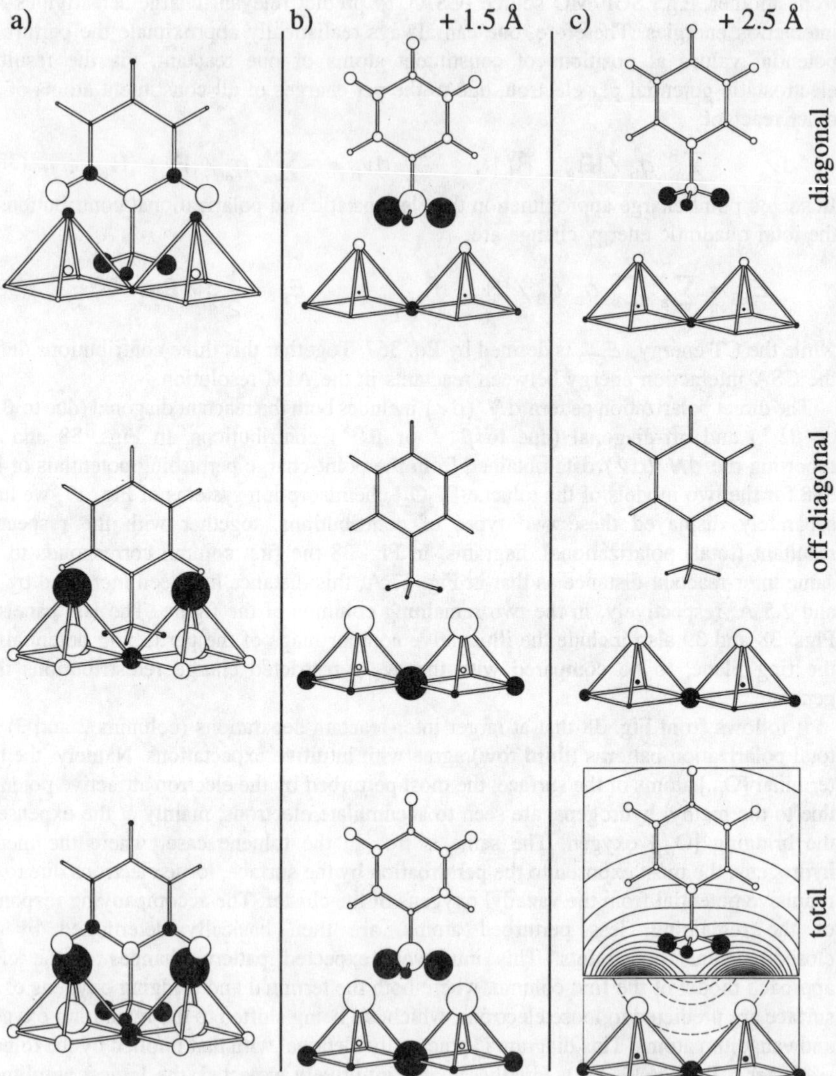

Figure 38. The AIM polarizational diagrams of reactants in the chemisorption system of Fig. 17 A (left column) and for the earlier stages of such a perpendicular approach of toluene to the bipyramidal cluster of V_2O_5. The same scale factor is adopted in all diagrams. The last panel of the third column additionally shows the contour maps of the perturbing potential (point-charge approximation) due to the other reactant.

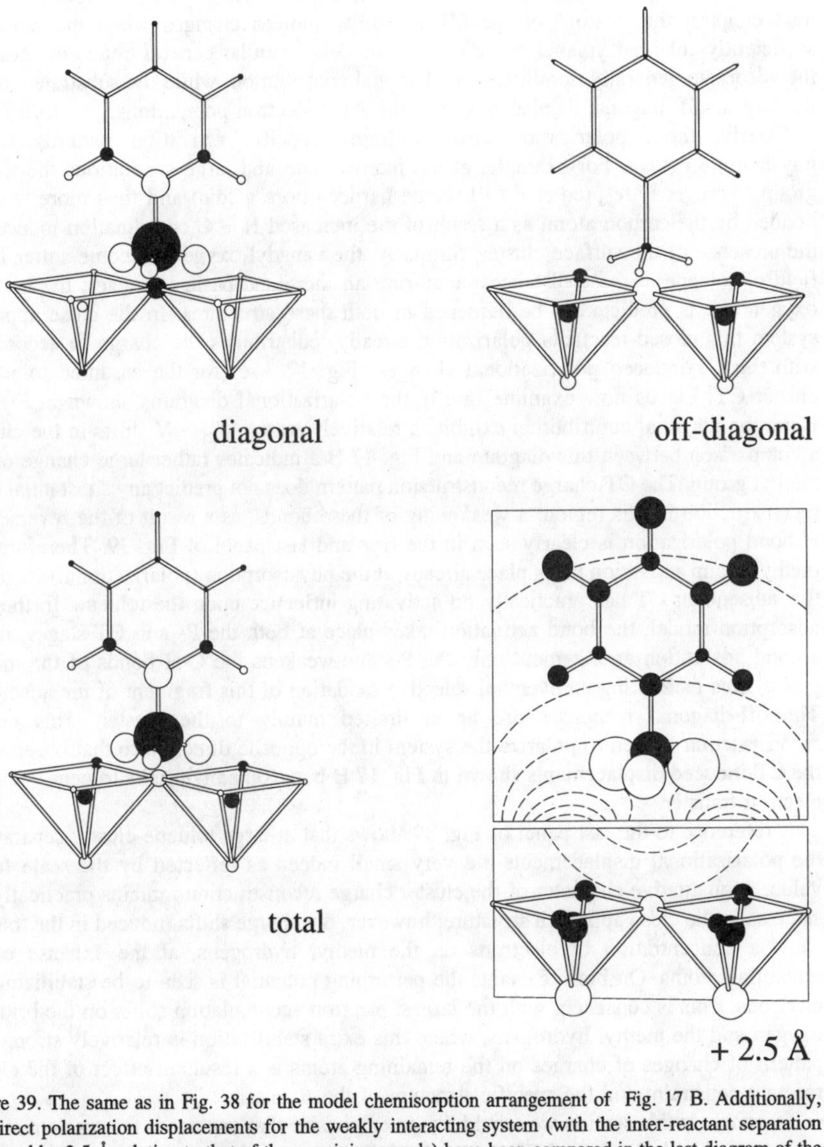

Figure 39. The same as in Fig. 38 for the model chemisorption arrangement of Fig. 17 B. Additionally, the direct polarization displacements for the weakly interacting system (with the inter-reactant separation increased by 2.5 Å relative to that of the remaining panels) have been compared in the last diagram of the figure with the contour map of the perturbing potential due to the other reactant; these AIM displacements should be multiplied by 0.07 to bring them into the common scale factor value with that of the remaining panels.

resolution of the *in situ* FF indices of Fig. 17 A, where both reactants exhibit a relatively large components in each of these contributions. However, as we shall demonstrate in the next chapter, this feature of the CT reactivity indices changes when the cluster is sufficiently enlarged; in such more realistic models a similar general trend emerges, that the adsorbate generally dominates the diagonal contribution, while the substrate exhibits the largest off-diagonal displacements of the AIM electron populations.

Clearly, these polarization changes imply specific activation patterns at the physisorption stage. For example, at the intermediate and large separations the methyl group hydrogens are predicted to become harder (more acidic) and thus more strongly bonded by the carbon atom, as a result of the increased H → C coordination induced by the presence of the surface cluster. Similarly, the vanadyl oxygens become softer in the field of toluene at these distances, acquiring an increased basic character; the bridging oxygen site is predicted to be hardened in both these structures. In the close approach system this closed-reactants polarization already redistributes the charge in accordance with the CT-induced polarizational changes (Fig. 17 A-c) for the assumed toluene → cluster CT. Let us now examine briefly the polarizational diagrams shown in Fig. 39. Here, the diagonal contribution exhibits a relatively strong $O_{(2)}$ → V shifts in the cluster; a comparison between this diagram and Fig. 17 B-a indicates rather large change on the methyl group. The CT-charge reconstruction pattern does not predict any substantial bond reconstruction in this region; a weakening of these bonds, as a result of the reverse C → H bond polarization is clearly seen in the first and last panel of Fig. 39. Therefore, the methyl group activation takes place already at the physisorption (polarizational) stage and the subsequent CT has practically no activating influence upon the toluene. In the first adsorption model, the bond activation takes place at both the P- and CT-stages; in the second adsorption arrangement only the P-stage weakens the C-H bonds of the methyl group, thus facilitating an eventual selective oxidation of this fragment of the adsorbate. The off-diagonal responses are again limited mainly to the cluster. This charge reconstruction is seen to polarize the system in the opposite direction to that observed in the CT-induced displacements shown in Fig. 17 B-b, accompanying the toluene → cluster electron transfer.

A reference to the last panel of Fig. 39 shows that at large toluene-cluster separations the polarizational displacements are very small indeed as reflected by the scale factor value. A qualitative character of the cluster charge reconstruction remains practically the same as in the close approach structure; however, the charge shifts induced in the toluene show a concentration of electrons on the methyl hydrogens, at the expense of all remaining atoms. On both reactants the perturbing potential is seen to be stabilizing for electrons. This is consistent with the largest electron accumulation shifts on the bridging oxygen and the methyl hydrogens, where this extra stabilization is relatively strong; the pattern of changes of charges on the remaining atoms is a resultant effect of the closed reactant constraint and the specific structure of the $\boldsymbol{\beta}$ matrix.

To conclude this section we would like to emphasize that one needs the reactant charge responses at both the P- and CT- stages, to fully diagnose the reactivity/activation trends in the system under consideration. The determination of a character and relative importance of the inter-reactant CT requires the information about the actual sign and magnitude of N_{CT}. This may be in principle determined within the CSA/EEM scheme itself, provided sufficiently adequate values of the AIM chemical potentials, perturbing external potentials, and of the hardness interactions are available.

9. Intersecting-State Model of Charge Transfer Processes

In this section we describe the *intersecting-state model* (ISM) of the N-restricted CT in the M = A(acid)---B(base) reactive system, defined in the reactant population space [32, 59, 61, 66]. The model follows the familiar approaches formulated in the nuclear position space [120-122], by generating an effective energy profile consisting of the parabolic segments of the effective charge displacement energy curves of both reactants. The model also identifies the representative *"transition state"* (TS) and the associated amount of CT, which allows one to relate this structure to both reactants, in the spirit of the Hammond postulate [123] formulated in the electron population space. The general case of the AIM/PNM-resolved ISM is summarized in the Appendix C.

Let N_{CT} correspond to the full, equilibrium amount of charge transfer, marking the transition from $M^+ = (A|B)$ to $M^* = (A¦B)$, and $z = x\,N_{CT}$ denotes the intermediate CT for the current reaction progress variable $x \in [0, 1]$. The N-constrained section $E_{CT}(z)$ through the general quadratic surface for the constant external potential due to both reactants at their current mutual separation and orientation,

$$\Delta E_{AB}(x, y) = \mu_A^+ x + \mu_B^+ y + \frac{1}{2}(\eta_{A,A} x^2 + 2\eta_{A,B}\, xy + \eta_{B,B}\, y^2)\,, \tag{390}$$

where $x = N_A - N_A^0$ and $y = N_B - N_B^0$, corresponds to $x = -y = z$:

$$\begin{aligned} E_{CT}(z) &= \left[\mu_A^+ z + \frac{1}{2}(\eta_{A,A} - \eta_{A,B})\, z^2\right] + \left[-\mu_B^+ z + \frac{1}{2}(\eta_{B,B} - \eta_{A,B})\, z^2\right] \\ &= \left[\mu_A^+ z + \frac{1}{2}\eta_A^M z^2\right] + \left[-\mu_B^+ z + \frac{1}{2}\eta_B^M z^2\right] \equiv \left[\bar{E}_A(z)\right] + \left[E_B(z)\right]. \end{aligned} \tag{391}$$

In the last equation the N-restricted CT energy is partitioned into effective reactant energy functions, by dividing equally the interaction contribution due to the coupling hardnesses. The reactant energies, $E_B(z)$ and $\bar{E}_A(z)$, represent the energy functions of a *charge withdrawal* from B and a *charge stabilization* in A, respectively. Since M^+ provides a reference state (before the inter-reactant CT) one has $E_B(0) = 0$; similarly, the charge stabilization curve should give the correct $E_{CT}(N_{CT}) \equiv E_A(N_{CT})$, which calls for a shift in the zero energy of the acid curve:

$$E_A(z) = \bar{E}_A(z) + E_B(N_{CT})\,. \tag{392}$$

The reactant energies $E_A(z)$ and $E_B(z)$ generate the CT energy profile for the CT process shown in Fig. 40 a. Their intersection at $z = z^{\ddagger}$ marks the representative TS, $M^{\ddagger}$, for which the activation $[E_B(z^{\ddagger})]$ and stabilization $[E_A(z^{\ddagger})]$ energies, associated with the transition $M^+ \rightarrow M^{\ddagger}$, are equal. The corresponding value of the CT progress variable, $x^{\ddagger}$, may be used as a measure of the donor-acceptor bond-order and position of the TS between M^+ and M^* along the CT coordinate.

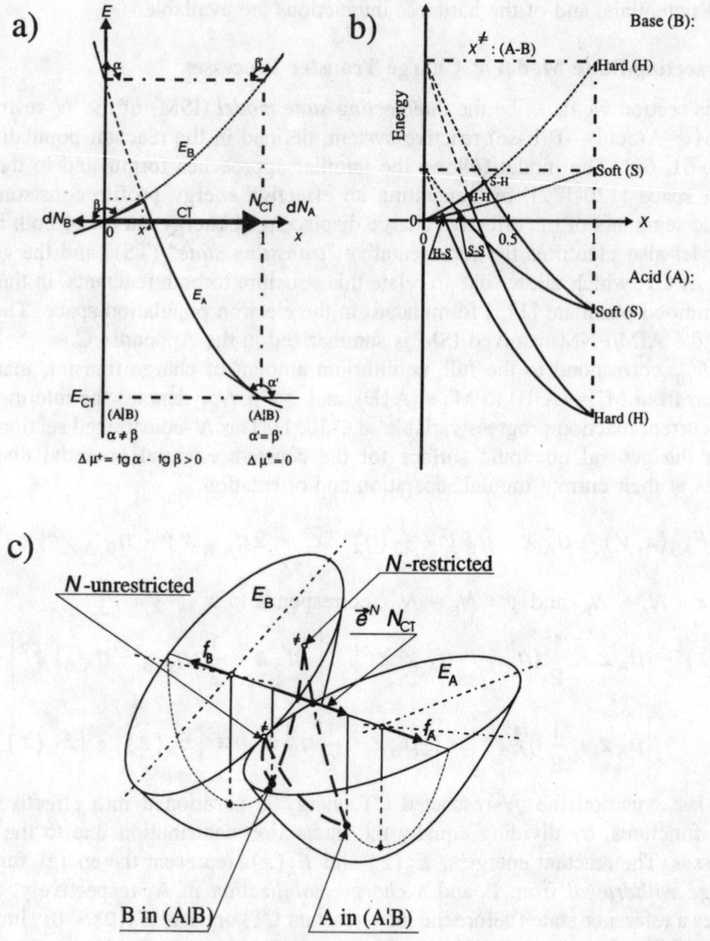

Figure 40. The ISM of the CT process: the ISM energy profile (a), corresponding to the N-restricted CT path; the effect of various hardness combinations of the HSAB principle on the TS activation barrier and the corresponding CT (bond-order) (b); panel c shows the perspective view of the intersecting energy paraboloids of A and B reactants, each with two populational degrees of freedom, and the corresponding N-restricted and N-unrestricted CT energy profiles (see Appendix C). The initial ($z = 0$) slopes of $E_A(z)$ and $E_B(z)$ in panels a and b reflect the (different) reactant chemical potentials in M^+; they are equalized in M^*, at $z = N_{CT}$. The effective CT energy profile consists of $E_B(z \in [0, z^\ddagger])$ and $E_A(z \in [z^\ddagger, N_{CT}])$, where $z^\ddagger$ corresponds to the intersection point (TS) value of the amount of CT.

The $E_B(z^\ddagger)$ measures the energy barrier in the exothermic direction (of forming the A-B bond); similarly, $E_B(z^\ddagger) - E_{CT}(N_{CT})$ represents the energy barrier in the endothermic direction (of breaking the A-B bond). The very construction of the present ISM in the electron population space of reactants indicates that this representative activation barrier should be proportional to the effective hardness of the basic reactant, η_B^M, and the TS CT value, $z^\ddagger$. Namely, the harder is B in M the more steep is the ascent along the base activation curve; also, the later is the barrier position the longer ascent along this curve before reaching the intersection point (Fig. 40 b). In the reactive system the hardness value of the isolated B, $\eta_{B,B}$, can be effectively reduced by the strength of the inter-reactant charge coupling, measured by $\eta_{A,B}$. Therefore, when this interaction is strong the activation process of B is relatively facile.

Illustrative applications to simple diatomic M have demonstrated that the ISM model satisfies the general Hammond postulate of an early ($x^\ddagger < 1/2$) barrier in the exoergic direction. The model also correctly predicts a general trend of shifting $x^\ddagger$ to an earlier position (closer to M^+) when the magnitude of the CT reaction energy increases (see Fig. 40 b). Qualitative (N-restricted) energy profiles of Fig. 40 b have been generated to satisfy the electronegativity equalization in the final stage of CT ($x = 1$). They predict the relative positions of the TS for various hard (H)/soft (S) combinations of the A-B reactants. The expected general ordering of $x^\ddagger$ values is: H-S < S-S <H-H < S-H.

The model resembles the ISM of Marcus [120] and similar semiempirical approaches [121, 122], all designed in the nuclear position space to identify and semiquantitatively estimate the most important thermodynamic, electronic, and geometrical factors influencing the CT reaction rates. The ISM defined in the electron population space or its mode- or AIM-resolved generalizations should provide a basis for examining how the electronic structure factors influence the CT reaction rate; also, they could help to establish general rules identifying the hardness (softness) matching of reactants, which is favorable for accelerating the CT process (*dynamical HSAB principle*). The AIM/PNM-resolved ISM (Appendix C) should enable one to follow the donor acceptor interactions in the mode-to-mode resolution. In the open A-B systems (Fig. 40 c), when reactants are coupled to an external electron reservoir, the N-unrestricted CT paths can be realized, which correspond to a lower activation energy. The relevant equations for determining this N-unconstrained TS are explicitly discussed in the Appendix C. The character of this lower pass between the reactant energy paraboloids in the electron population space provides a vital supplementary information about a role of the reactive system environment. For example, the deviation of the corresponding TS value of the global number of electrons from the initial value $N^0 = N_A^0 + N_B^0$, indicates how strong oxidizing/reducing action of the catalyst is required for the reactive system to follow this N-unrestricted, minimum energy path.

The quadratic energy function $\Delta E_{AB}(x, y)$ allows one to generate the model energy profile for any N-restricted or N-unrestricted charge reorganization process. Consider as an illustrative example the concerted charge flow involving the mutually closed (but interacting) A and B and their respective reservoirs, $[\mu_A(z) \mid \rightleftarrows A|B \rightleftarrows \mid \mu_B(z)]$, the chemical potentials of which are controlled in such a way that A gains and B looses exactly the specified number of electrons. Clearly, when $\mu_A(z) = \mu_B(z) = \mu_{AB}$

$= f_A^M \mu_A^+ + f_B^M \mu_B^+$, $z = N_{CT}$. Such a hypothetical system models the N-constrained CT process involving, e.g., two adsorbates A and B coupled to the isolated basic and acidic active sites, respectively. From the equalization of the reactant chemical potentials with those of their reservoirs one obtains:

$$\mu_A(z) = \mu_A^+ + z\eta_A^M \quad \text{and} \quad \mu_B(z) = \mu_B^+ - z\eta_B^M . \tag{393}$$

Hence, in this N-restricted CT process

$$\tilde{E}_A(z) = -\frac{1}{2}\eta_A^M z^2, \quad \tilde{E}_B(z) = -\frac{1}{2}\eta_B^M z^2, \quad \tilde{E}_{AB}(z) = \tilde{E}_A(z) + \tilde{E}_B(z) . \tag{394}$$

CHAPTER 5

ILLUSTRATIVE APPLICATIONS TO MODEL CATALYTIC SYSTEMS

One of the primary aims of the development described in previous chapters has been to formulate adequate two-reactant reactivity concepts and the underlying coordinate systems, which could be used to diagnose reactivity and selectivity trends in systems of very large donor/acceptor reactants, e.g., chemisorption structures. The CSA approach is both relevant and attractive from the chemist's point of view, since many branches of chemistry, and the theory of chemical reactivity in particular, consider responses of chemical species to the external potential and electron population perturbations. Clearly, the two-reactant approach is vital, since the charge response reactivity criteria and the interaction energy must involve measures of both the generalized polarizabilities of the perturbed reactants and the stimuli due to the presence of the other reactant. The CSA treats molecular systems as an interconnected ensemble of structural or functional units, with a flexibility to hypothetically open or to close any of its constituent parts, both mutually and relatively to an external reservoir. It adopts the thermodynamic-like description of the global and regional (constrained) equilibrium charge distributions, based upon the chemical potential (electronegativity) equalization principle. Such a perspective is indeed very close to that adopted in intuitive chemistry. Each reactivity problem requires an appropriate level of resolution (reactants, functional groups or other important structural units, local sites, e.g., AIM, etc.). The canonical AIM-division can be expected to be sufficient for most chemical problems, although for some subtle bonding effects the local or molecular orbital resolutions may be required. Both the function space (populational analysis) and physical space (topological approach) partitionings of the electronic density can be used to extract the required AIM information, e.g., the net charges. The AIM-type modelling has an *a priori* character, while the topological approach of Bader [7] represents an *a posteriori* extraction of the AIM information from the exact molecular calculations. When considering bond dissociation processes one has to take into account the chemical potential discontinuity [48].

The CSA in the AIM-resolution can be applied as a procedure supplementary to the standard *ab initio* SCF CI or DFT calculations, which may in principle provide the input AIM charges, but this would severely limit a range of the method applications. The real strength of CSA lies in its ability to fast diagnose the reactivity trends from rather crude, and limited information about the constituent atoms, including the hardness (valence electron repulsion) interaction between large reactants, parametrically dependent upon the system geometrical structure. Thus, even approximate AIM charges, e.g., those from the EEM [10], parametrized to reproduce the Mulliken populations of the STO-3G quality, or from the semi-empirical SCF MO calculations on small models, may be used as input information in the CSA calculations to quickly explore the reactivity possibilities in a system consisting of very large reactants.

Each physical phenomenon can be related to the system of coordinates in which it can be described in the most compact and simple way. Finding such an optimum reference frame is very important for interpretative purposes, since it allows one to identify the crucial elements of the mechanism and to separate them from the background information, irrelevant to the process in question, at least in the first approximation. We have shown previously that in reactive systems the highest degree of concentration of reactivity information involving charge displacements is obtained in the reactant collective populational coordinates, which reflect the inter-reactant charge couplings, e.g., the IRM. Each system of coordinates is probing different facets of molecular behavior. Some of them, e.g., MEC/REC, are closely related to the intuitive chemical thinking, *viz*., the inductive effect, charge relaxation accompanying localized reduction/oxidation, effects of a hypothetical or real CT between selected subsystems, etc.

Although most of the sensitivity criteria are closely related to the interaction energy, they are basically reactant charge responses, be it reflecting the reaction stimuli, i.e., the interaction between the two molecules. The mode resolved CSA may also be considered as supplementing the classical frontier orbital interaction approach of Fukui.

The AIM-resolved CSA starts with the information on isolated reactants, which it supplements with a realistic representation of the hardness interaction, interpolated from the reactant (AIM) data, and their current mutual geometrical arrangement in reactive system. Thus, the CSA provides a phenomenological treatment of molecular systems, which allows one to quickly predict their behavior in a changed environment, e.g., due to the presence of the other molecule. Clearly, this novel way of both thinking about and predicting chemical reactivity trends was possible due to the DFT, which provides a firm foundation for quantitatively implementing the chemical potential (electronegativity) equalization principle of Sanderson [18]. As we have demonstrated in the preceding chapters the CSA enables one to formulate new geometric concepts for describing charge rearrangements between complex reactants, including the alternative collective modes, resolutions of charge displacement patterns, etc.

It is a purpose of this chapter to summarize general observations on the CSA predicted CT and direct polarization patterns of large surface clusters, and to demonstrate a use of the MEC and IRM reference frames in diagnosing and interpreting the CT processes. In particular, we shall examine how the adsorbate – substrate interaction shapes the polarizational and *in situ* FF responses in catalytic systems. As we have emphasized in Section 8 of the preceding chapter both the P^+ and CT charge displacement information is required to fully diagnose the most probable reactivity trends. Therefore, these two complementary charge sensitivity components will be reported together with overall charge displacement diagrams for the three illustrative catalytic systems which have been extensively studied in the Cracow group.

Changes in atomic charges of reactants can be interpreted as the bond weakening/strengthening effects only approximately, since the overall bond order results from a subtle interplay between covalent and ionic components. A more direct, quantum-mechanical measures of the effective bond-multiplicity [4-8] are needed to predict reactivity/activation patterns in surface reactions involving large adsorbates. It has been recently demonstrated that useful absolute bond-order indicators can be generated by comparing bonding patterns in a molecule with those in the separated atoms limit, respectively [6, 124-126], generated within the one-determinantal (Hartree-Fock or Kohn-

Sham) difference approach, in the framework of the two-electron density matrix. In this chapter we shall supplement the CSA reactivity predictions with changes in such quantum-mechanical bond-order indices due to the chemisorption, for model chemisorption complexes involving allyl and the molybdenum oxide cluster. We have selected three systems which were the subject of recent theoretical and experimental studies [9, 32, 59, 88, 92, 93, 106, 107, 118, 127-131]. They include the water- [(110) rutile], toluene - [(010) vanadium oxide], and allyl - [(010) molybdenum oxide] chemisorption clusters. Since the one-reactant reactivity criteria may be inadequate in such complicated, prototype catalytic systems, we focus here on the two-reactant response criteria and related concepts introduced in previous chapters.

1. Rutile Surface Clusters and the Water-Rutile System

1.1. PNM Resolution of the AIM FF Indices

As we have already remarked previously, every external CT to/from a molecular system in question is accompanied by the conjugate CT-induced internal polarization. The CT component of molecular charge rearrangements should be dominated by the hardest, totally nonselective PNM, while the short range polarization of the softest PNM is expected to strongly contribute to the associated P component. This is indeed reflected by the numerical results [9, 32, 59]. However, when one is interested in the magnitude of the atomic population differentiation due to the external CT then the softest mode appears as the most important one; this is explicitly demonstrated in Fig. 41 for the $Ti_{49}O_{98}$ cluster of the (110)-rutile surface. A reference to the right column of Fig. 41 b shows that the softest mode $\alpha = 1$ indeed contributes the most to the FF diagram of panel a, reproducing the most important shifts in atomic populations, which accompany the global chemical reduction of the cluster. The observed minor differences between these two diagrams are to a large extent compensated by the contributions from the three harder modes included in the figure. For example, the modes $\alpha = 49$ and 103 are seen to correct the softest mode topology in the cluster artificial boundary region. Clearly, in the CT (w_α) hierarchy the most important is the hardest, non-selective mode, in which all atoms undergo practically the same, much smaller shifts in their atomic populations, $w_{i,\alpha} \approx w_\alpha/m$, whereas the site-differentiating effects of the external CT are due to the short-range polarization modes, which exhibit w_α values close to zero.

1.2. Use of the MEC in Diagnosing Surface Clusters

The MEC can be used to diagnose the extent of the populational responses in catalytic clusters, created by the localized test reductions/oxidations [9, 32, 57, 87, 91, 92]. This is illustrated in Fig. 42 a, which reports the external MEC for the surface system of Fig. 41. Both the hardness parameter Γ_k and the mode topology can be used in a qualitative diagnosis of the system responses to environmental changes. In Fig. 42 a the largest open circle identifies the primarily displaced atom k, $dN_K = 1$. All diagrams show that the appreciable populational displacements created by this perturbation practically do not extend beyond the second-neighbors of the primarily displaced site. The same effect has

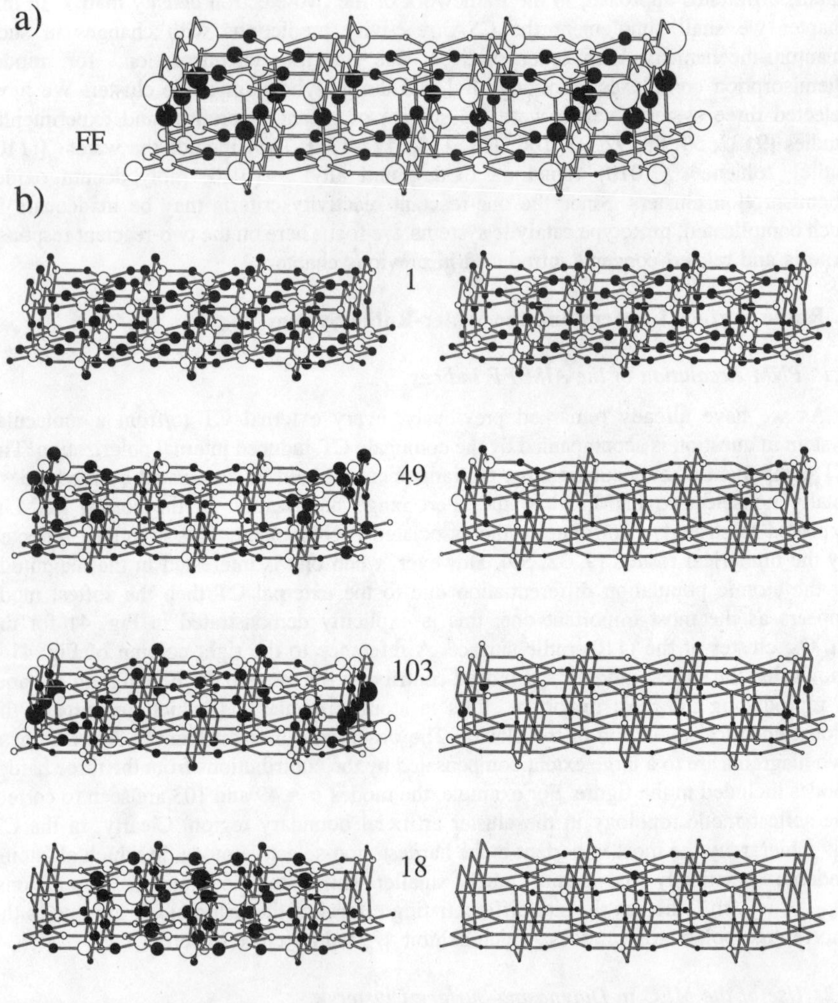

Figure 41. The AIM FF distribution in the two-layer, stoichiometric cluster model $Ti_{49}O_{98}$ of the (110)-rutile surface (panel a) and the four PNM (panel b), α = 1, 49, 103, and 18 (left column), which exhibit the largest magnitudes of contributions to the atomic FF (right column), measured by the mode maximum value of $|w_{i,\alpha}|$; a common scale factor values have been adopted in each column of part b.

been observed in the V_2O_5 systems [87]. This *a posteriori* validates the small, non-stoichiometric (50 atoms) cluster representation of this Ti active site and its effective environment shown in Fig. 42 b, which we have used in the previous study [129]. A relatively short range of the observed MEC charge shifts also suggests that the

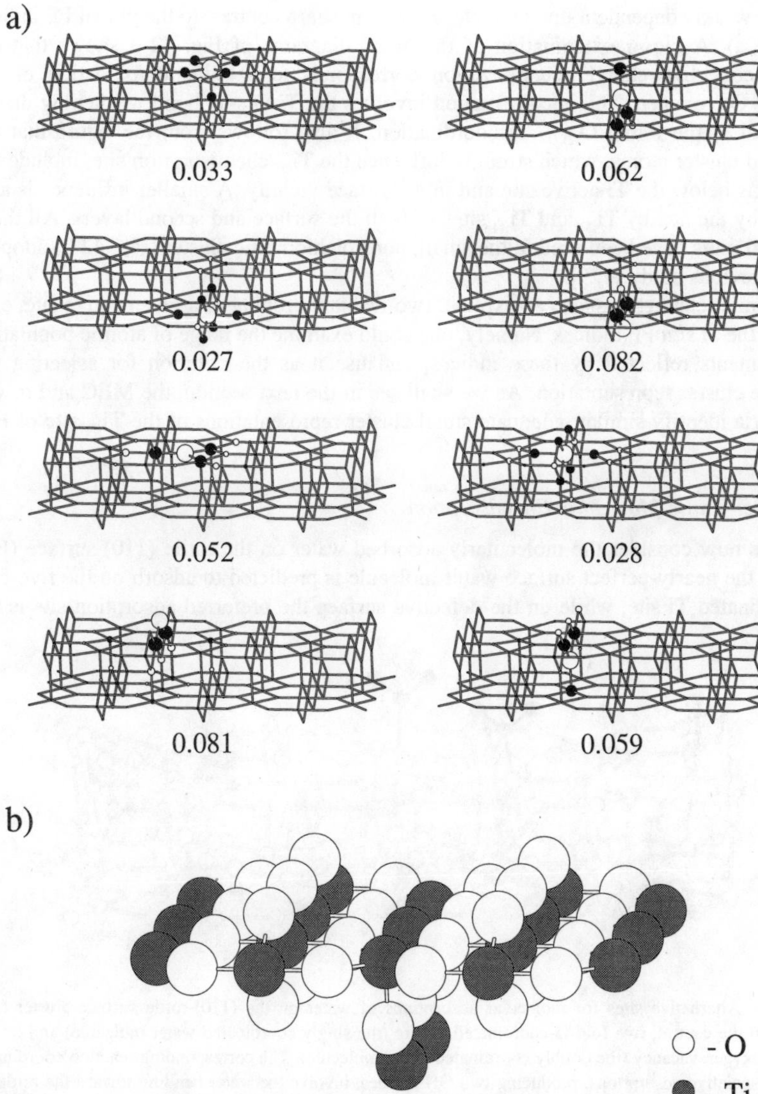

Figure 42. Selected external MEC (part a) of the surface cluster of Fig. 41 and the adequate cluster representation of the five O-coordinated Ti site (part b). The primarily displaced atoms of the MEC are identified by the largest open circle, and the mode hardness parameter Γ_k (a. u.) is reported below each mode diagram.

off-diagonal responses, which dominate the cluster contribution to the *in situ* FF indices, are only weakly dependent upon the cluster size, in sharp contrast to the global FF indices (Fig. 41 a). A closer examination of the MEC diagrams of Fig. 42 a shows that the hardest collective charge reorganization corresponds to a reduction/oxidation of the bridging $O_{(2)}$ oxygen. This reconstruction involves the Ti-$O_{(2)}$-Ti bond weakening due to a reversal of the initial $O_{(2)} \rightarrow$ Ti coordination. It also follows from the figure that the displaced cluster atoms, which strongly influence the $Ti_{(5)}$ chemisorption site, include the $O_{(3)}$ atoms below the Ti active site and in its surface vicinity. A smaller influence is also exerted by the nearby $Ti_{(5)}$ and $Ti_{(6)}$ sites of both the surface and second layers. All these structural units have been used in the small, nonstoichiometric cluster (Fig. 42 b) adopted in the previous study [129].

Of similar character is also an explicit, two-reactant probing of a given active site, e.g., through the *in situ* FF indices. Namely, one could examine the range of atomic population displacements reflected by these indices, and use it as the criterion for selecting the adequate cluster representation. As we shall see in the next section, the MEC and *in situ* FF criteria identify similar, adequate small cluster representations of the $Ti_{(5)}$ site of Fig. 42 b.

1.3. Water-Rutile Molecular Chemisorption

Let us now consider the molecularly adsorbed water on the rutile (110)-surface (Fig. 43). On the nearly-perfect surface water molecule is predicted to adsorb on the five-fold O-coordinated Ti site, while on the defective surface the preferred adsorption site is the

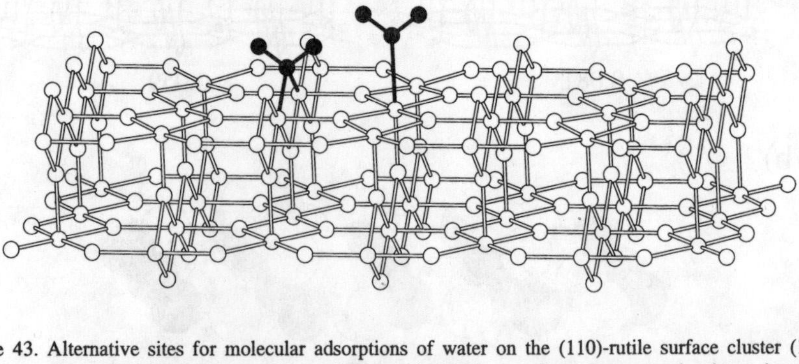

Figure 43. Alternative sites for molecular adsorptions of water on the (110)-rutile surface cluster (146 atoms): on the central, five fold O-coordinated Ti site (the singly coordinated water molecule) and on the bridging oxygen vacancy (the doubly coordinated water molecule). The corresponding reaction coordinates of the dissociative adsorptions, producing two OH species, involve the water bending toward the bridging oxygen of the singly coordinated adsorbate, and a rotation of the doubly coordinated water, respectively.

bridging O-vacancy, which allows the double σ-coordination of water (Fig. 43). The respective reaction coordinates, of the alternative adsorbed water dissociations, should involve the singly-coordinated water bending (toward a single $O_{(2)}$ site) and the doubly-coordinated water rotation (toward the two $O_{(2)}$ atoms in its vicinity) [9, 129, 130]. The first mechanism should involve much lower activation barrier, predicted to be around 5 kcal/mol [129].

Let us examine the FF quantities for the singly coordinated molecular adsorption cluster shown in Fig. 44. The panels a, b, and c, correspond to the closed chemisorption system (isoelectronic responses) and represent the diagonal, off-diagonal, and total CT FF indices, respectively, for the assumed water → rutile electron transfer, i.e., a direction reflecting the water → cluster coordination bond.

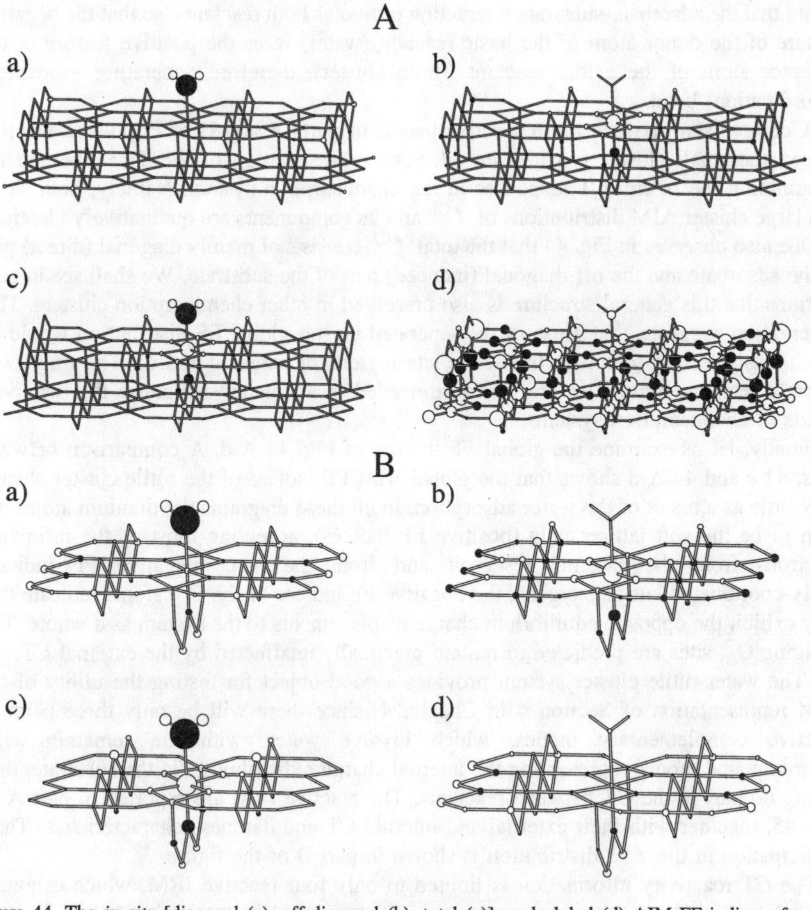

Figure 44. The *in situ* [diagonal (a), off-diagonal (b), total (c)], and global (d) AIM FF indices for the water – [(110)-rutile] chemisorption systems: large cluster (149 atoms, part A) and small cluster (53 atoms, part B). The large rutile cluster includes the bridging oxygen vacancy in the vicinity of the adsorption site, i.e., the five-fold O-coordinated, central Ti atom. The water → surface electron transfer is assumed in all the *in situ* FF diagrams.

Consider first the large cluster part of Fig. 44 A. In contrast to the $\boldsymbol{f}$ distribution of panel d, the $\boldsymbol{f}^{CT}$ (panel c) quantities are strongly localized on the adsorption site and its nearest vicinity. The phases of the charge displacements in panel c show that this internal CT is mainly from the water oxygen to the Ti site of the cluster, as expected during a formation of the coordination bond; however, the observed accompanying charge adjustments on the water hydrogens, diminishing the bond polarity relative to that in the isolated water, should weaken (lengthen) the O-H bonds, in accordance with the mapping relations of Section 5 in Chapter 4. Thus, one again observes in panels a and b of the figure that the adsorbate-substrate interaction promotes both reactants, so that the negative feature of the donor atom of the basic reactant (water) faces the positive feature of the acceptor atom of the acidic reactant (rutile cluster), therefore generating a covalent (coordination) bond.

A comparison between the respective panels of parts A and B of Fig. 44 shows that indeed the small cluster provides an adequate representation of the Ti active site for determining the *in situ* CT responses in the chemisorption system. Namely, both small and large cluster AIM distributions of $\boldsymbol{f}^{CT}$ and its components are qualitatively identical.

One also observes in Fig. 44 that the total $\boldsymbol{f}^{CT}$ consists of mainly diagonal (direct) part of the adsorbate and the off-diagonal (induced) part of the substrate. We shall see in next sections that this general structure is also preserved in other chemisorption clusters. This observation suggests, that a use of the separated cluster global FF distribution would be much less selective (compare the active site regions of panels c and d). The truly two-reactant response properties are therefore needed to adequately diagnose the reactivity trends of chemisorption systems.

Finally, let us examine the global FF indices of Fig. 44 A-d. A comparison between Figs. 41 a and 44 A-d shows that the global AIM FF indices of the rutile cluster change very little as a result of the water adsorption. In all these diagrams the titanium atoms are seen to be the soft lattice units (positive FF indices), accepting most of the inflowing electrons from the external reservoir and from the hard (negative FF indices) triply-coordinated lattice oxygens; the negative FF indices of the $O_{(3)}$ atoms indicate that they exhibit the opposite equilibrium charge displacements to the system as a whole. The bridging $O_{(2)}$ sites are predicted to remain practically unaffected by the external CT.

The water-rutile cluster system provides a good object for testing the utility of the IRM representation of Section 4 in Chapter 4, since there will be only three pairs of reactive complementary modes, which involve water, with the remaining 143 environmental modes representing the internal charge redistribution in the substrate, thus taking no part in the CT between reactants. The reactive IRM are reported in part A of Fig. 45, together with their external and internal CT and hardness characteristics. Their participation in the $\boldsymbol{f}^{CT}$ distribution is shown in part B of the figure.

The CT reactivity information is limited to only four reactive IRM, which originate from the two independent components on each reactant. The mode $\gamma = 1$ describes the electron transfer from water to cluster, with almost uniform distribution of the charge among the constituent atoms of each reactant. Its complementary mode, $\gamma = 149$, again with the "background", constant displacements on each reactant, also does not differentiate between local sites. The truly reactive channels $\gamma = (3, 147)$ generate the same reversal

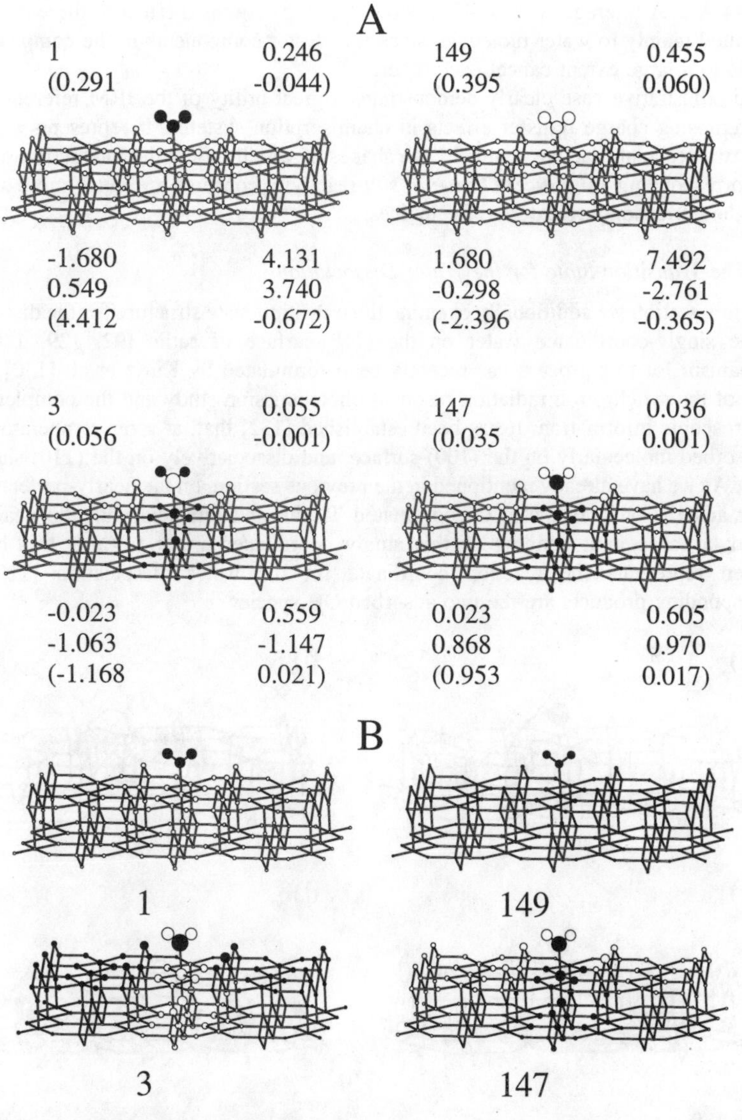

Figure 45. The complementary reactive IRM (part A), (γ and $m - \gamma + 1$), for the molecular chemisorption complex of Fig. 44, and their contributions to the total *in situ* FF indices (part B) of Fig. 44 A-c . The external and internal CT numerical data (a. u.) reported above and below each mode diagram are the same as in Fig. 30. The γ = (2, 148) reactive IRM have been omitted since, by the overall symmetry (only slightly distorted by the oxygen vacancy), they practically do not contribute to the inter-reactant CT.

of the original bond polarity in the adsorbed water, as that seen in the f^{CT} diagrams of Fig. 44 A-c. A reference to Fig. 45 B shows that the combined effect of these two modes is limited mainly to water molecule, since the cluster components in the complementary modes to a large extent cancel each other.

This illustrative case clearly demonstrates a great utility of the IRM reference frame in diagnosing charge transfer effects in chemisorption systems. It represents a set of a truly two-reactant reactive channels, the shapes of which reflect the interaction region of the corresponding FF indices. These *in situ* reactivity concepts are much more selective than the corresponding external FF indices.

1.4. The Transition-State for the Water Dissociation

In this section we additionally examine the transition-state structure for the dissociation of the singly-coordinated water on the (110)-surface of rutile [92, 129, 130]. The mechanism for this process has recently been formulated by Kurtz et al. [130], on the basis of the synchrotron radiation resonant photoemission study and the complementary spectroscopic information. It has been established [132] that, at room temperature, H_2O is adsorbed molecularly on the (100)-surface, and dissociatively on the (110)-surface of rutile. As we have already mentioned in the previous section, on the nearly-perfect surface water adsorbs on a five-fold O-coordinated Ti site and interacts with the nearby $O_{(2)}$ ligand; a subsequent bending of this singly coordinated water towards this bridging oxygen represents the reaction coordinate for the water dissociation [129]. The corresponding products are the two adsorbed OH species.

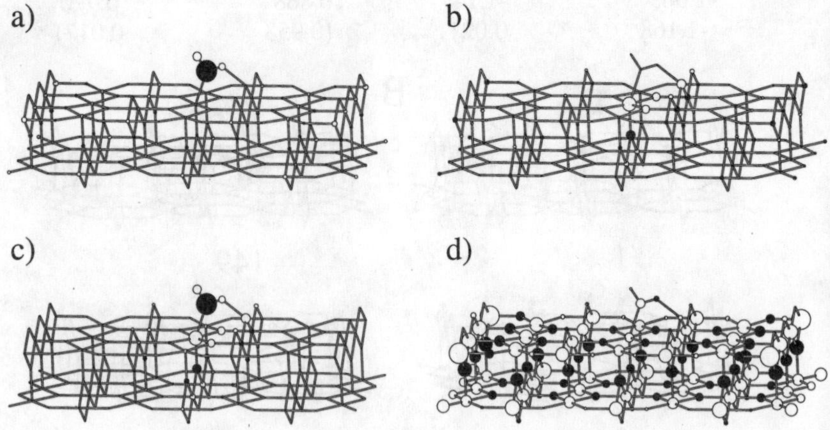

Figure 46. The AIM FF indices for the transition-state toward the water dissociation of the singly coordinated water molecule (with a tilt angle of 21° relative to the structure of Fig. 44). The panels a - d have the same meaning as in Fig. 44.

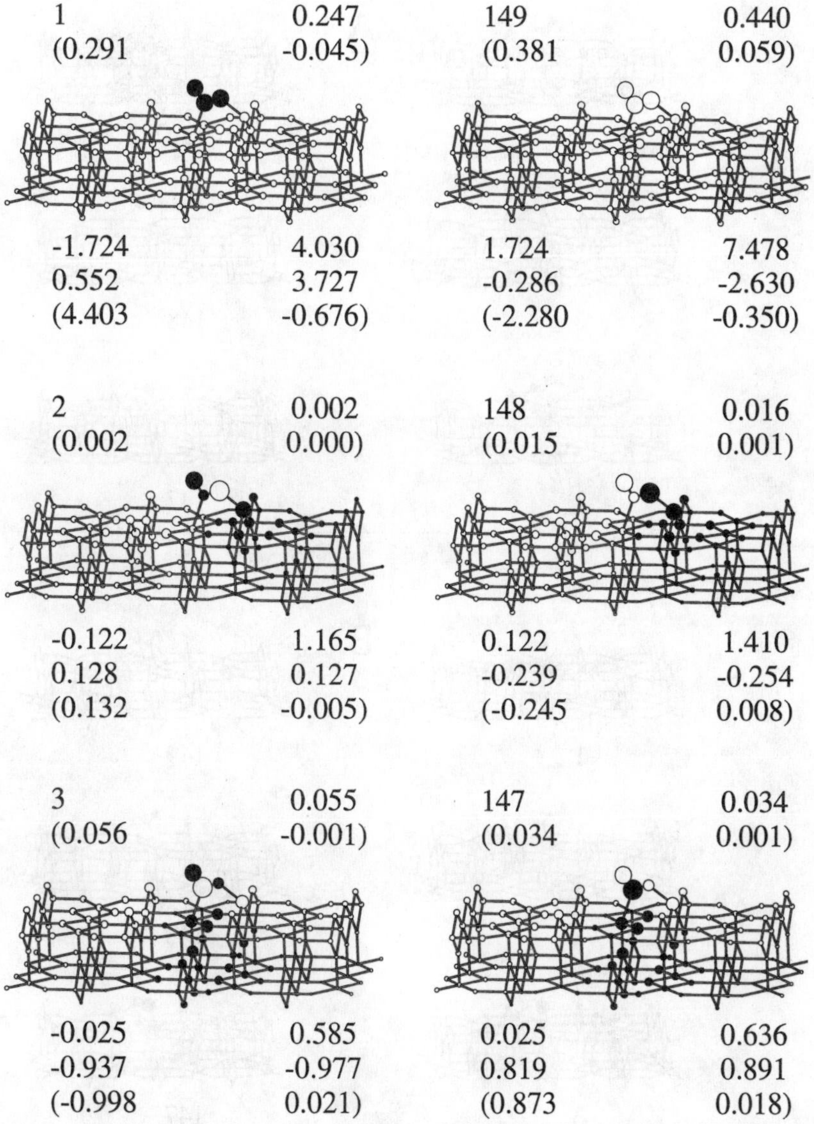

Figure 47. The complementary reactive (delocalized) IRM for the transition-state of the chemisorption complex of Fig. 46. The numerical data are reported in the same order as in Figs. 30 and 45 A.

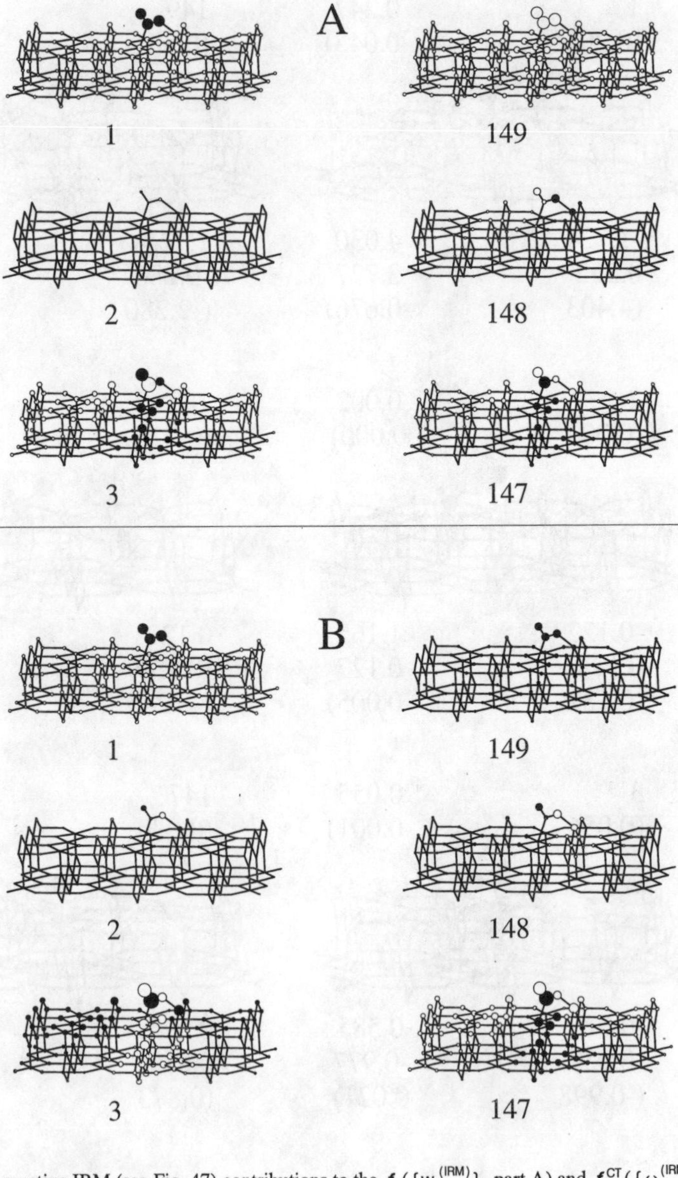

Figure 48. The reactive IRM (see Fig. 47) contributions to the $f\ (\{w_{i,\gamma}^{(IRM)}\}$, part A) and $f^{CT}(\{\omega_{i,\gamma}^{(IRM)}\}$, part B) FF distributions of the transition-state water - rutile chemisorption system.

The AIM global and *in situ* FF indices for the transition-state chemisorption system are shown in Fig. 46. The f^{CT} distribution is seen to be very similar to that reported in Fig. 44 A-c; however, the symmetry-breaking due to the adsorbate bending has created an increased off-diagonal response on the bridging $O_{(2)}$ and $O_{(3)}$ oxygens of the cluster (Fig. 46 b, c), a clear sign of the partial formation of the new OH bond. The overall structure of the total *in situ* diagram, dominated by the diagonal component of water and the off-diagonal component of rutile, remains unchanged in the transition-state complex.

A similar symmetry breaking changes are seen in Fig. 47, where we present the six CT-active complementary IRM. We have also included in the figure the modes γ = (2, 148), which should also contribute to f^{CT}. This is indeed seen in Fig. 48 B, where the modes 148 and 147 account for the total f^{CT} feature in the region of the newly formed bond.

The IRM resolution of the external FF shown in Fig. 48 A reflects the moderating effect of the electron reservoir (the cluster environment), which is missing in the isoelectronic, *in situ* indices. A comparison between the respective panels of parts A and B of Fig. 48 indicates that an opening of the chemisorption cluster strongly affects contributions from the specific inter-reactant CT-active IRM. Namely, the modes γ = 1, 2, 147 are seen to give rise to a similar charge rearrangements in both cases, while the remaining modes exhibit opposite phases in the closed and open chemisorption systems. We would like to recall, that Fig. 48 A refers to an inflow of a single electron to the whole chemisorption system ($dN = 1$), i.e., to an overall chemical reduction. Therefore, the opposite phases of the IRM contributions are predicted, when the whole system is oxidized ($dN = -1$).

1.5. Polarizational Patterns of the Closed Adsorbate and Substrate

The full reactivity diagnosis of the water-rutile system requires the complementary P^+ information, on how reactants polarize each other in M^+, $d\boldsymbol{N}^+_{A(B)}(d\boldsymbol{v})$, before the CT (physisorption stage), in addition to the P^* diagrams (CT-induced polarization) reported previously. Moreover, the actual signs and magnitudes of the f^{CT} displacements are determined by the direction and amount of CT between reactants, both being determined by the polarizational stage of the resultant charge reconstruction in the reactive system under consideration. The relevant AIM diagrams are shown in Fig. 49.

Let us first examine the molecular adsorption case shown in the left column of the figure. Here, the off-diagonal AIM displacements are very small indeed, so that the diagonal changes dominate the total polarizational displacements due to the presence of the other reactant. As seen in the last diagram of the first column the physisorption of water on the titanium site enhances the initial H $\rightarrow$ O polarization of bonds in the isolated water molecule, thus increasing (decreasing) its ionic (covalent) bond-order component, and generally strengthening the bonds. Hence, the physisorbed water hydrogens are predicted to become harder (increased acidic character) while the water oxygen becomes softer (increased basic character). This may facilitate the subsequent $O_{(water)} \rightarrow$ Ti and H $\leftarrow O_{(2)}$ coordinations, which lead to the water dissociation when the hypothetical barriers preventing the flow between reactants are lifted. Such flows are indeed seen in Fig. 44.

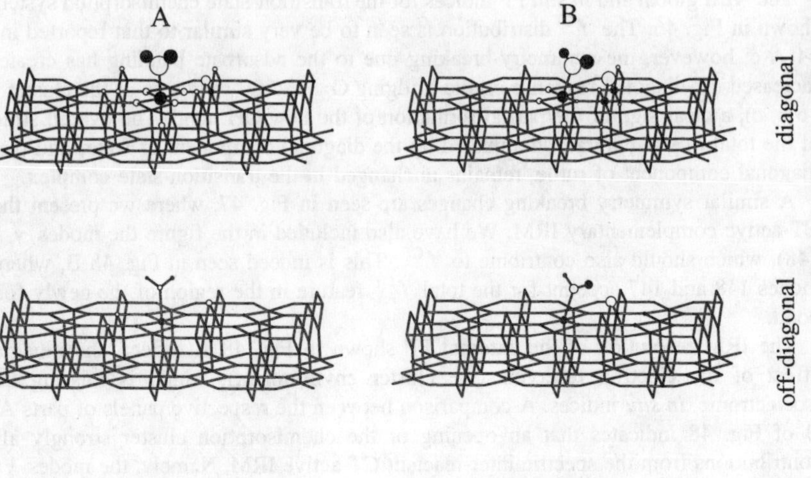

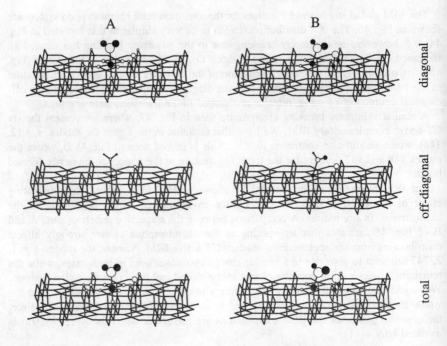

Figure 49. The AIM-resolved P^+ polarizational patterns, $dN^+_{A(B)}(dv)$, of the closed adsorbate and substrate reactants in the molecularly adsorbed water on the (110)-rutile surface (column A) and in the transition-state toward the water dissociation (column B). In each column the first, middle, and last diagrams show the diagonal, off-diagonal, and total polarizations, respectively (see Section 8 in Chapter 4).

At the same time one detects a reconstruction of the Ti-O bonds of the cluster, in the vicinity of the adsorption site. The initial O → Ti coordination is partly reversed, which should weaken these cluster bonds; the bridging oxygen, which eventually forms the bond with the water hydrogen in the dissociation reaction, becomes softer (increased basic character) after the water physisorption. Therefore, the P^+ stage is seen to promote the crucial atoms of both reactants toward the subsequent H_2O → Ti CT and the HOH ← $O_{(2)}$ donor/acceptor interactions.

The *in situ* FF diagrams of Figs. 44 and 46 have been generated for the assumed water → cluster electron transfer, in accordance with the CT direction determined from the semi-empirical SINDO1 calculations [131] (N_{CT} = 0.16). A comparison between Fig. 44 A-c and the bottom diagram of Fig. 49 A shows that the extra electron charge accumulated on water oxygen at the P^+ stage is subsequently transferred mainly to the Ti site at the CT stage of the chemisorption; one also observes that the extra positive charge of the water hydrogens from the P^+ charge redistribution is partly neutralized due to the interreactant CT. Moreover, the bridging oxygen $O_{(2)}$, the site of the eventual water dissociation, receives a part of the transferred electron charge. From this interplay between the P^+ $[(dN^+(dv)]$ and CT $[(dN^*(N_{CT})]$ components of the overall AIM electron population shifts, $[(dN(dv, N_{CT})]$, a picture emerges in which the spontaneous CT stage minimizes the displacements created by the primary (dv) perturbation, in accordance with the familiar thermodynamic analogy (Le Châtelier-Braun principle).

Let us consider next the transition-state P^+ diagrams shown in the second column of Fig. 49. The topology (nodal structure) of these polarizational charge displacements are similar to those for the molecular adsorption (left column). However, due to the symmetry breaking caused by bending the water molecule, there is a much stronger off-diagonal component, which enhances the main diagonal charge responses. In particular, one observes that the H---$O_{(2)}$ interaction is strongly promoted in the transition-state, via the increased $H^{\delta+}$ - $O^{\delta-}_{(2)}$ ionic interaction due to the extra charges $\delta+$ and $\delta-$, generated by the mutual polarization of reactants.

To summarize these P^+ plots, one already observes at the physisorption stage the extra (P^+-induced) inter-reactant ionic interactions: $Ti^{\delta+}$ - $O^{\delta-}_{(water)}$ and $H^{\delta+}$ - $O^{\delta-}_{(2)}$, which are the precursors of the two new bonds in the adsorbed dissociation products.

Finally, let us compare the transition-state CT $[dN^*(N_{CT})]$ and the P^+ $[dN^+(dv)]$ patterns of Figs. 46 c and 49 B (bottom diagram), respectively. Again, the same mechanism is observed as that detected for the molecular adsorption. The extra electrons accumulated on water oxygen at the P^+ stage are distributed in the CT stage to the nearby Ti site, hardened as a result of the P^+ polarization, while the extra charge of the cluster $O_{(2)}$ site goes to the approaching water hydrogen, also hardened at the P^+ stage. Thus, the above mentioned thermodynamic analogy (Le Châtelier-Braun principle) also holds at the transition-state structure of this chemisorption system.

2. Toluene–$[V_2O_5]$ System

2.1. Surface Cluster and Adsorption Arrangements

In previous illustrative applications of CSA to this chemisorption system (Chapters 3 and 4) we have used a very small, strongly non-stoichiometric cluster representation of the surface active site, which included two idealized VO_5 pyramids of the surface layer. In this section a relatively large, two-layer, nearly-stoichiometric cluster $V_{36}O_{98}$ shown in Fig. 50 has been used to examine the influence of both the surface cluster size and of the supporting layer, in a truly two-reactant description of charge responses. In Fig. 50 we present two views of the representative two-layer, nearly- stoichiometric cluster $V_{36}O_{98}$ of the real crystal: on the top (panel a) and bottom (panel b) layers, respectively. The corresponding elements of these top and bottom "surfaces" model alternative active-sites for the toluene adsorption; they are specified in panels c-j of the figure. The layer, which is not involved directly in the chemisorption bond(s), models the surface "support" in the crystal. The chemisorption systems in Fig. 50 c-g represent prototype perpendicular adsorptions of toluene, via the methyl group, on various oxygen sites and the vanadium atom: on the terminal, singly-coordinated (vanadyl) oxygen, $O_{(1)}$, of the top layer surface (panel c), on the bridging, doubly-coordinated oxygens, $O_{(2)}$, of the bottom (panel d) and top (panel e) layer surfaces, on the triply coordinated oxygen, $O_{(3)}$, of the bottom layer

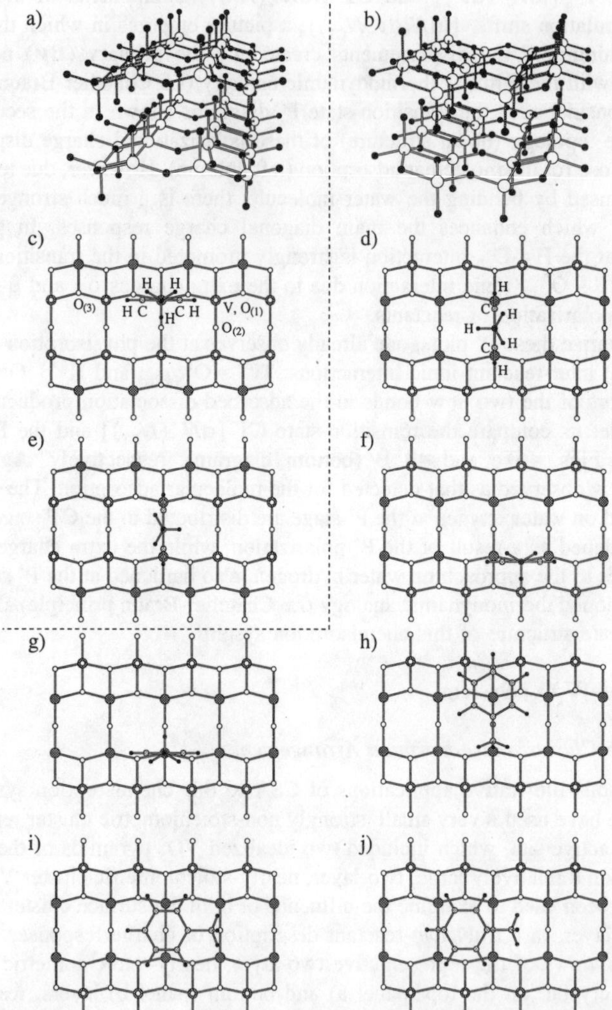

Figure 50. Perspective views from the top-layer side (panel a) and from the bottom-layer side (panel b) of the representative, nearly-stoichiometric surface cluster $V_{36}O_{98}$. The geometric structure of the V_2O_5 (010)-surface cluster involves the SINDO1 optimized intra-layer bond lengths as well as the crystallographic values of the bond angles and the inter-layer V-O bond length. The figure also shows the surface layer views of alternative perpendicular adsorptions of toluene on the vanadyl oxygen $O_{(1)}$ from the top layer side (c), on the bridging oxygen $O_{(2)}$ from the bottom (d) and top (e) layer sides, on the $O_{(3)}$ site from the bottom layer side (f), and on the vanadium atom from the bottom layer side (g); the three model parallel adsorptions on the bottom layer are shown in panels h-j. Unless specified otherwise the same adsorbate-substrate separations as in Fig. 36 have been used.

surface (panel f), and on the vanadium site of the bottom layer surface (panel g). For all perpendicular chemisorption structures other toluene conformations have also been examined, involving the 90° rotation of the toluene ring with respect to the methyl group, around the $C-CH_3$ bond. However, since practically no influence of this internal rotation on the predicted responses has been found, we report results for one rotamer only.

Figures 50 h-j correspond to the alternative parallel adsorption arrangements of the toluene benzene ring relative to the bottom layer surface. The first parallel complex (Fig. 50 h) corresponds to the central position of the benzene ring above the vanadium (positive electrostatic potential [91]) site, which should stabilize the ring π-electrons. The second parallel adsorption arrangement (Fig. 50 i) places the ring centre above the central $O_{(2)}$ atom; this brings the hydrogen atoms close to the four $O_{(3)}$ sites on the surface, and the ring electrons are located close to the two vanadium atoms. However, in this test structure hydrogens are also placed close to the vanadium atoms, which should give rise to a repulsive electrostatic interaction. Finally, in the third parallel complex (Fig. 50 j) the ring is above the "cage" defined by the four vanadium atoms connected through two pairs of the $O_{(2)}$ and $O_{(3)}$ bridges, respectively.

The remaining figures report the charge response patterns for the adsorption arrangements of Fig. 50. In Figs. 51 and 52 we present the resolutions of the P ($d\boldsymbol{N}^+$), CT ($d\boldsymbol{N}^*$), and overall ($d\boldsymbol{N}$) responses into the corresponding diagonal and off-diagonal contributions, for the illustrative perpendicular and parallel adsorptions, respectively. For all structures the total responses at the P, CT and (P + CT) levels of description are compared in Figs. 53 and 54. The arrow in the total CT diagrams shows the CSA determined direction of the inter-reactant CT. Finally, in Figs. 55 and 56 the global FF indices are reported.

In Figs. 53 (c, d), 54 [(a, b), (c, d), (e, f)], 55 (c, d), and 56 [(a, b), (c, d), (e, f)], two representative adsorbate-substrate distances are used, with the first panel in parentheses corresponding to a shorter distance [118], and the second panel representing an earlier stage of the adsorbate approach, with the corresponding inter-reactant separation increased by 1 Å relative to the first one. We have also carried out the CSA calculations for the increased (also by 1 Å) inter-reactant distance for the remaining perpendicular complexes of Fig. 50 c, d, f, g, although no charge reconstruction diagrams will be reported for these geometries. However, we shall discuss some qualitative energetical and the CT direction results for these early approach structures too.

In order to clearly see the reaction of the supporting layer to an adsorption on the surface layer, we have artificially separated the two layers of the vanadium oxide cluster in Figs. 51-56 by an arbitrary relative translation. These calculations have used the charges obtained from the semi-empirical SINDO1 [131] calculations. The geometric structure of the large cluster involves the SINDO1 optimized intra-layer bond lengths [87] for the assumed crystallographic values of the bond angles and the inter-layer V-O bond length.

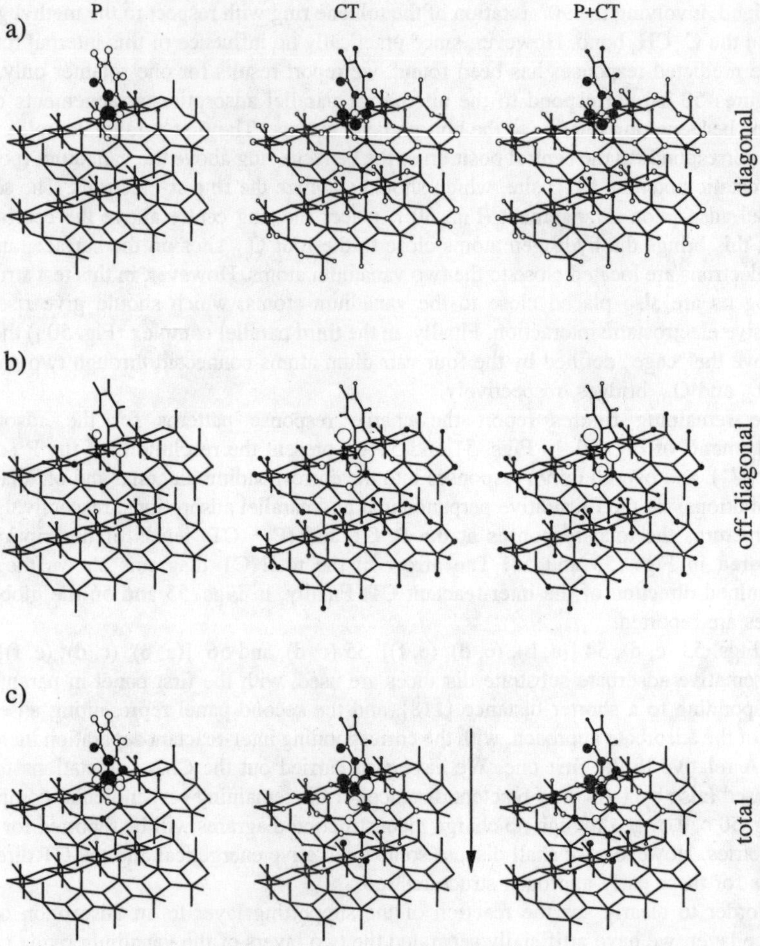

Figure 51. The representative resolution of the isoelectronic total (third row, c) polarizational (P, first column), charge transfer (CT, second column), and overall (P + CT, third column), electron population diagrams into the diagonal (first row, a) and off-diagonal (second row, b) contributions, for the perpendicular adsorption of toluene on the $O_{(2)}$ oxygen of the top layer (see Fig. 50 e). For reasons of clarity both layers have been shifted one relative to the other by an arbitrary mutual translation. The same scale factor has been used in all three panels of each column. The circle areas reflect the magnitude of populational shifts; the arrow in the total CT diagram shows the CSA predicted direction of the inter-reactant CT.

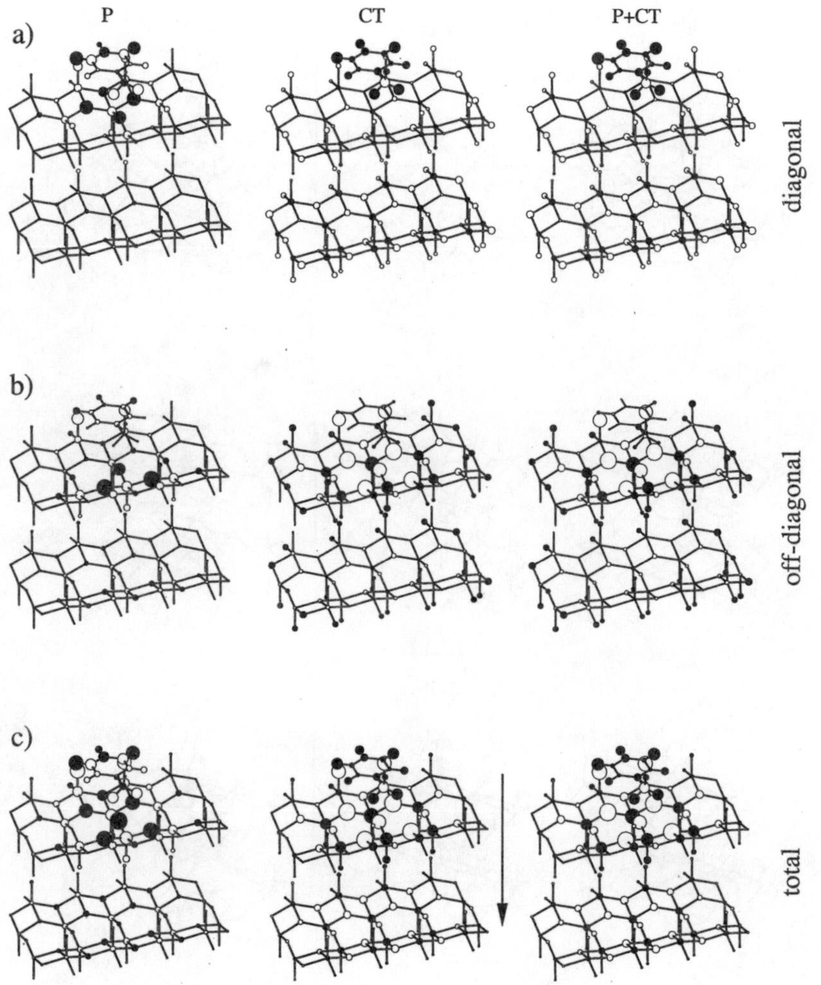

Figure 52. The same as in Fig. 51 for the representative parallel coordination of Fig. 50 h. The inter-reactant separation has been increased by 1Å relative to that in Fig. 36 i.

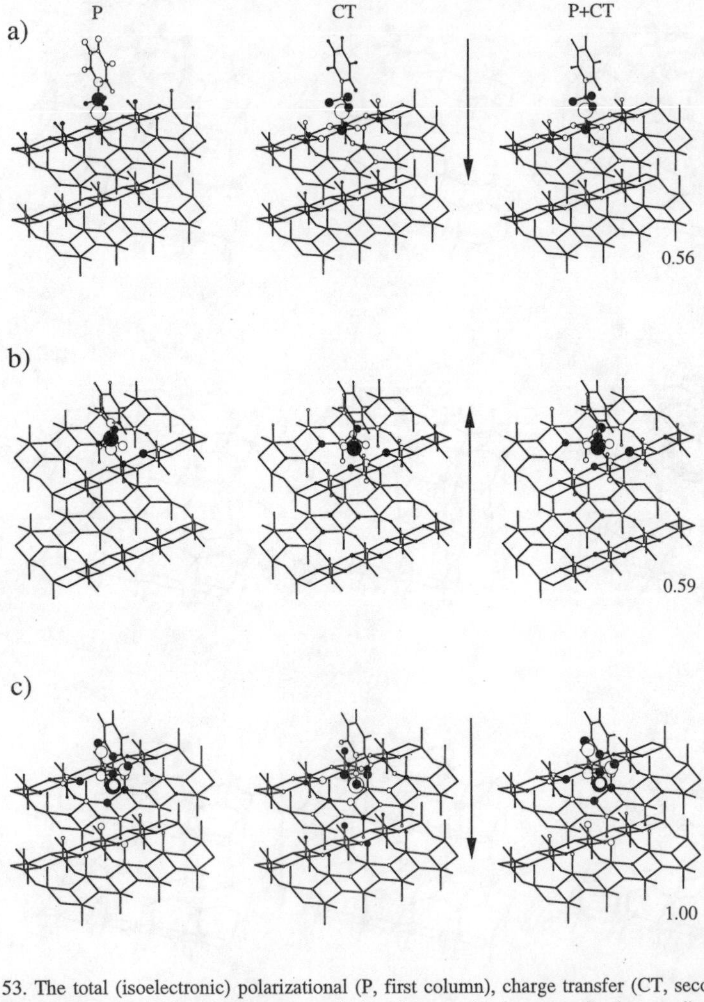

Figure 53. The total (isoelectronic) polarizational (P, first column), charge transfer (CT, second column), and overall (P + CT, third column), electron population diagrams for the perpendicular coordinations of Fig. 50 (c-g): parts a, b, (c, d), e, and f of the figure correspond to the perpendicular complexes c, d, e, f, and g of Fig. 50, respectively. Rows c and d display plots for two representative distances of toluene from the $O_{(2)}$ adsorption site in the chemisorption arrangement of Fig. 50 e, with the complex c corresponding to the same inter-reactant separation as in Fig. 36 e, and complex d representing an increase (by 1Å) in the inter-reactant separation. The numbers below the overall diagrams indicate the scale factor to be used to multiply the overall diagrams in the figure to bring them to the common scale of the row c.

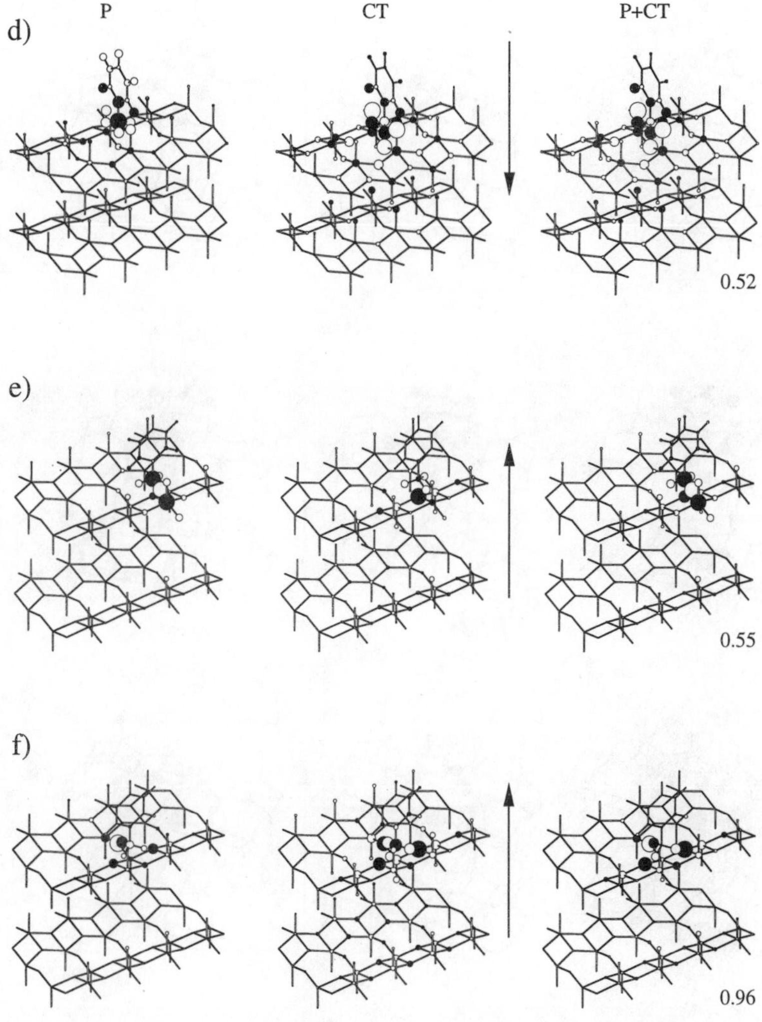

Figure 53. Continued.

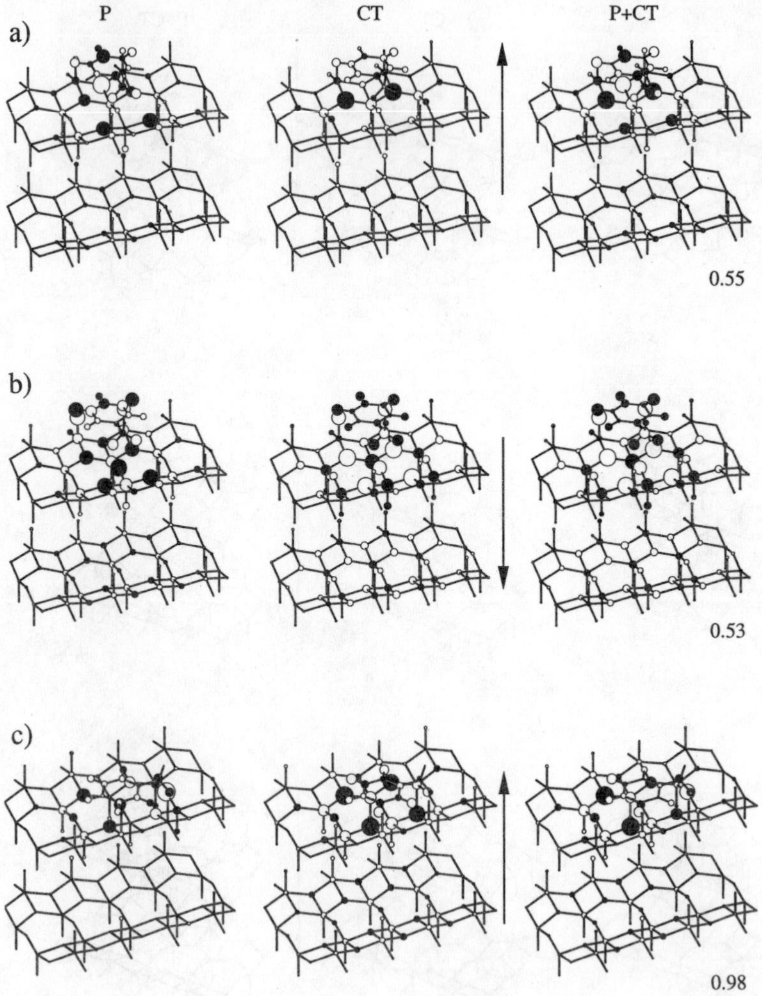

Figure 54. The same as in Fig. 53 for the parallel arrangements of Fig. 50, and the two representative adsorbate-substrate distances. Parts (a, b), (c, d), and (e, f) correspond to the parallel structures of Fig. 50 h, i, and j, respectively. The inter-reactant separation used in the first complex of each of the above pairs is the same as in Fig. 36 i; it has been increased by 1Å in the second arrangement of each pair of chemisorption complexes.

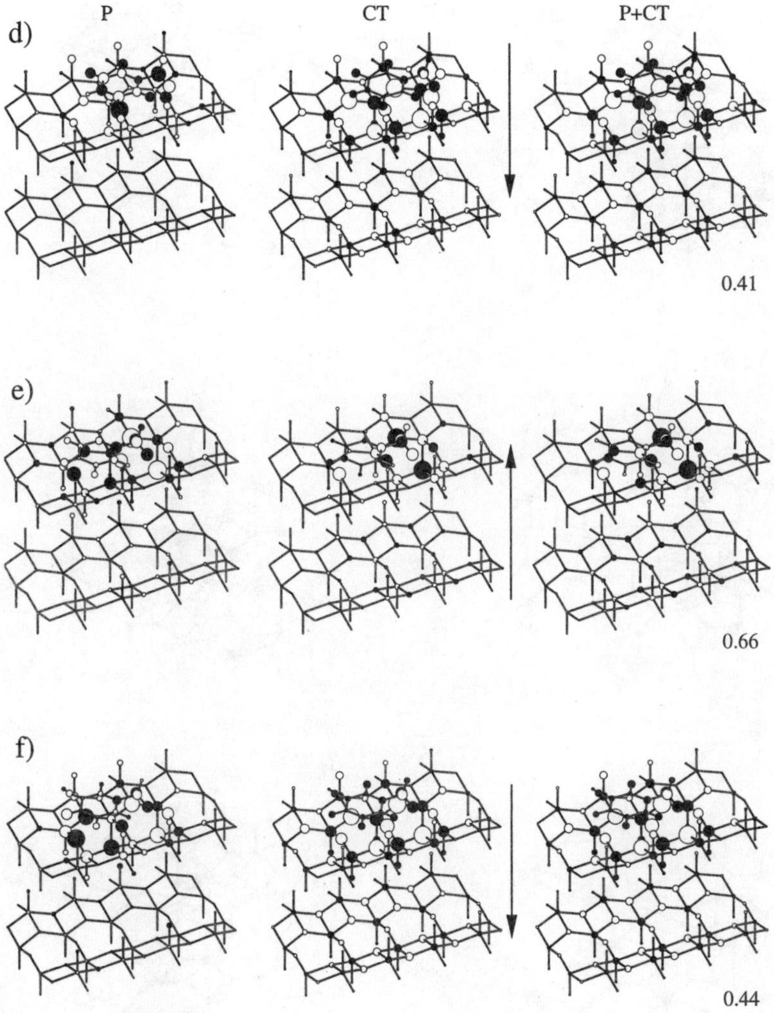

Figure 54. Continued.

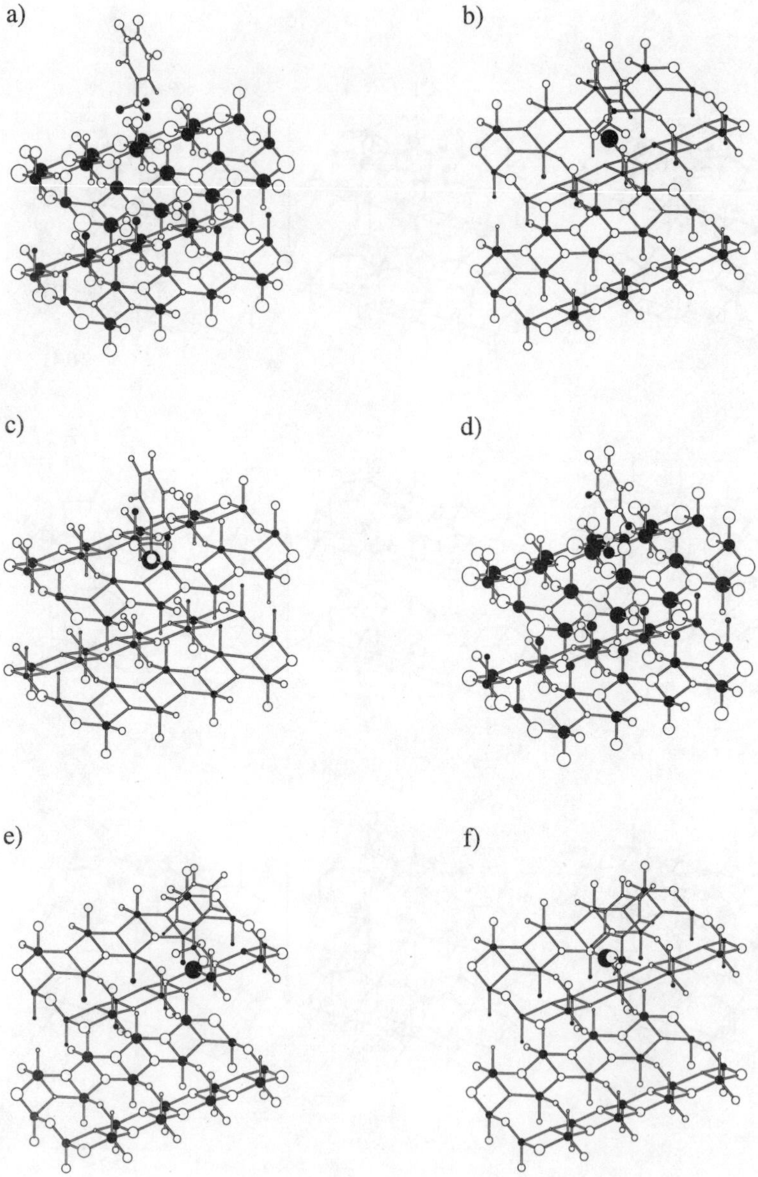

Fig. 55. The global (open system) FF indices for the perpendicular chemisorption arrangements of Fig. 53; for reasons of clarity for each panel a separate value of the diagram scale factor has been adopted.

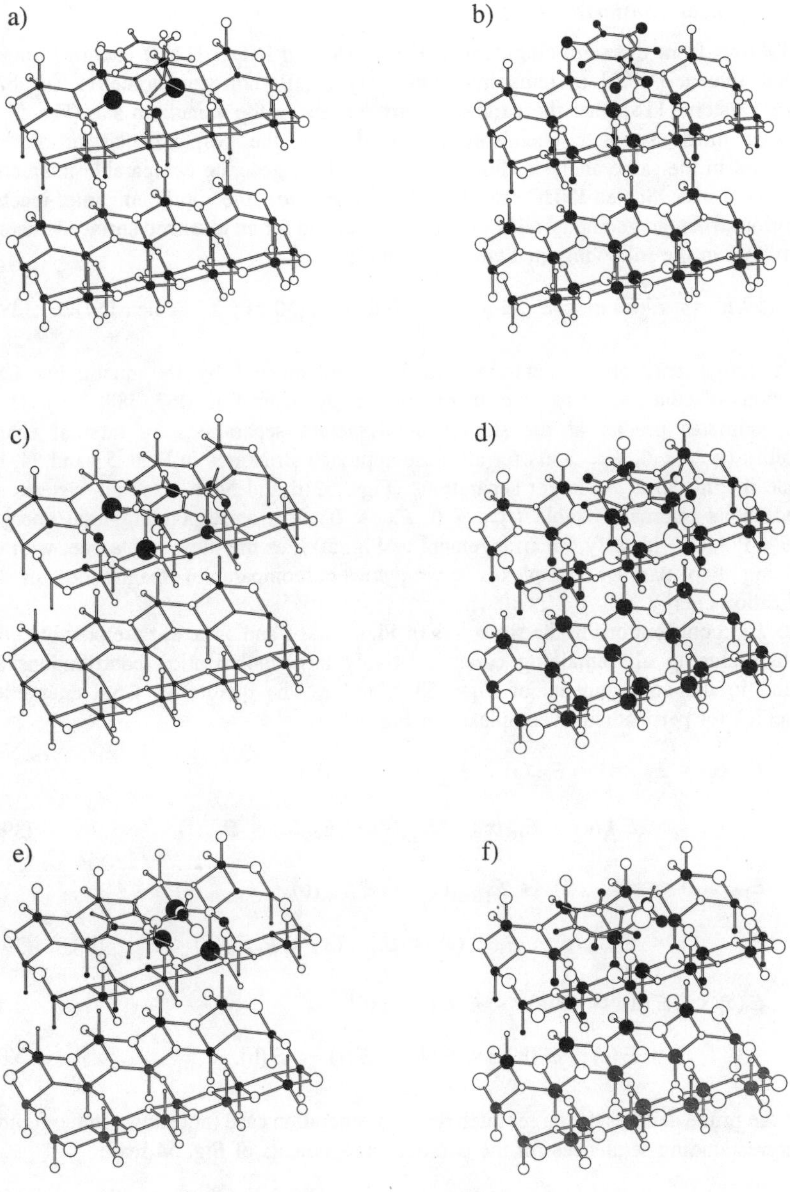

Fig. 56. The same as in Fig. 55 for the parallel chemisorption structures of Fig. 54.

2.2. *Energetical Considerations*

It follows from the semi-empirical SCF MO (Scaled-INDO [119]) energies, for the selected toluene-[V_2O_5] systems involving very small, non-stoichiometric, idealized surface clusters [118], that the parallel adsorption above the vanadium site (Fig. 50 h) generates much stronger chemisorption bond than the perpendicular adsorptions considered in the same analysis (Fig. 50 c-e, g), which generate comparable interaction energies. These Scaled-INDO results, determined for the optimum inter-reactant separation, when no geometry relaxation is allowed, and for an idealized surface structure, are ordered in the following qualitative sequence:

$$E(50\text{ h}) << E(50\text{ e}) < E(50\text{ g}) < E(50\text{ d}) < E(50\text{ c})\,; \qquad \text{Scaled-INDO.} \quad (395)$$

This rough energetical hierarchy can be supplemented by the qualitative CSA predictions of contributions to the interaction energy (from Eqs. 367, 388, 389). These crude estimates predict at the shorter inter-reactant separations an internal charge instability ($\eta_{CT} < 0$, $E_{CT} > 0$) for all close-approach structures in Figs. 53 and 54, but the one in Fig. 53 a. At larger separations (Figs. 53 d and 54 b, d, f) all systems are diagnosed as internally stable ($\eta_{CT} > 0$, $E_{CT} < 0$); the corresponding total energies E(ES + P + CT) identify the arrangement of Fig. 50 h as the most stable one, with the remaining two parallel complexes giving lower (comparable) magnitudes of the stabilization energy.

The E_{ES} contributions in the structures of Figs. 53 d-f and 54 a, e, f are positive; they are stabilizing in all remaining cases; relatively large polarization contributions are detected in the arrangements of Figs. 53 c and f. The qualitative CSA energetical hierarchies for perpendicular complexes in Fig. 53 are:

$$E_{ES}(\text{c}) < E_{ES}(\text{b}) < E_{ES}(\text{a}) < E_{ES}(\text{a}') \sim E_{ES}(\text{b}')$$
$$< E_{ES}(\text{e}) \sim E_{ES}(\text{e}') \sim E_{ES}(\text{f}') \sim E_{ES}(\text{d}) < E_{ES}(\text{f}), \quad (396)$$

$$E_{ES+P}(\text{c}) << E_{ES+P}(\text{f}) << E_{ES+P}(\text{e}) << E_{ES+P}(\text{b}) < E_{ES+P}(\text{a})$$
$$< E_{ES+P}(\text{d}) < E_{ES+P}(\text{a}') < E_{ES+P}(\text{e}') < E_{ES+P}(\text{f}') < E_{ES+P}(\text{b}'), \quad (397)$$

$$E(\text{c}) << E(\text{d}) << E(\text{a}) << E(\text{e}') < E(\text{f}')$$
$$< E(\text{e}) \sim E(\text{b}') << E(\text{a}') < E(\text{f}) < E(\text{b}), \quad (398)$$

where the prime denotes the larger inter-reactant separation case (not shown in the figure). The corresponding sequences for the parallel arrangements of Fig. 54 read:

$$E_{ES}(\text{c}) < E_{ES}(\text{b}) \sim E_{ES}(\text{d}) < E_{ES}(\text{f}) < E_{ES}(\text{e}) < E_{ES}(\text{a})\,, \quad (399)$$

$$E_{ES+P}(a) < E_{ES+P}(e) < E_{ES+P}(c) \sim E_{ES+P}(b) < E_{ES+P}(d) < E_{ES+P}(f)\,, \quad (400)$$

$$E(b) \ll E(d) \sim E(f) \ll E(a) \ll E(e) \ll E(c). \quad (401)$$

Of interest in the theory of chemical reactivity also is a qualitative comparison of the energetical preferences at a comparable separation between reactants. At the earlier stage of the adsorbate approach one obtains the following sequences of the total CSA interaction energies (a.u., in parentheses) (see Eqs. 398 and 401):

(i): for the perpendicular structures of Fig. 53:

$$d(-1.145) \ll e'(-0.386) < f'(-0.337) < b'(-0.237) < a'(-0.196); \quad (402)$$

(ii): for the parallel structures of Fig. 54:

$$b(-1.235) \ll d(-0.461) \sim f(-0.456). \quad (403)$$

At the close approach structures the corresponding orderings are:

$$\text{(i):} \quad c(-2.572) \ll a(-0.754) \ll e(-0.249) < f(0.075) < b(0.198); \quad (404)$$

$$\text{(ii):} \quad a(-0.198) < e(0.325) < c(1.166). \quad (405)$$

Thus, the structures 53 c and 54 b are predicted as the CSA energetically most favourable perpendicular and parallel adsorption arrangements, respectively. This qualitatively agrees with energies from the Scaled-INDO calculations on much smaller clusters, which we have summarized at the beginning of this section. Also, at an earlier approach, the parallel structure of Fig. 54 b is predicted to exhibit a lower interaction energy than the most stable perpendicular arrangement of Fig. 53 d.

2.3. Relative Role of the Diagonal and off-Diagonal Isoelectronic Responses

A reference to Fig. 51 shows that the total charge response patterns for the perpendicular adsorption of toluene on the bridging $O_{(2)}$ site of the top layer (Fig. 50 e) have a common structure at the P, CT, and P+CT stages of the charge reconstruction. Namely, the total patterns in the figure (third row) are dominated by the respective diagonal contributions of the adsorbate (first row) and the off-diagonal contributions of the substrate (second row). The same general rule has been observed in the remaining perpendicular coordinations. At the larger separation between reactants the internal CT is from the toluene (B) to cluster (A), due to a relatively weak overall charge coupling between reactants. At the overall description level (third column) the resultant effect of the P and CT displacements is shown; these contributions may enhance each other, e.g., in the ring fragment in the vicinity of the methyl group or in the two V-$O_{(1)}$ fragments of the surface in the chemisorption region, or they may reduce one another, e.g., on the methyl hydrogens and the remaining, peripheral part of the benzene ring. The total P and CT shifts are seen to act mainly "in phase", with the exception of the methyl group, where the dominating CT pattern is partly reduced by the P component. A comparison between the respective panels in the three columns of the figure reveals that the toluene → cluster CT-induced charge displacements dominate the overall (P+CT) trends.

There is a strong localization of the *in situ* responses in Fig. 51, which exhibit the largest magnitudes of the AIM charge displacements in the chemisorption region, with only marginal participation of the supporting layer and peripheral atoms of the surface layer. As also seen in the figure, only the $O_{(1)}$ sites of the second layer are more strongly involved in moderating the primary charge shifts in the chemisorption region of the surface layer.

Similar conclusions follow from Fig. 52. For this parallel arrangement (Fig. 50 h, larger distance between reactants) the second layer plays quite a substantial role in the cluster reconstruction, particularly at the CT stage. Again, the total patterns of the third row are roughly the combinations of the diagonal charge shifts in the toluene and the off- diagonal charge displacements in the surface layer of the cluster, at all stages of description. A general similarity of the respective panels in the CT and (P + CT) columns indicates that the toluene → cluster CT component is the dominant one in this particular chemisorption arrangement.

2.4. *Direction of the Inter-Reactant Charge Transfer*

It follows from Eq. 368 that the adsorbate-substrate interaction at finite separations may change the sign of N_{CT}, relative to the corresponding separated reactants estimate, since both the chemical potential difference ("force") and the CT hardness ("stiffness modulus") in this equation are interaction dependent. There is always a strong relaxational influence of the cluster upon the toluene chemical potential, while the opposite influence is relatively weak. For example, in the parallel structure of Fig. 50 h (Fig. 54 b) the CSA predicted levels (a.u.) of the polarized reactant chemical potentials are: $\mu^{(+)}_{toluene}$ = -0.0807 and $\mu^{(+)}_{cluster}$ = -0.2532; these values can be compared with the separate reactant estimates from the same canonical $\{\mu^{0}_{X}$ and $\eta^{X,X}\}$ input data: $\mu^{0}_{toluene}$ = -0.2542 and $\mu^{0}_{cluster}$ = -0.2524 (the chemical potential discontinuity [48] has not been included).

A reference to Figs. 51-54 shows that indeed both directions of the CT are predicted. For example, in the parallel arrangements of toluene (Fig. 54), the strongly interacting complexes (at shorter distance between reactants, panels a, c, e) exhibit the cluster (B) → toluene (A) CT, reversed in comparison with that found in the weakly interacting complexes (at larger separation, panels b, d, and f).

In the perpendicular structures of Fig. 50 c-g at the larger inter-reactant separation a common toluene → cluster CT direction is predicted from the CSA calculations. For the close approach complexes of Fig. 53 a-c, e, f, for which the inter-reactant perturbation becomes stronger, this direction is reversed in panels b, e, and f.

2.5. Relative Role of the P and CT Components in the Overall Isoelectronic Responses

A reference to Fig. 53 shows that in most cases the CSA predicted responses are indeed dominated by the CT component. However, there are important exceptions to this rule at some close approach structures. For example, an inspection of the three panels of Fig. 53 c shows that it is the direct polarization, before the CT, which determines the gross features of the overall charge reconstruction pattern. It should be observed, that at the larger internuclear separation for the same chemisorption arrangement (Fig. 53 d) the CT component alone generates qualitatively correct charge reconstruction diagram. Additional examples of a dominant role of the reactant mutual polarization can be found in Figs. 53 e and f.

Similar cases can be found in Fig. 54, although in most cases, particularly at larger separations, it is the CT component that determines the most important features. An example of a strong P contribution to the overall pattern is found in Fig. 54 a, where both components enhance each other on atoms undergoing the largest charge displacements. When the distance increases (Fig. 54 b) both components act with opposite phases on most of the cluster sites in the chemisorption region. Similarly, in Fig. 54 f, the opposite displacements on the two pairs of the $O_{(3)}$ surface sites in the P diagram have only a minor effect of differentiating these sites in the chemisorption area; this is a clear sign of a strong CT dominance of the overall charge reconstruction pattern due to the chemisorption.

2.6. Mutual Polarization of Reactants

In this section we shall examine the charge polarization patterns of the left (P) columns in Figs. 53 and 54; the bond weakening/strengthening implications of these polarizational changes will also be summarized.

The P diagram of Fig. 53 a predicts a flow of electrons from the methyl group to the benzene ring in the toluene and the accompanying (bond weakening) $V \rightarrow O_{(1)}$ polarization of the vanadyl adsorption site. These trends are further enhanced at the CT stage, as shown in the middle diagram of Fig. 53 a, via the polarization induced by the predicted toluene → cluster CT. The ring carbons are only weakly affected in the P diagram during the adsorption on the $O_{(1)}$ site, with the exception of the methyl substituted ring carbon, which acquires relatively more electrons; this should strengthen the hyperconjugation π-component between this atom and the methyl group.

It follows from Fig. 53 b that the direct polarization diminishes the electron population of the methyl carbon; the electrons are shifted mainly to its nearest neighbours. The $O_{(2)}$ adsorption site increases its negative charge at the expense of the nearby V and $O_{(3)}$ sites; this should weaken the $V-O_{(2)}-V$ bonds, since such a polarization reverses the original $O_{(2)} \rightarrow V$ donation of these coordination bonds. Again, the polarizational influence upon the ring is predicted to be rather small. It should be observed that the next (CT) diagram predicts a reversal of the main P trends in the surface chemisorption region, as a result of the subsequent cluster → toluene CT.

A reference to the P diagram of Fig. 53 c shows that in this case the polarizational influence is more extended. Namely, in addition to the $V \rightarrow O_{(1)}$ bond polarization (weakening), due to the presence of the *o*-hydrogens of the ring, the $O_{(2)}$ adsorption site is hardened by the presence of the softer methyl hydrogens, which acquire electrons from the methyl carbon and the benzene ring. Such a polarization promotes the subsequent $(C)H_3 \rightarrow O_{(2)}$ and $O_{(1)} \rightarrow H_{(o)}$ flows accompanying the toluene → cluster CT (see the next diagram).

In Fig. 53 d one detects an extra accumulation of electrons on the methyl hydrogens and in the peripheral part of the toluene, at the expense of the $(H_3)C-C$ carbons and the *o*-hydrogens. This polarization of toluene weakens the C-H bonds in the coordinating methyl group, since it partly reverses the H → C polarization in the isolated adsorbate. One also observes in the same diagram that the $O_{(1)}$ and $O_{(2)}$ atoms on the surface acquire electrons from the coordinating metal sites (weakening of the $V-O_{(1)}$ and $V-O_{(2)}$ bonds). Again, this mutual polarization creates favourable conditions for the subsequent transfer of electrons from the methyl hydrogens to these oxygen sites, as indeed detected in the CT diagram of Fig. 53 d.

A comparison of the P diagrams of structures c and d of Fig. 53 shows that the pattern of polarizational changes is qualitatively similar, with the exception of the $O_{(2)}$ site, which exhibits different displacements for the two inter-reactant separations.

The next P diagram of Fig. 53 e refers to the toluene coordination on $O_{(3)}$. Here the toluene polarization is practically limited to the methyl group; it is seen to shift electrons mainly from the methyl carbon toward hydrogens facing the vanadium sites. The surface undergoes a complementary V → O charge reconstruction in the most perturbed VO_5 "pyramid". The corresponding CT diagram reveals that these changes are further enhanced at the cluster → toluene CT stage.

Finally, the CSA calculations for the toluene adsorption on vanadium site (Fig. 53 f) also predict strongly localized polarizational responses at the physisorption stage. The range of these changes covers the methyl group (shifting electrons away from the hydrogen facing the nearby $O_{(2)}$ site), and the two surface pyramids linked through the $O_{(2)}$ bridge (shifting electrons to the bridging oxygen, i.e., weakening the $V-O_{(2)}$ bonds). One therefore observes a degree of differentiation of both the methyl hydrogens and lattice oxygens, which should create favourable conditions for the $O_{(2)} \rightarrow$ methyl $\leftarrow O_{(3)}$ coordination at the CT stage of the chemisorption process.

Let us examine now the P diagrams of Fig. 54. In the first structure of part a the electrons are accumulated at the *m*-hydrogens, *o*-carbons and on the two methyl hydrogens facing a pair of the $O_{(3)}$ sites. The active site region of the surface undergoes the following polarizational shifts: the $O_{(3)}$ sites below the toluene hydrogens lose electrons, which are shifted toward the vanadium atom below the ring centre (strengthening of these $V-O_{(2)}$ bonds) and the $O_{(3)}$ atom below the C-H *p*-fragment accumulates electrons at the expense of the nearby vanadium sites (weakening of these $V-O_{(3)}$ bonds); these charge reconstruction displacements, creating favourable conditions for a selective oxidation in the *p*-position and on the methyl group, are seen to be enhanced at the CT stage (see the next diagram in Fig. 54 a).

This direct polarizational pattern changes with the increase in the toluene-surface separation (Fig. 54 b). The main displacements in the physisorbed toluene now involve the increase of the electron charge on the *m*-carbons and the two methyl hydrogens directed toward the $O_{(3)}$ surface sites. In the cluster the two VO_5 "pyramids" located below toluene exhibit a strong enhancement of the original O → V coordination bonds (bond strengthening) and the two vanadyl fragments in the vicinity of the *m*-hydrogens of the ring accumulate electrons. These terminal $O_{(1)}$ oxygens are predicted to further receive electrons at the CT stage, while most of the other surface polarization trends will be reversed in the toluene → cluster CT.

In the close-approach parallel complex of Fig. 54 c the ring atoms mainly gain electrons from the methyl group, but the changes are relatively minor in comparison with the surface charge shifts. The latter exhibit a strong accumulation of electrons on the bridging oxygen below the methyl group, which should facilitate a selective oxidation of the methyl group by this lattice oxygen, and a decrease of the electron population of the $O_{(2,3)}$ sites below the opposite peripheral toluene fragment. We would like to emphasize the reverse displacements exhibited by the $O_{(3)}$ surface sites facing the *o*- and *m*-hydrogens of the toluene ring, respectively.

When the distance between reactants is increased (see Fig. 54 d) a relatively strong ring and methyl group polarizations are predicted. In the ring they should strengthen the initial C–H and $C_{(o)}-C_{(m)}$ bonds, while weakening the remaining bonds. Similarly, in the methyl fragment the C–H bonds are weakened as a result of a partial reversal of the original (isolated toluene) bond polarization. The complementary polarization of the cluster surface layer is now more delocalized; the main displacements represent a shift of electrons from the bridging $O_{(2)}$ sites to the nearby oxygen and vanadium atoms. At this P stage, therefore, no substantial release of the above bridging oxygen sites is predicted, since they become more strongly bonded to the coordinating vanadium atoms as a result of the reactant mutual polarization.

Finally, let us examine the P plots for the parallel complexes of Fig. 54 (e, f). A comparison of these two diagram reveals that in this case the P pattern remains qualitatively unchanged, when the inter-reactant separation increases. The main polarization trends of toluene involve the two C–H bond strengthening in the methyl fragment, and an overall accumulation of electrons in the ring, with the exception of the substituted carbon. The surface exhibits relatively larger polarizational flows, which shift electrons mainly to the $O_{(2,3)}$ sites below the methyl side of the toluene, from the $O_{(2,3)}$ locations below the other end of the adsorbate. The electron gaining surface oxygens are therefore weakly bonded by the coordinating vanadium sites (a partial reversal of the original coordination), and thus become more accessible for oxidizing the adsorbate.

2.7. *Charge Transfer Induced Displacements*

In this section we shall examine in some detail the CT-induced charge reconstructions shown in the middle columns of Figs. 53 and 54. Consider first *the perpendicular adsorption on the vanadyl oxygen of the top layer surface* (Figs. 50 c and 53 a). The phases of populational displacements in the region of the surface bond are similar to those observed at the P stage, with the methyl group hydrogens losing and the $O_{(1)}$ atom gaining electrons, respectively, as expected in the $-CH_3$ → $O_{(1)}$ coordination. No appreciable participation of the ring and the second-layer atoms is predicted for this inter-reactant CT process. The predicted surface reconstruction displacements involve a substantial $V \rightarrow O_{(1)}$ polarization (further weakening of this bond) and a slight increase in the electron population of remaining oxygen sites in the chemisorption region. The effective range of appreciable charge reorganization created on both reactants by this CT practically does not extend much beyond the methyl group and the coordinating VO_5 "pyramid".

The opposite CT charge reorganization is observed in the middle panel of Fig. 53 b, corresponding to the *perpendicular adsorption on the bridging oxygen of the bottom layer surface* (Fig. 50 d). This is due to the reverse, cluster → toluene CT. Again, the cluster is seen to respond only on the surface, in the vicinity of the adsorption site. Both the $O_{(2)}$ adsorption site and the nearby four $O_{(3)}$ atoms donate electrons to the methyl hydrogens and the vanadium atoms, thus becoming more strongly bonded. It should be observed that this charge displacement on $O_{(2)}$ partly reverses that observed at the P stage in the same complex. The overall *polarizational matching* of the AIM population shifts in the chemisorption region of the CT panel of Fig. 53 b, with the donor atoms of the basic reactant (cluster) facing the acceptor atoms of the acidic reactant (toluene), indicates a relatively "soft" (facile) charge reconstruction.

Next two CT diagrams of Fig. 53 (parts c and d) correspond to the *perpendicular adsorption on the bridging oxygen of the top layer surface* (Fig. 50 e). It should be noted that in both cases the same toluene → cluster CT direction is predicted. At both distances the same $-(C)H_3 \rightarrow O_{(2)}$ charge shift (*"forward chemisorption donation"*) is detected. At the close approach arrangement of Fig. 53 c one also finds the $(VO_{(1)}) \rightarrow H_{(o)}$ *"backward chemisorption donation"*, which is reversed at the larger distance between reactants in Fig. 53 d. The other observed difference is the $C(ring) \rightarrow C(H_3)$ bond polarization in Fig. 53 d, which weakens this C–C bond by reducing its original polarization in the isolated toluene and partially reversing the hyperconjugation donation from the methyl group to the π-bond system of the ring. In Fig. 53 c both these carbon atoms are predicted to gain electrons during the toluene → cluster CT, so that this C–C bond is not predicted to be appreciably affected by the toluene → cluster coordination.

It should be observed that at the larger separation between reactants the methyl group hydrogens are placed closer to the vanadyl oxygens of the surface, so that in Fig. 53 d they effectively coordinate to both $O_{(1)}$ and $O_{(2)}$ sites. A formation of partial $O_{(1)} \leftarrow H_{(o)}$ bonds implies an effective removal of electrons from the (C–H) σ-bond region, which weakens (strengthens) the covalent (ionic) bond component. This also suggests a relatively facile activation mechanism of a possible breaking of the methyl C–H bonds and forming the O–H bonds in this chemisorption arrangement. In Figs. 53 c, d the layer-bridging oxygens are seen to be involved more strongly than in previous perpendicular complexes, even at the larger separation between reactants.

The *perpendicular adsorption on the triply coordinated oxygen of the bottom layer surface* (Fig. 50 f) generates the CT diagrams of Fig. 53 e, in which the surface donates electrons to the toluene. It should be observed that, similarly to the CT pattern of Fig. 54 b, the oxygen adsorption site donates electrons to the nearby vanadium sites (bond strengthening) and to the methyl hydrogens. This polarization is qualitatively similar to that seen at the P stage (left diagram). The $O_{(3)}$ adsorption site is therefore more tightly bonded on the surface; thus, its subsequent participation in the toluene oxidation is less probable. Also, a formation of the surface bond with the methyl hydrogens is seen to have only a minor effect upon these C–H bonds in this CT process. The net increase in the electron populations of the "bridging" hydrogens in this perpendicular complex prevents a weakening of the C–H bonds, while providing at the same time an extra "coordination" bond component $O_{(3)} \rightarrow H$ responsible for the chemisorption bond.

When toluene is perpendicularly adsorbed on the vanadium atom (Figs. 50 g and 53 f), the methyl hydrogens acquire electrons from the $O_{(2)}$ and $O_{(3)}$ surface sites and the methyl carbon; these CT-induced charge shifts stimulate a dissociation of the C–H bonds of the methyl group, and a formation of the surface O–H bonds.

Let us examine next the CT diagrams of Fig. 54, corresponding to the *parallel adsorption arrangements on the bottom layer surface* (Fig. 50 h-j). The two probing inter-reactant distances, representing the close and earlier (more distant, by 1 Å) stages of the parallel approach of the toluene ring to the VO_5 "pyramid" bases, are considered in parts (a, c, e) and (b, d, f), respectively.

Consider first the chemisorption structures of Fig. 50 h (Fig. 54 a, b). A comparison between these CT panels indicates, that even at the larger distance both toluene and cluster are strongly affected. These two diagrams are qualitatively different, though, due to the reverse inter-reactant CT they correspond to. At the closer approach (Fig. 54 a) electrons are removed mainly from the two $O_{(3)}$ sites below the *o*- and *m*- hydrogens of the ring, and shifted to the *o*- and *m*-CH fragments of toluene and to vanadium surface sites; therefore, this pattern implies a strengthening of the V–O bonds in the adsorption region of the surface. The net inflow of electrons to the ring (Fig. 54 a) must populate the antibonding π^* orbitals, thus facilitating the ring destruction. This chemisorption exerts practically no influence upon the methyl fragment of toluene.

At the larger distance (Fig. 54 b) all ring bonds are also predicted to be weakened by an effective removal of electrons from all atoms, including mainly the bonding electrons of the π C–C bonds. This can also promote the total degradation of the ring structure, and - ultimately - the nonselective, total oxidation of the toluene [118]. The major toluene reconstruction trends now also include a substantial $H \rightarrow C$ bond strengthening polarization of the methyl substituent; hence the methyl part of the adsorbate is not predicted to be activated at the CT stage in this parallel complex. On the surface one detects a strong charge reconstruction in the adsorption region, involving various degrees of the $V \rightarrow O$ back donation, which weakens the $O \rightarrow V$ coordination bonds. Therefore, these $O_{(2,3)}$ lattice oxygens become accessible for a subsequent oxidizing of toluene. The increase of the positive charge of the toluene hydrogen atoms and the accompanying increase in the negative charge of the surface oxygen sites should effectively strengthen the overall electrostatic (ionic) component of the resultant toluene---cluster bond. This is indeed reflected by the position of this structure in the total energy sequence of Eq. 401. The second layer reaction to the chemisorption is similar to that observed in the surface layer, but much weaker.

In the second mutual arrangement of the "parallel" oriented reactants (Figs. 50 i and 54 c, d) the CT diagrams predict a strengthening of the V–O bonds in the chemisorption region in the close-approach case (Fig. 54 c) and their weakening for the larger inter-reactant separation (Fig. 54 d). The toluene ring patterns also exhibit a general change in the phases of atomic displacements, when this distance increases. Obviously, this is caused by the reverse directions of the inter-reactant CT predicted for both these complexes. An inspection of the two CT diagrams also shows that the $C_{(o)}$–$C_{(m)}$ and C-H (*o, m*)-bonds in the ring are weakened in Fig. 54 c, due to a removal of electrons and the reversal of the initial bond polarization, respectively. In Fig. 54 d electrons are withdrawn from the ring, similarly to the CT pattern in Fig. 54 b; this also implies a general bond weakening in this fragment of the adsorbate. Moreover, as in the previous parallel arrangement, the methyl substituent practically does not participate in the inter-reactant CT in the short-separation complex, while its polarization at the earlier approach stage strengthens the C–H bonds. The above activation trends again suggest a tendency toward the ring degradation and a possible involvement of the lattice oxygens in the toluene oxidation in the structure of Fig. 54 d.

Let us finally examine the two CT diagrams of Fig. 54 (e, f) corresponding to the third parallel arrangement of Fig. 50 j. The main features of the CT pattern in Fig. 54 e are the methyl bond weakening reconstruction in toluene, and the surface V–O bond strengthening in the surface layer of the cluster, in the vicinity of the methyl group; both these trends should facilitate a selective oxidation of the methyl part of toluene by the lattice $O_{(2,3)}$ oxygens. With the increase in the substrate-adsorbate separation (Fig. 54 f) the CT pattern changes: the range of the surface reconstruction increases, with the dominant V → O bond weakening polarization, the electrons are removed from the ring (general bond weakening) and the CH methyl bonds are strengthened. Therefore, only at the close-approach structure the selective oxidation of toluene should be expected in this particular parallel arrangement.

2.8. *Toluene Activation and Surface Reconstruction Trends from the Overall Diagrams*

In this section we briefly comment on the implications of the P+CT diagrams of Figs. 53 and 54, by identifying the dominant component of these overall plots. The CT component dominates the overall charge reorganization patterns in Figs. 53 b, d and 54 b, c, d, e, f, while the P component is dominant in the overall diagrams of Fig. 53 c, e, f; in the remaining structures of Fig. 53 a and 54 a both components are comparable and they generally display the same phases of the main AIM displacements. With an increase in separation between reactants the polarization component decays faster than the CT one. Therefore, one generally observes the CT domination at the earlier stage of the adsorbate approach to the surface, while the P component contributes more strongly at shorter distances. Clearly, the implications for the bond-order toluene activation/surface reconstruction of the dominant component, described in detail in the preceding two sections, determine the resultant, overall charge reconstruction effects (see Appendix D).

Consider first the perpendicular adsorption systems. The weakly bonded $O_{(1)}$ lattice oxygen in Fig. 53 a is expected to form a bond with either the methyl hydrogen or carbon, since the C–H bond dissociation is strongly promoted in this adsorption arrangement. In Fig. 53 b, where the V–$O_{(2)}$–V bonds are strengthened and the C–H methyl bonds are weakened, the dissociation in the coordinating methyl group can be expected. The forward ($H_{(o)} \rightarrow O_{(1)}$) and back ($O_{(2)} \rightarrow H_3$) donations are clearly seen in Fig.

53 c. This promotes the dissociation of the C–H bonds in methyl and the *o*-ring positions. The $O_{(1)}$ lattice oxygens participation in the selective oxidation is also highly probable, due to their being less strongly bonded to the vanadium atoms in the chemisorption system. These trends partially change in Fig. 53 d, where the two forward donation channels, $H_{(o)} \rightarrow O_{(1)}$ and $H_3 \rightarrow O_{(2)}$, are detected. Therefore, in this particular case both these lattice oxygen sites become accessible for the toluene oxidation; also, since the methyl hydrogens are seen to donate their electrons mainly to the surface oxygen sites, the C–H bond breaking at both the *o*-positions in the ring and in the methyl group are highly probable in this energetically favoured chemisorption complex. In Fig. 53 e only the methyl substituent is activated (bond weakening). Finally, in Fig. 53 f the two methyl hydrogens are involved in the $O_{(3)} \rightarrow H$ back donation, while the third hydrogen donates electrons to the surface bridging oxygen: $H \rightarrow O_{(2)}$. Both these channels may lead to the C–H bond dissociation; however, only the $O_{(2)}$ site is chemically more accessible for oxidizing the toluene, since the $O_{(3)}$ atoms are predicted to be more strongly bonded to the coordinating vanadium sites.

Let us similarly summarize the main reactivity trends implied by the P+CT column in Fig. 54. In the first panel of the figure the surface reconstruction "releases" the $O_{(3)}$ oxygen below the *p*-CH bond , thus facilitating its use in oxidizing the *p*-carbon of the ring; the latter is also seen to weaken its bonds with the neighbouring *m*-carbons, due to its diminished electron population. The other two $O_{(3)}$ sites, coordinated by the vanadium atom below the ring centre, are seen to be more strongly bonded in the chemisorption system. When the distance between the adsorbate and substrate increases (Fig. 54 b) all oxygen sites of the surface chemisorption region are more loosely bonded by the metal sites, the methyl C–H bonds are strengthened, while the ring electron population is diminished throughout most of its constituent atoms, particularly on the *m*-hydrogens in the vicinity of the two $O_{(1)}$ sites. These changes imply an increased tendency towards ring oxidation, particularly in the *m*-position.

The next parallel arrangement (Fig. 54 c, d) in the close approach complex generates some tendency toward the bond breaking on the methyl substituent, due to the forward $H \rightarrow O_{(2)}$ donation; one also detects in Fig. 54 c the back donation from the four $O_{(3)}$ sites, all more tightly held by the vanadium atoms, to the *o*- and *p*-hydrogens in the ring. For the increased separation between reactants a stronger participation of the CT charge reorganization component and the reverse CT direction produce together a pattern in which the electrons are partially withdrawn from all constituent atoms in the ring, the methyl bonds are strengthened, and all nearby $O_{(2,\,3)}$ surface sites are more accessible (less tightly bonded to the vanadium atoms). This relative "availability" of the lattice oxygens and a general tendency towards weakening of all bonds in the ring create favourable conditions for a nonselective ring oxidation (destruction).

Similar conclusions follow from the final pair of overall diagrams (Fig. 54 e, f) for the third parallel arrangement. Namely, at the shorter separation the ring activation is more selective, due to a stronger P component, and the surface bridging oxygen sites coordinate more electrons to neighbouring vanadium atoms and the methyl hydrogens; at the larger distance the CT component and the reverse CT direction create a delocalized pattern of bond weakenings in the ring, a strong "release" of the lattice oxygens, and a strengthening of the C–H bonds in toluene. Hence, a selective oxidation at the *p*-carbon in the ring and on the methyl substituent is likely in the overall charge reorganization of Fig. 54 e, while a nonselective ring oxidation (degradation) is more probable in the complex of Fig. 55 f.

The energetical considerations (see sequences of Eqs. 403 and 405) indicate that the larger separation between cluster and toluene gives rise to an increased stabilization energy; at this distance the overall pattern of charge displacements in the benzene ring is practically nonselective. This strongly suggests that the parallel structures at this earlier stage of the toluene approach indeed promote the total ring oxidation [118].

These general conclusions about the selective character of the toluene activation in the perpendicular complexes and the nonselective ring destruction in the parallel chemisorption arrangements qualitatively agree with the previously reported conjectures from the semi-empirical SCF MO calculations on selected toluene-[V_2O_5] model systems involving very small non-stoichiometric surface clusters [118].

2.9. *External Charge Transfer Effects*

The global FF indices for the chemisorption structures of Figs. 53 and 54 are shown in Figs. 55 and 56, respectively. They represent how an inflow of a single electron (a test, normalized chemical reduction of the system as a whole) from an external reservoir, redistributes in the system under consideration, when no constraints on the internal flows are imposed. These diagrams can therefore be used to probe the effect of an environmental CT upon the chemisorption, by comparing them with the respective isoelectronic patterns of Figs. 53 and 54. In systems, where the main *in situ* displacements in the chemisorption region are similar to those observed in the global FF diagrams, the overall system reduction ($dN > 0$) promotes the chemisorption (strengthens the adsorbate-substrate CT); similarly, in systems where both these diagrams exhibit opposite phases, the overall system oxidation ($dN < 0$) is required to promote the internal CT. Accordingly, a qualitative similarity between the isoelectronic and global CT patterns in the adsorption region indicates that the system chemical oxidation hinders the inter-reactant CT and facilitates the associative (molecular) desorption of the adsorbate; the same chemisorption hampering influence of the overall chemical reduction is therefore predicted for the chemisorption systems, in which the respective CT and FF patterns in the chemisorption region have opposite phases.

Such an overall similarity of the global FF diagrams in the adsorption region to their isoelectronic analogs is detected in Figs. 55 a, b, d-f and 56 a-f. In these systems $dN > 0$ promotes chemisorption (strengthens the chemisorption bond) and $dN < 0$ hinders the chemisorption (weakens the chemisorption bonds, promotes desorption). In Fig. 55 c the trends exhibited by the main AIM displacements in the CH_3 ---$O_{(2)}$ region are opposite in the closed and open system cases, so that the above dN influences will be reversed in this particular complex.

Another general conclusion following from a comparison between the global and isoelectronic CT diagrams is that the latter are relatively localized and selective, probing mainly a region of a strong inter-reactant charge coupling. In the global FF plots the inflowing charge generates AIM displacements throughout the system; the only exception to this generally delocalized character of the global FF diagrams is detected in Fig. 56 c, where the chemisorption region exhibits the increased AIM charge shifts. Therefore, as already stressed elsewhere [9, 57, 106, 108], the two-reactant, *in situ* diagrams constitute much more subtle and selective diagnostic tools for probing $(dN, d\boldsymbol{v})$- reactivities, than the corresponding global FF data of the reactive system as a whole.

2.10. Concluding Remarks

These illustrative results demonstrate that the inclusion of the external potential perturbation and all adsorbate-substrate charge couplings is necessary for an adequate probing of chemical reactivity trends in large systems, within a truly two-reactant approach. The CSA in the AIM resolution provides a simple and efficient scheme for generating charge responses of reactants, including both $d\boldsymbol{N}^{+}(d\boldsymbol{v})$ (P) and $d\boldsymbol{N}^{*}(N_{CT})_{\boldsymbol{v}^0+d\boldsymbol{v}}$ (CT) components of the overall shifts in the AIM populations, $d\boldsymbol{N}(d\boldsymbol{v}, N_{CT})$ (P + CT), due to the presence of the other reactant. This feature is particularly important for the heterogeneous catalysis, since both the adsorbate activation and surface reconstruction trends can be quickly explored using these responses, before and after the inter-reactant CT. These three types of populational (charge) displacements in reactive systems have been demonstrated to constitute much more selective reactivity criteria than the global FF indices of the chemisorption system as a whole. However, the latter allow one to additionally diagnose the effect of the external CT upon the adsorbate-substrate bond(s).

A remarkable localization of the isoelectronic charge responses in the catalytic systems, limited mainly to the chemisorption region, *a posteriori* validates the cluster approximation in this analysis, since the responses in the missing crystal remainder can be safely extrapolated as marginal. In all cases the role of the supporting layer has been found to be relatively unimportant, due to a relatively weak interlayer charge coupling; only in the parallel adsorption arrangements, for short inter-reactant separations, it was predicted to be relatively more pronounced, due to stronger mutual perturbations of the adsorbate and substrate.

3. Allyl - [MoO_3] System

The chemisorption systems examined in this section include allyl (adsorbate) and the [MoO_3] (010)-surface cluster. This choice of the third illustrative system has been made to study the catalytic reaction of a selective oxidation of allyl to acrolein [108]. As shown by Grzybowska *et al.* [133] MoO_3 is a highly efficient and selective catalyst for this process. It has also been found experimentally that the (010)-surface represents the cut of the crystal which is active in this reaction [134, 135]. The allyl oxidation has also been postulated as the second stage of the selective oxidation of propylene on molybdenum oxide catalysts [136, 137]. Large two-layer surface clusters will be used in the CSA calculations, to include effects due to the crystal environment of a small active site region.

The CSA calculations of the CT effects depend strongly on the predicted levels of the substrate and adsorbate chemical potentials. In cluster representations of the surface the cluster stoichiometry is crucial for obtaining realistic CSA values of the substrate electronegativity. To test this sensitivity in the MoO_3 case we have examined in Fig. 57

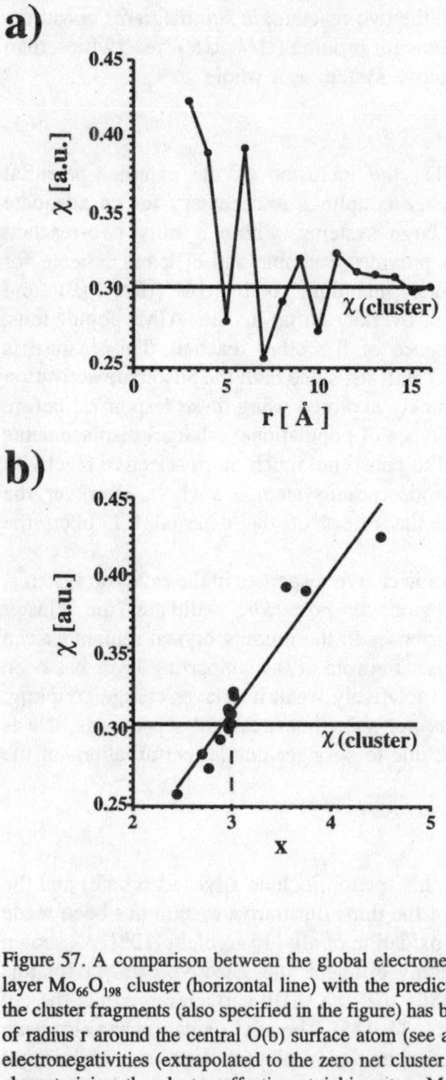

r [Å]	Subsystem or cluster	Effective stoichi ometry MoO_x
3	Mo_2O_9	$MoO_{4.5}$
4	Mo_4O_{15}	$MoO_{3.75}$
5	$Mo_{10}O_{27}$	$MoO_{2.70}$
6	$Mo_{11}O_{39}$	$MoO_{3.55}$
7	$Mo_{20}O_{49}$	$MoO_{2.45}$
8	$Mo_{28}O_{81}$	$MoO_{2.89}$
9	$Mo_{32}O_{97}$	$MoO_{3.03}$
10	$Mo_{43}O_{119}$	$MoO_{2.77}$
11	$Mo_{50}O_{151}$	$MoO_{3.02}$
12	$Mo_{56}O_{166}$	$MoO_{2.96}$
13	$Mo_{62}O_{184}$	$MoO_{2.97}$
14	$Mo_{64}O_{192}$	$MoO_{3.00}$
15	$Mo_{66}O_{196}$	$MoO_{2.97}$
16	$Mo_{66}O_{198}$	$MoO_{3.00}$

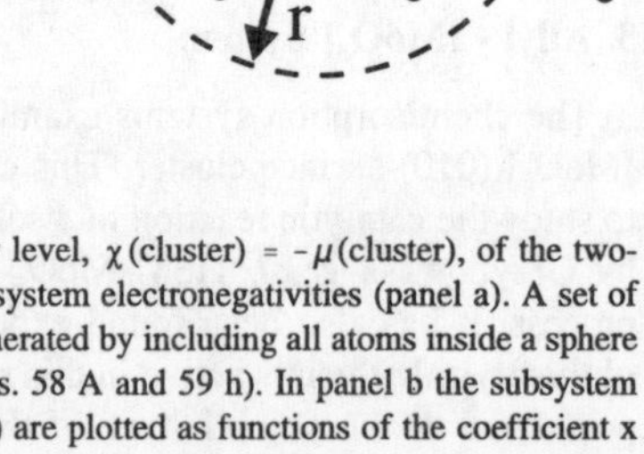

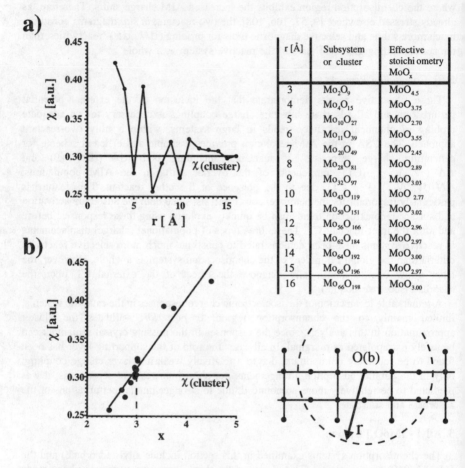

Figure 57. A comparison between the global electronegativity level, χ (cluster) = −μ(cluster), of the two-layer $Mo_{66}O_{198}$ cluster (horizontal line) with the predicted subsystem electronegativities (panel a). A set of the cluster fragments (also specified in the figure) has been generated by including all atoms inside a sphere of radius r around the central O(b) surface atom (see also Figs. 58 A and 59 h). In panel b the subsystem electronegativities (extrapolated to the zero net cluster charge) are plotted as functions of the coefficient x characterizing the cluster effective stoichiometry: MoO_x. It is seen in panel b that for all subsystems exhibiting x close to that of the cluster as a whole (x = 3) the predicted values of χ (subsystem) are close to χ (cluster).

a dependence of the CSA cluster electronegativity $\chi = -\mu$ (extrapolated to the same cluster charge, $Q = 0$) upon the cluster size and its effective stoichiometry. Namely, we have generated a set of subsystems of the $Mo_{66}O_{198}$ cluster used in the present study [108], corresponding to the increasing cluster radius around the central O(b) surface oxygen. It follows from this analysis, that the global electronegativity (or the chemical potential) is very sensitive to the cluster stoichiometry, while its size has a lesser effect upon the CSA predictions. This is well illustrated in the panel b of the figure, which shows that all subsystems of effective stoichiometry close to that of the cluster as a whole exhibit very close electronegativity values. This result indicates that for adequate predictions of the CT properties a special care should be taken to preserve the infinite crystal stoichiometry in the cluster representations of catalytic surfaces.

Properties of the cluster representations of the (010)- and (100)-faces of the MoO_3 crystal have recently been a subject of theoretical (DFT) investigations [138, 139]. The semiempirical SCF-MO (ZINDO [140]) calculations on the two-layer, nearly stoichiometric $Mo_{20}O_{68}H_{16}$ cluster (Fig. 58 A), modelling the (010)-surface of MoO_3, have been performed to generate the prototype charges of crystallographically non-equivalent sites, to be translationally propagated in a larger, stoichiometric cluster, $Mo_{66}O_{198}$, used in the subsequent CSA calculations. Both clusters use the crystallographic bond lengths and angles [141]; in the smaller $Mo_{20}O_{68}H_{16}$ cluster hydrogen atoms have been added at the cluster boundary to saturate valences of peripheral oxygen atoms, in accordance with the saturation scheme established in previous *ab initio* (DFT) calculations [138]. The CSA input charges of the large stoichiometric cluster $Mo_{66}O_{198}$ have been obtained by translating the four source atomic charges from $Mo_{20}O_{68}H_{16}$ calculations; they included charges of the central Mo atom and those of its three geometrically nonequivalent oxygen neighbors (Fig. 58 A).

The CSA calculations in the AIM resolution, performed on both a smaller allyl-$[Mo_{20}O_{68}H_{16}]$ (Fig. 58) and large allyl-$[Mo_{66}O_{198}]$ (Figs. 59-63) chemisorption complexes, have been undertaken to evaluate a relative role of the intra-reactant polarization and the inter-reactant charge transfer effects in an overall mechanism of the catalytic allyl oxidation. One of the goals also was to probe the importance of the specific components of the P and CT AIM charge reconstruction patterns; more specifically, the total P and CT diagrams have again been decomposed into the corresponding diagonal and off-diagonal contributions (Figs. 58 and 59), which explicitly expose a relative role of the reactant interaction. The CT diagram for one orientation, which has been identified as the most important for the reaction mechanism [6], is also interpreted in the IRM framework in a search for the most important reactive channels.

In order to further verify the qualitative CSA predictions of the overall changes in the charge distribution and bonding patterns of the adsorbed allyl we shall also report the independent LSDA DFT calculations of density difference diagrams for the small chemisorption system shown in Fig. 64 a [6]. It corresponds to the LSDA energy minimum of the energy profile along the rigid reactant approach. The small cluster approximation used in these calculations has in fact been validated by the CSA predictions, where both P and CT cluster reconstruction trends were indeed found to be localized in such a small surface fragment.

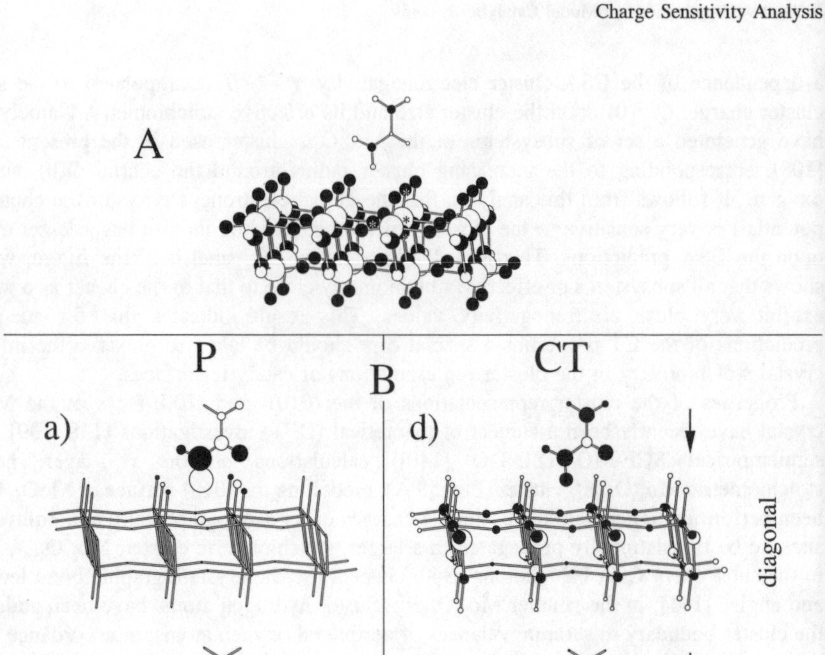

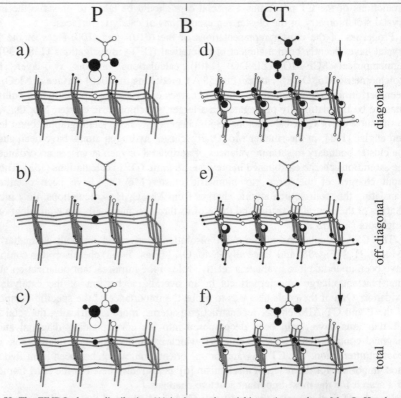

Figure 58. The ZINDO charge distribution (A) in the nearly stoichiometric, two-layer $Mo_{20}O_{68}H_{16}$ cluster modeling the (010) - surface of MoO_3 crystal. The terminal hydrogens have been added to saturate valences of the oxygen atoms at the artificial cluster boundary. The charges marked with an asterisk have been translationally propagated, when constructing the larger cluster used to obtain the charge response characteristics reported in Figs. 59-63. Part A also reports the isolated allyl charge distribution obtained in the same ZINDO approximation. In part B the polarizational (P^+, left column) and CT (*in situ* FF, right column) diagrams for the chemisorption arrangement of Part A are also reported (panels c and f), together with their diagonal (panels a and d) and off-diagonal (panels b and e) components. The arrows indicate the CSA predicted direction of the inter-reactant CT.

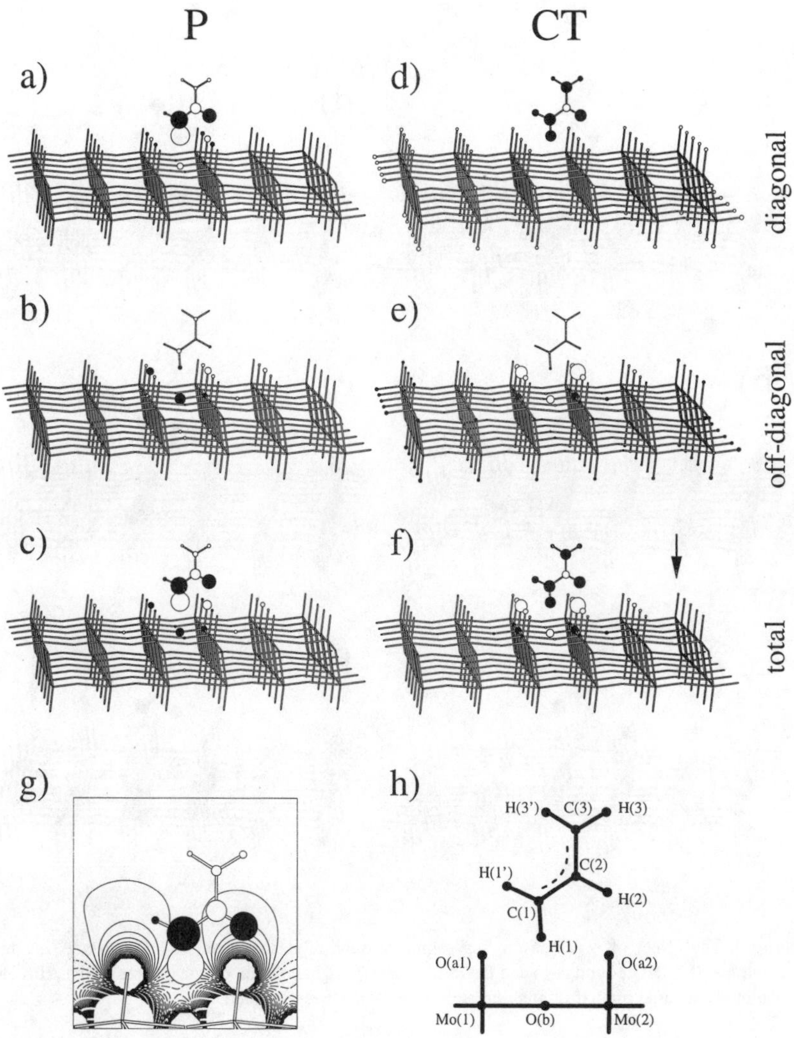

Figure 59. The P (left column, panels a-c) and allyl → $[MoO_3]$ CT (right column, panels d-f) diagrams for a representative allyl-[(010)-MoO_3] chemisorption complex; the first three rows show the diagonal, off-diagonal and total charge responses, respectively. A separate scale factor values are used in the P and CT plots, common to all three panels in question. In panel g the allyl polarization is shown with the background map of electrostatic potential (due to the AIM charges of the substrate) in the allyl plane. The atom designations on allyl and in the surface active site region are shown in panel h.

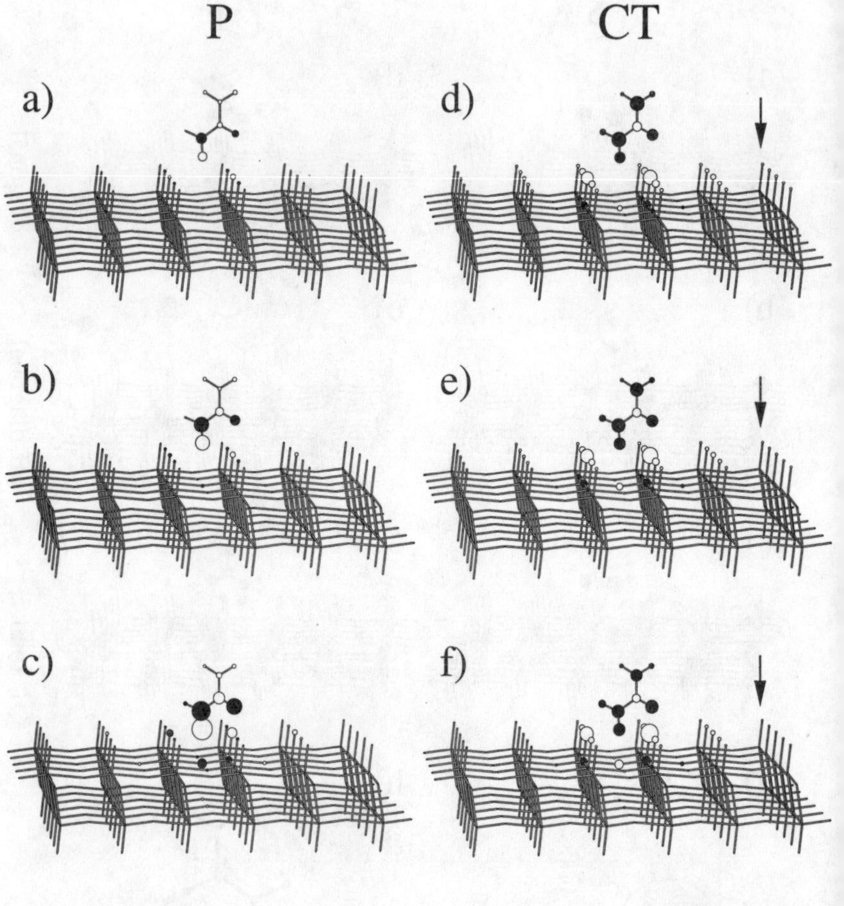

Figure 60. The effect of a decreasing separation between allyl adsorbate and the surface, on the total polarizational (P^+, left column, a-c) and the *in situ* FF (CT, right column, d-f) responses of the AIM electron populations; the same mutual orientation as in Fig. 59 has been adopted.

Changes in atomic charges of reactants can be related to bond weakening/strengthening effects only approximately, since the overall bond-order is a result of a subtle interplay between covalent and ionic components. A more direct, quantum-mechanical measures of the bond-multiplicity are needed to predict reactivity/activation patterns in surface reactions involving large adsorbates. It has been recently demonstrated that useful absolute bond-order indicators can be generated by comparing quantum-mechanical density matrices in a molecule and in the separated atoms limit, respectively, generated within the

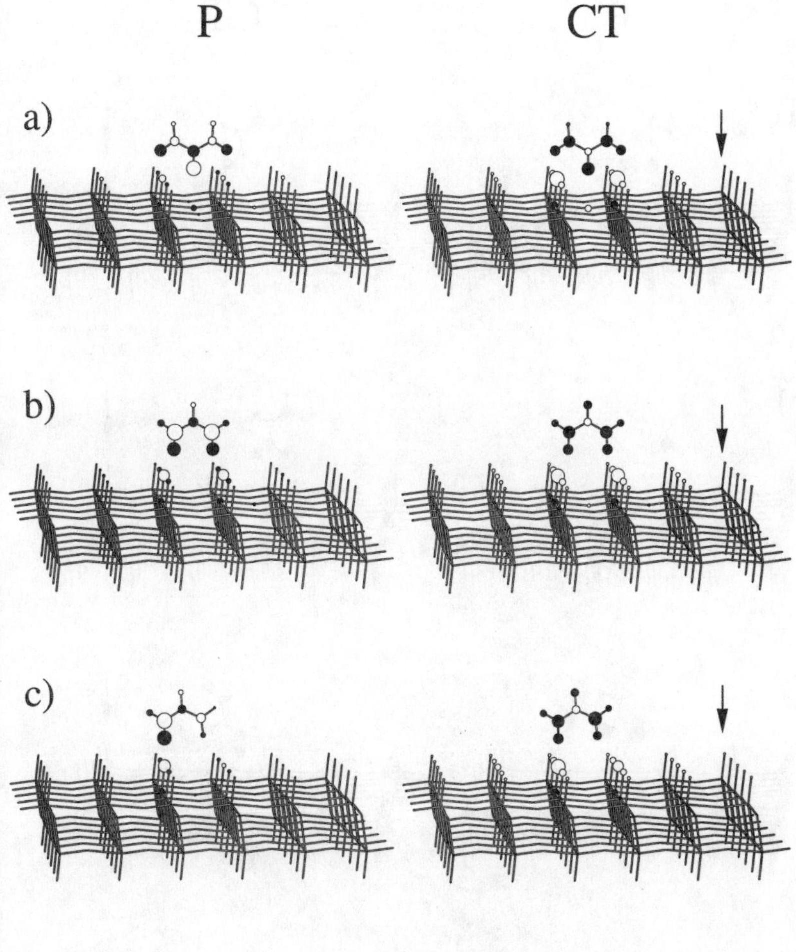

Figure 61. A comparison of the total P (left column) and CT (right column) charge displacement diagrams for the six alternative mutual arrangements of allyl relative to the [(010)-MoO_3] surface.

one-determinantal (Hartree-Fock and Kohn-Sham) difference approach (see Appendix D) [6]. From such absolute bond-orders one can predict relative changes in the bonding pattern due to the chemisorption, by subtracting the bond-orders in the separated-reactants limit (substrate and adsorbate) from those in the chemisorption complex. However, these supplementary calculations can be performed only for much smaller (saturated) clusters

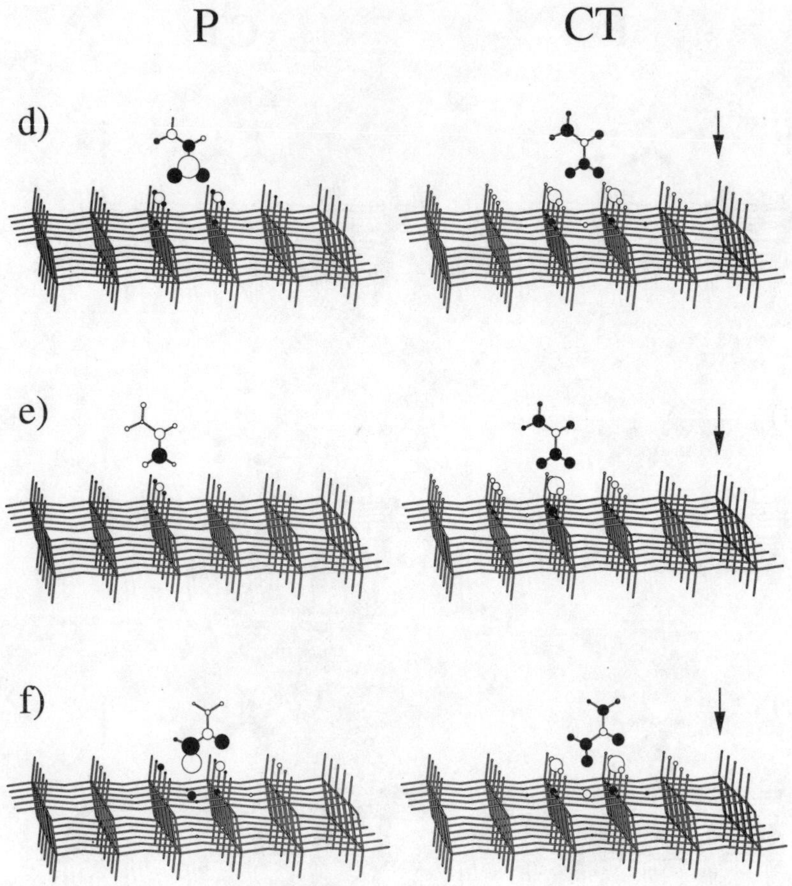

Figure 61. Continued.

involving only the active site region of the surface. The bond-order (valence) data have been obtained from the KS calculations on $Mo_2O_{11}H_{10}$ cluster, which has been shown to provide an adequate representation of the surface in the adsorption region [138]. It also involves a saturation of artificial, terminal oxygen valences by the four hydrogen atoms simulating four broken "single" Mo-O bonds, and the six hydrogens simulating three broken "double" Mo-O bonds at the cluster boundary. In Fig. 65 we report the calculated DFT (LSDA) bond-order estimates obtained within the difference approach to the hemical valence and bond multiplicity. All this CSA and bond-order reactivity ıformation is subsequently used to form conjectures about the most probable mechanism f the allyl catalytic oxidation to acrolein (Fig. 66).

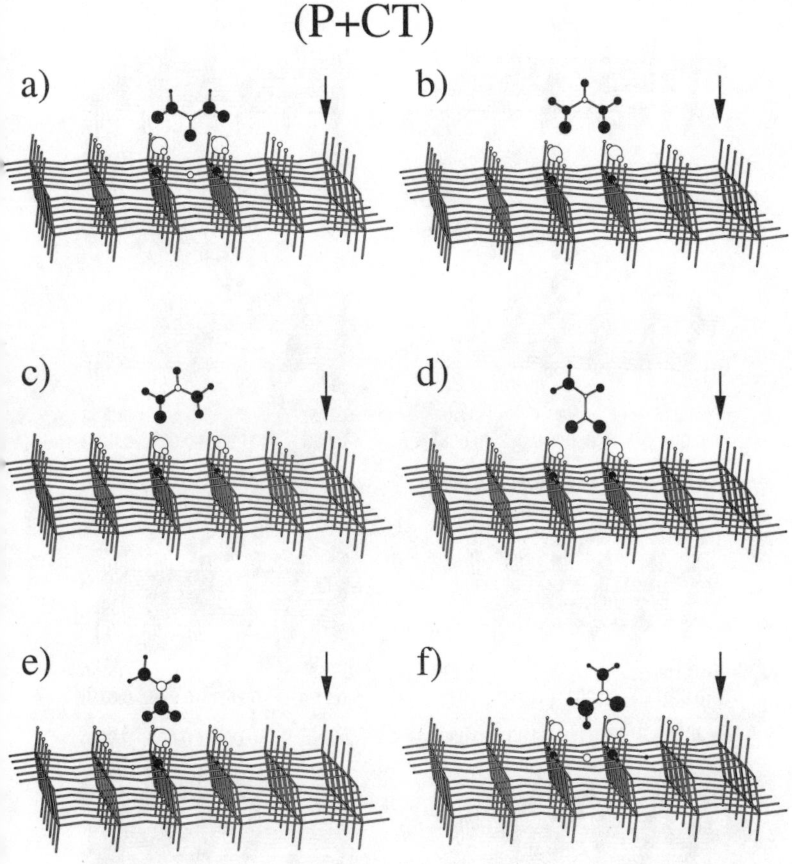

'igure 62. The overall charge reconstruction diagrams for the chemisorption structures of Fig. 61, epresenting the sum of the corresponding P and CT contributions shown in Fig. 61. The same scale factor as been adopted in all panels of the figure.

The six probing mutual arrangements of the adsorbate and substrate in the allyl-[MoO_3] eactive system, representing comparable inter-reactant separations, are shown in Fig. 61; hey have been used in both CSA and bond-multiplicity calculations. In the bond-order nalysis we have additionally considered yet another adsorption complex shown in Fig.

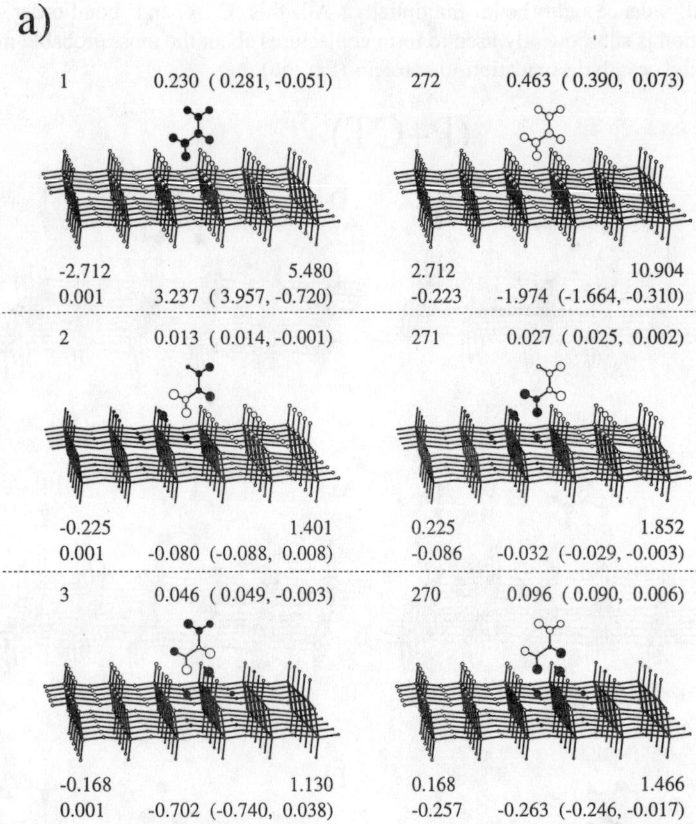

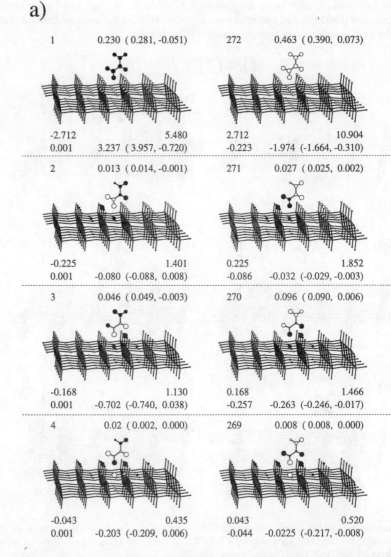

Figure 63. The complementary reactive (delocalized) IRM (part a) for the chemisorption structure of Fi 59. The numerical data are reported in the same order as in Figs. 30, 45 and 47. The CT parameters a calculated for the allyl (B) → [MoO] (A) electron transfer. The part b of the figure reports a decompositic of the *in situ* FF diagram (see Fig. 61 f - CT) in the IRM representation; only 8 modes, exhibiting t largest magnitudes of the AIM contributions are included; the sum of their respective contributions to t electron transfer amounts to 92 % (cluster) and 100 % (allyl).

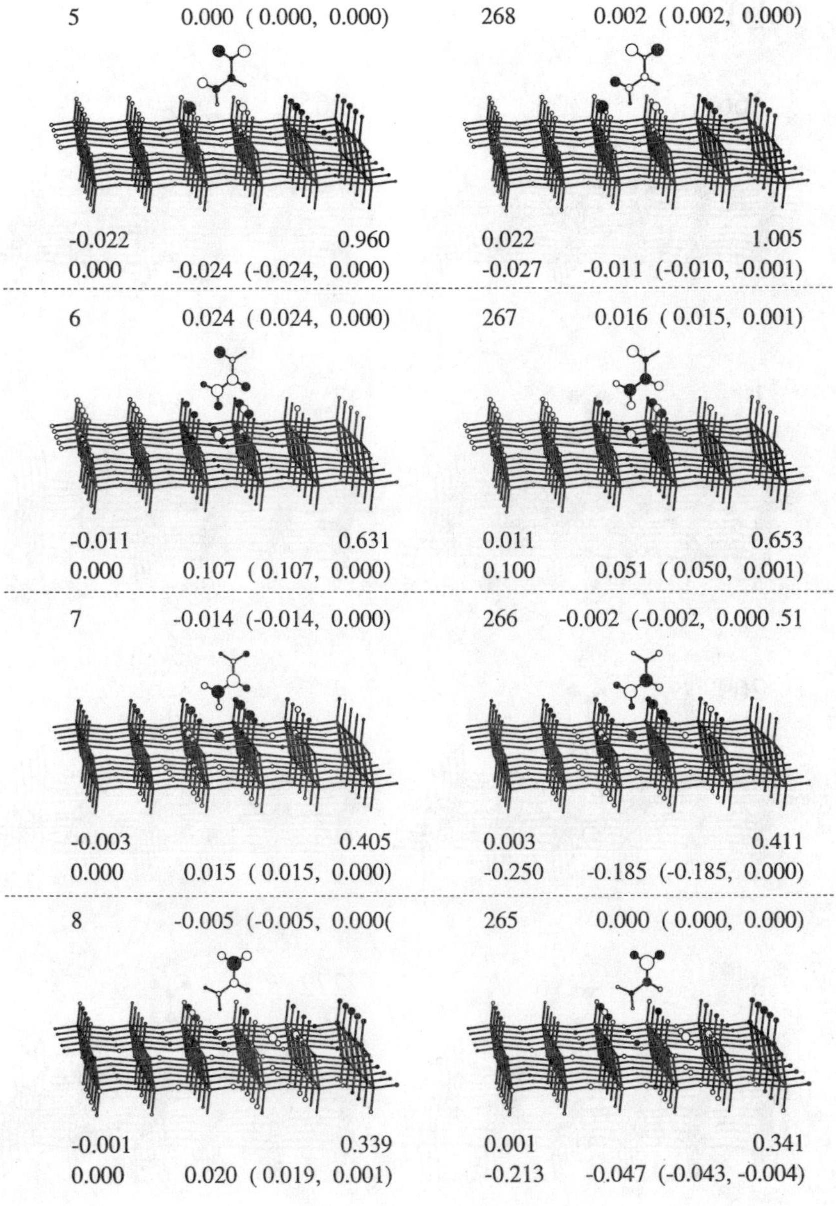
5
0.000 (0.000, 0.000)
-0.022
0.960
0.000
-0.024 (-0.024, 0.000)
268
0.002 (0.002, 0.000)
0.022
1.005
-0.027
-0.011 (-0.010, -0.001)
6
0.024 (0.024, 0.000)
-0.011
0.631
0.000
0.107 (0.107, 0.000)
267
0.016 (0.015, 0.001)
0.011
0.653
0.100
0.051 (0.050, 0.001)
7
-0.014 (-0.014, 0.000)
-0.003
0.405
0.000
0.015 (0.015, 0.000)
266
-0.002 (-0.002, 0.000 .51
0.003
0.411
-0.250
-0.185 (-0.185, 0.000)
8
-0.005 (-0.005, 0.000(
-0.001
0.339
0.000
0.020 (0.019, 0.001)
265
0.000 (0.000, 0.000)
0.001
0.341
-0.213
-0.047 (-0.043, -0.004)

Figure 63. Continued.

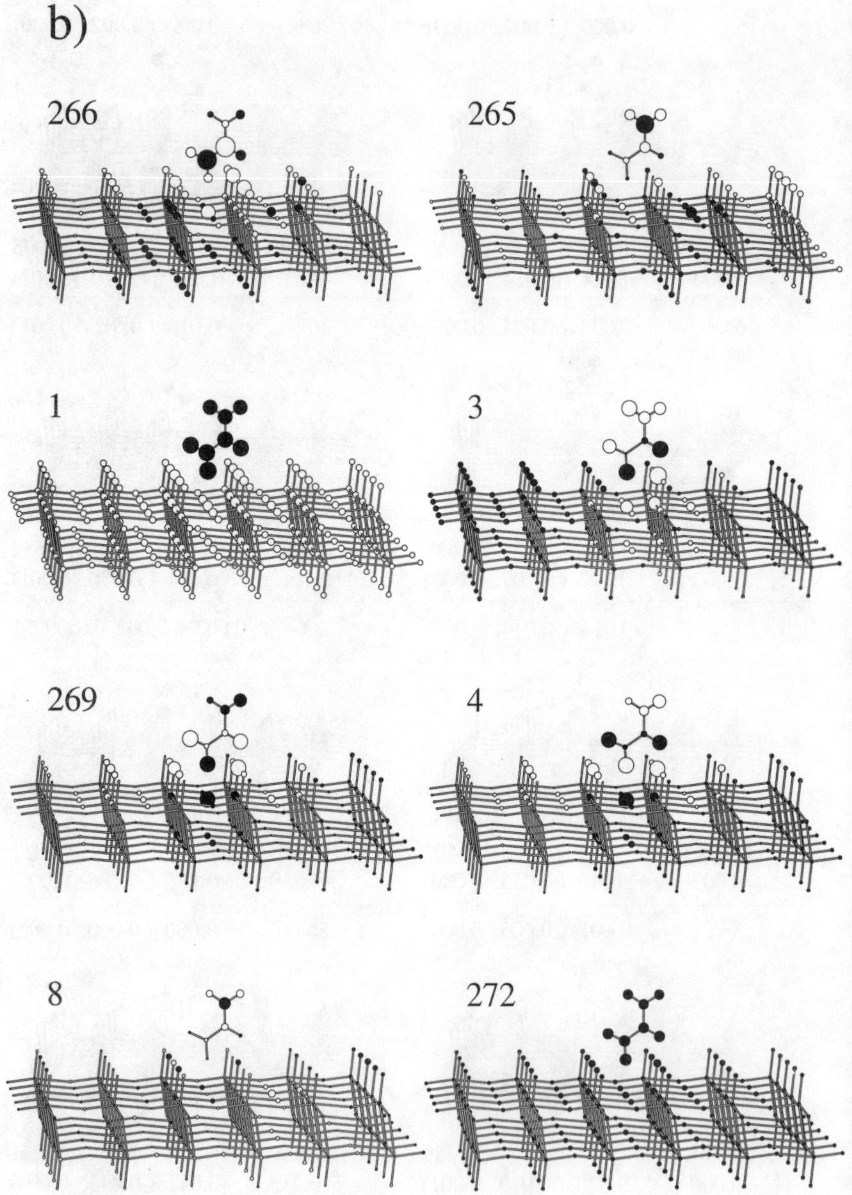

Figure 63. Conclusion.

a)

b)

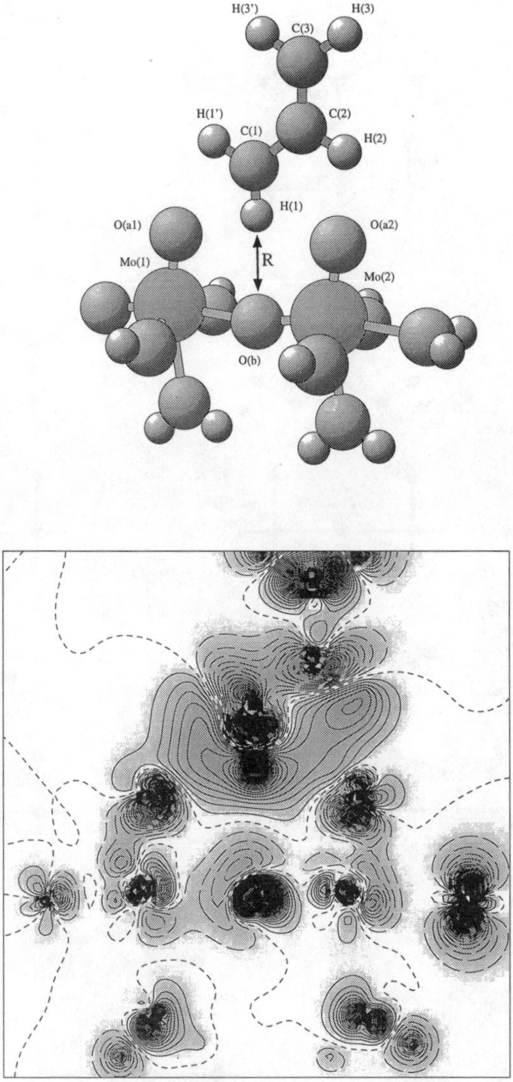

Figure 64. The small chemisorption system (a) including allyl and the $Mo_2O_{11}H_{10}$ surface cluster used in the LSDA DFT calculations and the resulting density difference diagram (b). The diagram of panel b is the difference between the molecular (chemisorption system as a whole) and the sum of reactant (isolated allyl and cluster) densities at the current separation R = 2.6 Å [6]. The broken (solid) lines correspond to the electron outflow (inflow) regions, relative to the nonpolarized reactant densities.

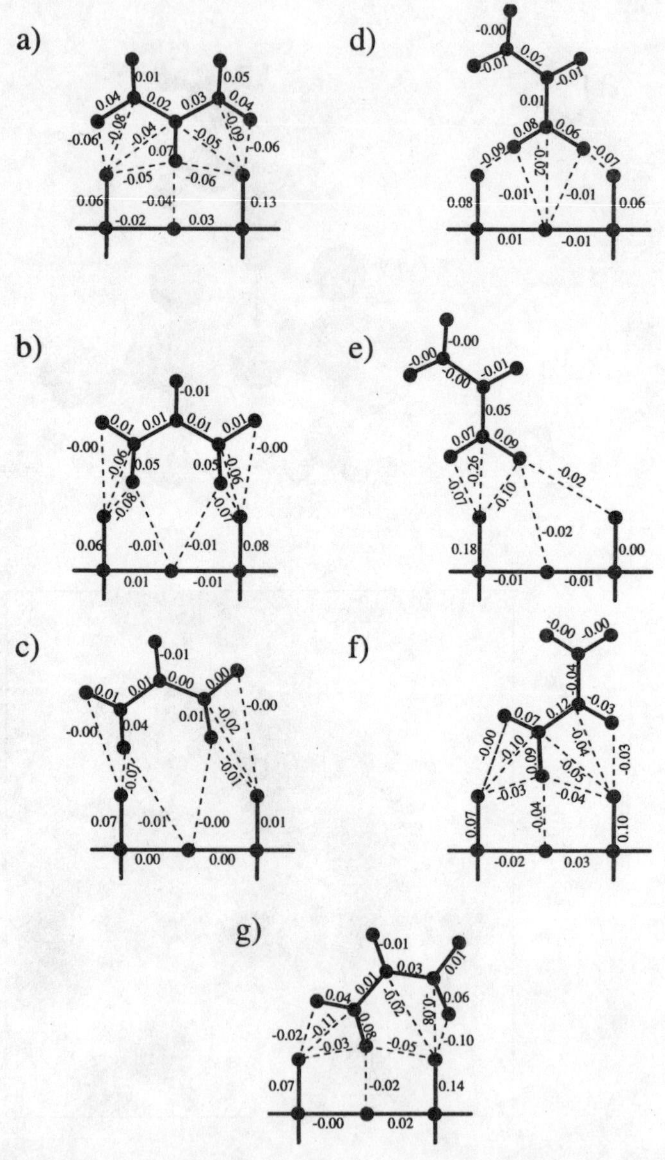

Figure 65. The bond-multiplicity patterns from the valence indices [one-determinantal (KS) difference approach] for the six adsorption arrangements of Fig. 62 and the additional chemisorption complex shown in panel g.

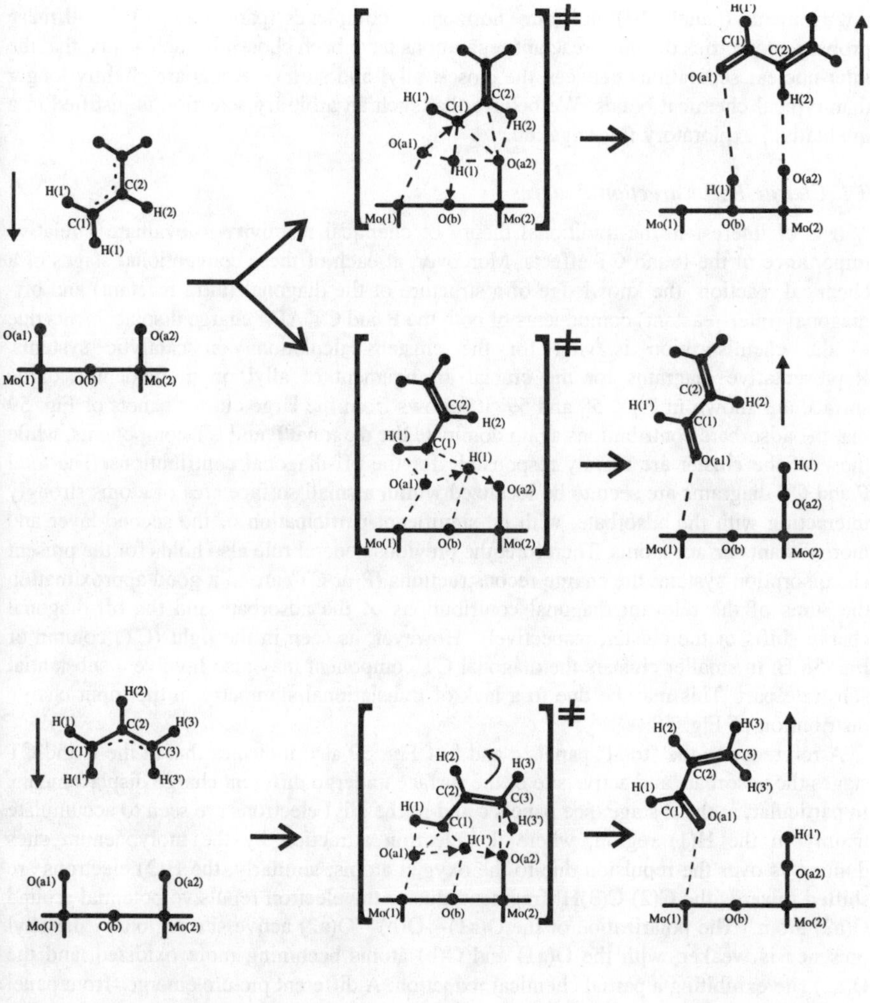

Figure 66. Schematic diagrams illustrating the two alternative concerted reaction mechanisms for the selective allyl oxidation to acrolein.

65 g. All these orientations can be physically rationalized on the purely electrostatic ground. Namely, in all these structures the positive (electron stabilizing) regions of the allyl electrostatic potential (in the vicinity of the molecular plane) are overlapping with the negative (proton stabilizing) regions of the surface electrostatic potential (around the active site oxygens). This selection of adsorption arrangements includes the "vertical" allyl

arrangements (panels d-f) and four "horizontal" complexes (panels a-c, g). In all these probing geometries the inter-reactant separations have been chosen in such a way, that the inter-nuclear separations between the closest allyl and surface atoms are slightly longer than typical chemical bonds. We believe that such an arbitrary selection is justified in a qualitative, exploratory investigation.

3.1. Charge Reconstruction Patterns

It is of interest in the traditional theory of chemical reactivity to evaluate a relative importance of the P and CT effects. Moreover, at each of these conventional stages of a chemical reaction the knowledge of a structure of the diagonal (intra-reactant) and off-diagonal (inter-reactant) components of both the P and CT AIM charge displacements due to the chemisorption is vital for the cluster calculations on catalytic systems. Representative diagrams for the crucial arrangement of allyl on molybdenum oxide surface are shown in Figs. 58 and 59. It follows from the large cluster panels of Fig. 59 that the adsorbate contributions again dominate the diagonal P and CT components, while those of the cluster are mainly responsible for the off-diagonal contributions. The total P and CT diagrams are seen to be localized within a small surface area of atoms strongly interacting with the adsorbate, with no significant participation of the second-layer and more distant surface atoms. Therefore, the previous general rule also holds for the present chemisorption system: the charge reconstructions (P or CT) are to a good approximation the sums of the relevant diagonal contributions of the adsorbate and the off-diagonal charge shifts of the cluster, respectively. However, as seen in the right (CT) column of Fig. 58 B, in smaller clusters the diagonal CT component may also involve a substantial substrate part. This may be due to a lack of translational symmetry in the input charge distribution of Fig. 58 A.

A reference to the "total" panels c and f of Fig. 59 also indicates that at the P and CT stages the adsorbate and active site of the surface undergo different charge displacements. In particular, at the P stage (see panels c and g) the allyl electrons are seen to accumulate mainly in the H(1) region, where the electron attraction by the molybdenum sites dominates over the repulsion due to the oxygen atoms; similarly, the H(2) electrons are shifted towards the C(2)-C(3)H_2 fragment, due to the electron repulsive potential around O(a2) atom. The polarization of the O(a1)---O(b)---O(a2) active site region by the allyl presence is weaker, with the O(a1) and O(b) atoms becoming more oxidized, and the O(a2) site exhibiting a partial chemical reduction. A different picture emerges from panel f, where all active site atoms are chemically reduced due to the allyl → substrate CT, determined from both the CSA and DFT calculations; on the allyl side a majority of atoms is predicted to be oxidized by such an inter-reactant CT, with the exception of the C(2) site.

In the left column of Fig. 60, corresponding to the large cluster representation, we investigate the effect of an increasing distance between reactants upon the P diagrams. A reference to panels a-c of the figure shows that the nodal structure of these polarizational displacements, with the scaled down magnitudes of charge responses, remains qualitatively the same when this distance increases. A comparison between Figs. 58 B-c and 60 c also shows that an enlargement of the cluster has practically no effect upon these polarizational responses, just additionally validating the smaller cluster representation of Fig. 58. The only minor difference observed in Fig. 60 c is an increased hardening influence at the O(a1) position. The R dependence of the total FF plot is illustrated in the right column of Fig. 60. One observes that, due to the constant normalization to $N_{CT} = 1$, these plots are only weakly dependent upon R; however, we would like to stress at this point that the real CT component in the reaction mechanism is obtained by scaling these normalized FF diagrams to the actual value of the amount of CT, which is R dependent [9 b].

In Fig. 61 we have shown the total P and CT diagrams corresponding to the six alternative adsorption arrangements at comparable inter-reactant separations. In all these structures the allyl → substrate CT is predicted by the CSA calculations, with the amounts of CT in the range of N_{CT} = 0.64 - 0.83 for the structures a-f. The same direction of the inter-reactant CT follows from the LSDA calculations on small chemisorption systems [108], although the population analysis results predict much smaller magnitudes of N_{CT}. The energetical hierarchies for these structures, determined from the rough energy estimates of Eqs. 367, 368, 388, and 389 are: d < b < c < a < f < e (ES), f < d < b < c < a < e (ES + P), and a < f < e < b < d < c (ES+P+CT). Therefore, the arrangement of Fig. 59 indeed appears as the preferred "vertical" one, while the structure 61 a is predicted as the energetically most favourable "horizontal" complex, of comparable total energy to that of the 61 f geometry. As we shall argue in the last section, the structure 61 f and 65 g are expected to play crucial roles in the inferred concerted mechanisms of the allyl oxidation to acrolein on MoO_3.

The resultant charge rearrangements (shown in Fig. 62) depend upon a relative participation of the CT and P components in the overall (P + CT) displacements. An inspection of the corresponding charge redistribution diagrams of Figs. 61 and 62 indicates, that the CT component plays a dominant role in the overall diagrams of Fig. 62. Only in arrangements d and f one detects increased P contributions, particularly on the adsorbed $-CH_2$ fragment of allyl. As a rule all constituent AIM of the adsorbate, with the exception of the middle allyl carbon, are donating electrons to the surface in the isoelectronic allyl → cluster CT process. The P diagrams of the adsorbate exhibit more diversified patterns; one detects a relatively large number of nodes in these diagrams, with the neighboring atoms acting with opposite phases, which signifies a soft charge rearrangement with the dominant intra-adsorbate polarization and a small inter-reactant CT component. Another general trend emerges from an examination of the charge displacement phases of the most strongly interacting pairs of constituent atoms on the adsorbate and substrate. Namely, it can be seen that they are opposite in most cases with the allyl atoms donating electrons to the neighboring surface oxygens.

It follows from the P (physisorption) column of Fig. 61 that the allyl polarization of the $-CH_2$ fragment(s), induced in arrangements a-d, is in the H → C direction, consistent with the original bond polarization: $H^{+\delta}$ — $C^{-\delta}$; such a charge reconstruction should strengthen the C-H bonds on terminal carbons. The middle CH fragment is seen to be strongly polarized in the two energetically preferred structures a and f; in panel a one detects a weakening of the C-H bond on the middle carbon atom, associated with the reverse induced bond polarization C → H, while in panel f this bond is predicted to be strengthened at the P stage of chemisorption. Of interest also is the C-C bond polarization relative to the isolated allyl pattern: $C^{-\delta}$ — $C^{+\delta}$ — $C^{-\delta}$. In complexes a-d this C-C polarization is increased (bond strengthening), while the reverse trend (bond weakening) is observed in the two remaining structures a and f at the P stage of adsorption.

In all CT diagrams of Fig. 61 the middle C-H bond in allyl is extra polarized in the original direction, thus giving rise to a slight bond strengthening. The coordinating terminal -CH_2 groups are loosing (bonding) electrons, which should result in the C-H bond weakening. A comparison of the P and CT diagrams for a given geometry indicates that in structures a-d the carbon chain is polarized in opposite directions at these two stages of adsorption; in panels e and f they are similar, thus enhancing one another. In general, the CT diagrams are relatively insensitive to the adsorbate arrangement on the surface, in comparison to the corresponding P diagrams. Since they dominate the overall trends of Fig. 62 the same adsorbate bond activation and surface reconstruction predictions follow from the charge reconstruction at the P + CT stage (chemisorption). This general conclusions are further supported by the quantum mechanical bond multiplicity predictions reported in Fig. 65.

3.2. *IRM Representation of the Inter-Reactant CT Effects*

There are 16 reactive IRM in the allyl-[MoO_3] chemisorption system (Fig. 63 a), involving 8 independent components of both the adsorbate and substrate, regardless of the surface cluster size; they can be arranged in the complementary pairs, each grouping the in-phase and out-of-phase combinations of these reactant components. These pairs are shown in Fig. 63 a; they can be identified by the same magnitude of the mode hardness, $| \eta_\alpha |$. In Fig. 63b we have ordered the IRM diagrams in accordance with diminishing of the maximum magnitude of the AIM contribution $\omega_{i,\gamma}^{CT}$ to the AIM *in situ* FF diagram (shown in panel f of Fig. 60), in order to limit the figure to the most important charge reorganization channels. Among the remaining modes there are 8 reactive modes carrying negligible AIM displacements on allyl, and a large number of non-reactive modes, with the exactly vanishing allyl component. The latter represent the pure intra-surface modes, of no direct importance for interpreting the inter-reactant CT.

It follows from Fig. 63 b that the complementary third-row modes produce mainly a change in the cluster, since their allyl components approximately cancel each other; in the active site region of the surface the terminal surface oxygens O(a1, a2) receive electrons and the bridging oxygen O(b) looses electrons via this reactive channel. The second pair of the $\eta = \pm 2.712$ complementary modes $\gamma = 1$ and 272 represents uniform (background) changes of populations on both reactants, with practically zero polarizational (site differentiating) components on each reactant; together they describe an outflow of electrons from allyl and an inflow of electrons to the surface. As such this pair of modes is of little importance for interpreting the reactant activation patterns and the reaction mechanism. The topology of these modes explains their extreme hardness, since a simultaneous removal/addition of electrons from/to many neighboring atoms must involve large energy changes. The third complementary pair of the soft IRM, $\eta = \pm 0.001$, $\gamma =$

8 and 265, exhibits mostly a polarization inside the -C(3)H_2 allyl fragment, with the opposite phases of relatively small cluster components; therefore, they too are expected to be only marginally involved in the CT reaction mechanism for this particular arrangement.

This leaves the two non-complementary modes γ = 266 ($\hbar$ = 0.003) and γ = 3 ($\hbar$ = -0.168) as the most important IRM channels for interpreting the inter-reactant CT in this particular adsorption complex. Indeed, these modes are seen to produce relatively large displacements on atoms in the chemisorption region. They represent the main independent collective charge components of the surface atom exchange process summarized in Fig. 66. Namely, the γ = 3 ($\hbar$ = -0.168) IRM facilitates an anchoring of the H(1) and H(2) allyl hydrogens on the O(b) and O(a2) sites of the surface. Similarly, the γ = 266 ($\hbar$ = 0.003) mode represents a reactive channel through which the C(1)-O(a1) bond can be realized; one also notices an accumulation of electrons on both O(b) and H(1) atoms *via* this channel; this should also promote a stronger covalent component of this surface bond.

This illustrative application of the IRM framework supports our earlier conclusion that this particular set of collective charge displacements provides an adequate and attractive reference system for interpreting the inter-reactant CT processes.

3.3. Density Difference Diagram

The density difference diagram of Fig. 64 b reports the effective charge reconstruction due to chemisorption for a reduced size of the surface active site. It involves the two neighboring MoO_6 distorted octahedra sharing a common bridging oxygen, and a saturation of artificial, terminal oxygen valences by ten hydrogen atoms (four to simulate broken "single" Mo-O bonds and the remaining six - to simulate three broken "double" Mo-O bonds) at the cluster boundary. As was demonstrated previously [138] such an approximation provides a realistic representation of the surface in the adsorption area. The electron density changes in Fig. 64 b generally agree with the CSA charge redistribution pattern of Fig. 62 f. More specifically, in the density difference map there is an accumulation of charge between the pairs of atoms [C(1) and O(a1)] and [H(2) and O(a2)], responsible for oxidizing the allyl terminal -CH_2 group and for the anchoring of the middle -CH group on O(a2). There is also a clear indication (charge accumulation) of the H(1)-O(b) partial bond being formed. This contour map also shows a slight increase in electron density between C(2) and C(3), and a decrease of electron density in the bond region between C(1) and C(2). Such changes are indeed expected in the final (acrolein) stage of this catalytic reaction, in which these atomic pairs exhibit a single and double bonds, respectively.

3.4 Bond-Order Analysis

In order to further verify the above qualitative CSA predictions of changes in the charge distribution and bonding patterns in the chemisorption region, we have performed independent LSDA DFT calculations of the recently developed bond-multiplicity indices [6, 124-126] (see Appendix D), for all adsorption arrangements of allyl on the small $Mo_2O_{11}H_{10}$ cluster (Fig. 64 a) representing the surface active site. In Fig. 65 we have collected bond-multiplicity diagrams for a selection of seven adsorption arrangements, where the main chemisorption and cluster/allyl activation bond-order shifts are shown.

A reference to this figure shows changes in diatomic bond-multiplicities [6] induced by the chemisorption, i.e., the bond-order displacements, relative to the corresponding values in the separated reactants limit. The negative (positive) values signify a bond strengthening (weakening) shifts [6]. Let us first examine general trends in the active site reconstruction. In all adsorption complexes the terminal (molybdenyl) oxygens O(a1) and O(a2) are more weakly bonded to the Mo atoms after the chemisorption. This effect is particularly strong in panel e of the figure, representing a "vertical" adsorption of allyl on a single terminal oxygen; in complexes a and f an intermediate degree of such a disengaging of the surface sticking oxygens is observed. It also follows from panels c and e that the asymmetrical adsorption structures result in a strong differentiation between the Mo-O bond activation of the two molybdenyl oxygens. This surface activation by the adsorbate is crucial for the proposed mechanism of the selective oxidation of allyl to acrolein. The Mo-O bonds involving the bridging oxygen O(b) are seen to be weakly affected by the chemisorption in all arrangements. The cluster parts of the bond-order diagrams a, f, and g exhibit almost identical values of the bond-order displacements.

Changes in the total cluster bond-order ($\Delta \mathcal{V}^A$), representing a sum of all diatomic bond multiplicities in this subsystem, do indeed indicate that there is an overall bond weakening in the surface active site region; the largest effect is detected for structures a, g, and e. The smallest bond activation of the surface cluster is found in the non-symmetrical arrangement c, which is distinctly lower than that corresponding to its symmetrical analog of panel b. A comparison between these total cluster quantities reveals a rough correlation between the total cluster reconstruction bond-order $\Delta \mathcal{V}^A$ and the interaction energy, with the two structures exhibiting the largest shift in the cluster bond-order giving rise to the positive (destabilization) interaction energies.

The bonding patterns inside allyl exhibit a relatively strong dependence on the geometry of the chemisorption, as intuitively expected. In panel a all allyl bonds are slightly weakened with the strongest effect being observed on the middle C-H bond, involving the hydrogen appreciably bonded to the three oxygens of the active site. For this structure the overall bond weakening in allyl ($\Delta \mathcal{V}^B$) is the highest among all seven complexes considered, being roughly comparable to those corresponding to panels e and g. It should be recalled at this point that these three arrangements also correspond to the largest intra-cluster bond weakening; this provides the bond-multiplicity rationalization of the unfavourable interaction energies in these three cases. It is also seen in Fig. 65 e that only three bonds involving the carbon atom of the coordinating -CH_2 group are activated (weakened) as a result of such a "vertical" adsorption on the O(a1) surface atom; one also observes a weakening of the C(1)-C(2) bond for this arrangement, a clear demonstration of the competition effect between this and C(1)-O(a1) bonds for the valence electrons of C(1). This influence is weaker in panel d ["vertical" adsorption on O(b)] so that only C(1)-H bonds are appreciably weakened in such a complex, due to the interaction with the terminal oxygens of the active site.

A similar concentration of the bond activation pattern is observed in the three related "horizontal" adsorption structures shown in panels b, c, and g. In the asymmetrical arrangement of panel c only one C(1)-H bond is visibly affected, while all the remaining allyl bonds are practically unchanged by the chemisorption. In complexes b and g one detects a similar bond weakening in both C(1)-H and C(3)-H bonds. The geometries b, c, and d exhibit a relatively low total bond-order reorganization inside allyl, and the chemisorption system as a whole; this nicely correlates with favourable interaction energies predicted for these complexes, comparable to that found for panel f. The total allyl bond reorganization and the total intra-reactant bond-order shifts in panel g are comparable to those in panels a and e; in all these cases the net destabilization energy is predicted for the frozen geometry of separated reactants.

The panel f is of particular importance for the selective oxidation of allyl to acrolein [108]. It exhibits the allyl bond activation pattern, which facilitates the concerted process of the C(1)-O(a1) bond formation and the C(1)-H bond weakening, leading to acrolein (Fig. 66). The C-C bonds are also changed in the direction of the bonding pattern in the acrolein product, with the C(2)-C(3) bond-multiplicity being increased from the initial approximate value of -1.5 in allyl, toward the intuitively expected bond-order of -2 in acrolein. This "vertical" adsorption also corresponds to relatively small total intra-cluster and intra-allyl bond reconstructions, thus giving rise to an overall stabilization of the chemisorption complex relative to the separated reactants.

Finally, let us examine the interaction part of Fig. 65. The numbers reported in panel a reflect a formation of partial bonds between allyl carbons and the molybdenyl oxygens; the three coordinating allyl hydrogens are predicted to form chemisorption bonds with the three surface oxygens. This structure generates the largest inter-reactant bond-order ($\mathcal{V}^{A-B}$), defined as the sum of all inter-reactant diatomic contributions. Also, the magnitude of a change in the overall bond-multiplicity ($\Delta\mathcal{V} = \Delta\mathcal{V}^{A} + \Delta\mathcal{V}^{B} + \mathcal{V}^{A-B}$) is the largest in the complex a, despite a relatively large intra-reactant bond weakening. A similar trend is detected in complex g. For the structure f one obtains a comparable value of the net chemisorption bond-order, but at reduced overall value of the intra-reactant bond weakening (approximately by a factor of 2). This and the observed differences in the interaction energies indicate that the arrangement f is indeed very important for explaining the known selectivity of the allyl oxidation to acrolein. One also concludes that the "horizontal" complexes b and c give almost identical overall bond-order changes despite a large difference between the total interaction valence ($\mathcal{V}^{A-B}$) values, which compensates the opposite trend exhibited by the overall intra-reactant bond-order displacement, $\Delta\mathcal{V}^{A} + \Delta\mathcal{V}^{B}$. The LSDA interaction energies for structures b and c differ only by 0.3 and 2.4 kcal/mole, respectively, from that obtained for panel f; therefore, transitions b $\rightarrow$ f and c $\rightarrow$ f require little activation, even in this approximation of the frozen reactant geometries. The observed pattern of the allyl-cluster diatomic bond-orders in Fig. 65 b reveals the bonding interactions between the two coordinating hydrogens and terminal carbons of allyl, and the molybdenyl oxygens of the surface, respectively; a comparison with panel a of the figure indicates that the C---O (H---O) interactions in complex b are relatively weaker (stronger). In complex c only one -CH_2 fragment of allyl forms appreciable partial bond with the nearby molybdenyl oxygen.

The main chemisorption bond-orders in the "vertical" complexes d and e of Fig. 65 are practically limited to the coordinating -CH_2 fragment and the two (panel d) or a single (panel e) terminal oxygens of the surface. In the arrangement d only allyl hydrogens form a relatively strong bond with the O(a1) and O(a2) atoms; a "vertical" coordination to a single molybdenyl oxygen in panel e leads to a similar H---O and a very strong C---O bonding interactions. These two geometries give rise to relatively large overall chemisorption bond-orders, comparable to that predicted for complex f. However, there is a difference in the interaction energy between the comparable value of the stabilization energy obtained for structures (d, f) and a relatively large destabilization energy predicted for complex e. This result clearly indicates that, for all practical purposes, the latter structure can be excluded from the list of potentially important chemisorption structures in this system. An additional argument for disregarding it in a search for the mechanism of the allyl oxidation to acrolein is that this structure cannot assure a synchronous process of the O(a1) attachment and the hydrogen removal in allyl, since the whole -CH_2 group coordinates strongly to a single surface oxygen only in such an arrangement.

Let us finally examine the chemisorption bond-orders in the two remaining panels f and g. The complex f produces quite large magnitudes of both $V^{A\text{-}B}$ and ΔV; this extra bonding is achieved at relatively low value of the total bond weakening inside adsorbate and substrate. The large chemisorption bond-order predicted for structure g (second largest, after that found for complex a) is partly reduced by the medium level of the total intra-reactant bond weakening, giving as a result the second largest value of $|\Delta V|$, identical to that predicted for structure f.

3.5. Implications for the Mechanism of Catalytic Allyl Oxidation to Acrolein

The calculated LSDA energies for the chemisorption structures of Fig. 65 show that all geometries but complex e have the chemisorption energies differing by less than 10 kcal/mole. This indicates that transitions from one such a structure to another are relatively easy. For the selective oxidation process of allyl to acrolein the two structures f and g are the most important, since they provide an efficient anchoring of allyl to one molybdenyl oxygen, while exposing the terminal -CH_2 fragment to an oxidation by another molybdenyl oxygen, as schematically shown in Fig. 66. These two structures also facilitate the concerted mechanism of the oxygen attachment and a simultaneous hydrogen removal in allyl, leading to the *trans*- and *cis*- conformers of acrolein, as shown in the figure. In the first case (structure f and the first mechanism of Fig. 66) there is only a small net bond-order change in allyl, and a relatively large covalent component of the chemisorption bond. In the second case (structure g and the second mechanism of Fig. 66) there is a larger bond weakening both inside allyl and cluster, and the ionic component of the chemisorption bond dominates the interaction [108]. This indicates that the overall covalent bond-order in the chemisorption complex g roughly conserves the sum of the overall covalent bond-orders of the separated reactants.

The first mechanism of Fig. 66 involves a movement of the O(a1) atom to the terminal carbon of allyl, and a simultaneous transfer of the H(1) hydrogen to either O(b) or O(a2) surface oxygens, thus producing the surface hydroxyl, an oxygen vacancy, and the acrolein product. On the basis of the present results we cannot rule out any of these possibilities. In both transition states the atom exchange processes should involve little if any activation energy, since a breaking of bonds in reactants is accompanied by a formation of new bonds in products. This and efficient anchoring of allyl on the active site, which exposes the terminal $-CH_2$ group to the action of the crucial O(a1) site, explains in our opinion the observed high selectivity of this catalytic reaction.

The second mechanism of Fig. 66 refers to the chemisorption complex g, in which there is a very strong anchoring of the allyl through the C(3) carbon and the H(1′), H(3′) hydrogens to the O(a2) terminal oxygen of the surface. This again facilitates a strong bonding interaction between C(1) and O(a1); the transition state will involve a further strengthening of this bond, simultaneous transfer of H(1) towards O(a2), and a return of the C(3)-H(3′) bond to its original strength associated with a "raising of the anchor" on the other molybdenyl oxygen. One can also envisage alternative mechanism in which H(1′) replaces the H(3′) atom in the $-CH_2$ group of the acrolein product, with a simultaneous abstraction of the H(3′) hydrogen by the O(a2) molybdenyl oxygen leading to a formation of the surface hydroxyl.

4. Conclusion

We have intended in this chapter to demonstrate how various computational methods can be applied to study specific catalytic systems. The AIM-resolved CSA have been used to determine gross features of the adsorbate/substrate charge reconstructions due to the chemisorption. Such analysis can be performed on very large surface clusters, thus providing a simple tool for testing the range of perturbations created by the chemisorption, both on the surface and in the supporting layer(s). We have also stressed in our analysis that the consistent, unbiased two-reactant CSA approach is required to adequately predict both the P and CT components of the chemisorption bond. This formalism has enabled us to further investigate the role of the intra-reactant (diagonal) and inter-reactant (off-diagonal) charge reconstruction patterns, and to formulate more general observations concerning their participation in the chemisorption interactions. Finally, we have demonstrated how the collective modes of charge displacements which reflect the adsorbate---substrate interactions, *viz.* the IRM, can help to provide the mode-resolved picture of the inter-reactant CT.

Since CSA-generated atomic charge diagrams may in some cases fail to diagnose the detailed bond weakening/strengthening patterns, we have additionally used the finite-difference bond-order analysis, based upon the KS calculations on small (saturated) chemisorption clusters. The predicted changes in bond-multiplicities, both inside adsorbate/substrate and of the newly formed diatomic chemisorption interactions, offer a sound basis for conjectures concerning the molecular mechanisms of catalytic reactions.

CHAPTER 6
CHARGE SENSITIVITIES IN KOHN-SHAM THEORY

It has already been remarked in Chapters 1-3 that the electronegativity equalization (EE) and the derivative quantities of the CSA originate from the HK DFT. It is the purpose of this chapter to summarize the basic CSA quantities, local and nonlocal, in the standard Kohn-Sham (KS) theory [15], which represents the practical, numerical implementation of the HK DFT, determining the g.s. electronic density, energy and other properties. This problem has been examined in the literature by several researchers [12, 26, 27, 29, 40, 42, 46, 47, 55, 79, 80, 84, 109]. The recent exposition by Cohen [109], specifically addressing the reactivity theory in the global equilibrium state will be used as the basis for the survey presented in Sections 1-5. In Section 6 of this chapter we additionally examine the local charge sensitivities and basic relations in the constrained equilibrium case (multicomponent KS approach).

We shall begin with some rudimentary remarks on the HK DFT and its KS realization, comparing the DFT approach with the conventional (Schrödinger) quantum mechanics. Next, the explicit expressions for the local softness and FF will be examined and the basic kernels identified. The reactivity kernels of the open and closed systems will be summarized and the nuclear reactivity indices, which closely follow their electronic FF analogs, will also be expressed in terms of the basic KS derivative properties. The problems of the softness (hardness) spectrum and the charge stability of molecular systems will also be discussed.

Next, we shall specifically address the two-component description of the reactive system $M^+ = (A^+|B^+)$. We shall use the relevant Euler equations to derive the non-local (kernels and matrices) and local (vectors and functions) quantities, as well as the condensed charge sensitivities of molecular fragments, both in the local and AIM descriptions. The hardness equalization principle in such constrained-equilibrium state will be examined in some detail; the joint transformations between perturbations and responses (in quadratic approximation) of reactive systems will be derived in the local theory and its AIM discretization. We shall also briefly describe the relevant KS scheme for such a two-component description of the polarized reactants, before the inter-reactant CT.

1. Conventional and Density Functional Quantum Mechanics

The basic variational principle of the wavefunction theory in the Born-Oppenheimer approximation ("frozen" external potential $v(\mathbf{r})$ due to the nuclei in their fixed positions, $\mathcal{X} \equiv \{R_\alpha \equiv \vec{R}_\alpha\}$, $\alpha = 1, \ldots, m$, involves the minimization of the system electronic energy $E_v[\Psi] = \langle\Psi|\mathrm{H}|\Psi\rangle$, the expectation value of the hamiltonian $\mathrm{H}(N, v)$ including all terms but the nuclear kinetic energy, subject to the wavefunction normalization constraint (Eq. 2):

$$E_{g.s.} = \min_{\Psi} E_v[\Psi]_{\langle\Psi|\Psi\rangle=1} = E[N, v]$$
$$\text{or} \quad \delta\{E_v[\Psi] - E_{g.s.}(\langle\Psi|\Psi\rangle - 1)\}_{\Psi_{g.s.}} = 0 . \tag{406}$$

It is equivalent to the familiar electronic Schrödinger equation, $H\Psi_{g.s.} = E_{g.s.}\Psi_{g.s.}$, which implies that the local energy, $E_v(\boldsymbol{x}) \equiv H\Psi/\Psi$, is equalized throughout the whole (spin-position) configurational space of electrons, $\boldsymbol{x} \equiv \{\sigma_i, \mathbf{r}_i\} = \{\mathbf{x}_i\}$, $i = 1, ..., N$, at $\Psi = \Psi_{g.s.}$:

$$E_v^{g.s.}(\boldsymbol{x}) \equiv H\Psi_{g.s.}/\Psi_{g.s.} = E_{g.s.} . \tag{407}$$

As demonstrated by Hohenberg and Kohn [14], $E_{g.s.}$ is alternatively the minimum of the energy functional $E_v[\rho] \equiv E[\rho, v] = \int v\rho\, d\mathbf{r} + Q[\rho]$ of the electronic density, $\rho(\mathbf{r}) = \langle\Psi|\hat{\rho}|\Psi\rangle = N\int|\Psi|^2\, d\boldsymbol{x}_1'$, for the constrained number of electrons, $N[\rho] \equiv \int\rho\, d\mathbf{r} = N$ (Eq. 1),

$$E_{g.s.} = \min_{\rho} E_v[\rho]_{N[\rho]=N} = \min_{\rho}\{\min_{\Psi\to\rho} E_v[\Psi]\} = E[N, v]$$
$$\text{or} \quad \delta\{E_v[\rho] - \mu(N[\rho] - N)\}_{\rho_{g.s.}} = 0 ; \tag{408}$$

here $\hat{\rho}(\mathbf{r}) = \Sigma_i\delta(\mathbf{r} - \mathbf{r}_i)$ denotes the density operator, $\boldsymbol{x}_1' = (\sigma_1, \{\boldsymbol{x}_{i>1}\})$, $Q[\rho] = T_e[\rho] + V_{ee}[\rho]$ (Eq. 12), and $\Psi \to \rho$ stands for any trial wavefunction giving the specified density. The chemical potential (electronegativity) equalization (Eq. 6) at the g.s. density immediately follows from the last equation: $\delta E_v[\rho]/\delta\rho(\mathbf{r})|_{\rho_{g.s.}} = \mu = (\partial E[N, v]/\partial N)_v = -\chi$; this equalization principle (in physical space $\mathbf{r}$) replaces the local energy equalization (in the configurational space $\boldsymbol{x}$) of the wavefunction theory.

The problem of a practical implementation of the DFT variational principle has been solved by Kohn and Sham [15], who demonstrated that it can be formulated in terms of the one-electron equations, similar to those of the familiar Hartree and Hartree-Fock approximations. Consider the independent-electron problem resulting from the following partitioning of the electronic energy [*Local Density Approximation* (LDA)]:

$$E_v[\rho] = T_{KS}[\rho] + \int v_H\rho\, d\mathbf{r} + E_{xc}[\rho] . \tag{409}$$

where $T_{KS}[\rho]$ is the kinetic energy of non-interacting electrons, distributed in accordance with the specified density $\rho(\mathbf{r})$, $v_H(\mathbf{r}) = \phi(\mathbf{r})$ denotes the Hartree (electrostatic) potential, $v_H(\mathbf{r}) \equiv v(\mathbf{r}) + \int\rho(\mathbf{r}')/|\mathbf{r} - \mathbf{r}'|\, d\mathbf{r}'$,and $E_{xc}[\rho]$ stands the remaining *exchange-correlation* energy which also includes the relevant correction to $T_{KS}[\rho]$. This partitioning leads to equations determining the optimum spin-orbitals $\{\psi_i\}$ determining the density:

$$\rho(\mathbf{r}) = \sum_i \sum_\sigma n_i\,|\psi_i(\mathbf{x})|^2 = \rho_\alpha(\mathbf{r}) + \rho_\beta(\mathbf{r})\,, \qquad \mathbf{x} = (\mathbf{r}, \sigma)\,,$$

$$\left(-\frac{1}{2}\Delta + v_{KS}\right)\psi_i \equiv \mathrm{H}_{KS}\,\psi_i = \varepsilon_i\,\psi_i\,, \qquad \varepsilon_i \le \varepsilon_{i+1}\,,$$

$$v_{KS}(\mathbf{r}) = v_H(\mathbf{r}) + v_{xc}(\mathbf{r})\,, \quad 0 \le n_i \le 1\,, \qquad \sum_i n_i = N\,; \tag{410}$$

here $(\rho_\alpha, \rho_\beta) \equiv \{\rho_\sigma\}$ stands for the densities for specific spins (spin-densities) and the occupation numbers $\{n_i\}$ are integer (fractional) for nondegenerate (degenerate) ground states. The effective KS potential, $v_{KS}(\mathbf{r})$, includes the exchange-correlation potential, $v_{xc}(\mathbf{r}) = \delta E_{xc}[\rho]/\delta\rho(\mathbf{r})$, as the correction to the classical Hartree potential.

Thus, given sufficient knowledge of $E_{xc}[\rho]$, accurate total energy and other properties, themselves functionals of ρ, can be determined using the KS selfconsistent-field equations. This extremally successful algorithm is as simple as that of the Hartree theory, but it realistically accounts for both exchange and correlation effects, in contrast to the Hartree-Fock theory, which takes into account the exchange exactly and neglects correlation completely.

In the spin-polarized formulation [*Local Spin Density Approximation* (LSDA)] $E_{g.s.} = E_v[\{\rho_\sigma\}] = \int v\rho\, d\mathbf{r} + F[\{\rho_\sigma\}]$, where again the universal functional $F[\{\rho_\sigma\}] = T_e[\{\rho_\sigma\}] + V_{ee}[\{\rho_\sigma\}]$ can be formally defined using the Levy constrained search of Eq. 12, as the infimum of the expectation value of the sum of the electronic kinetic and repulsion energies, with the search being carried out over all wavefunctions (or density operators), which integrate to the prescribed spin densities $\{\rho_\sigma\}$. The effective potentials for specific spins, determining the KS equations for the spinorbitals $\psi_i \equiv \psi_i^\sigma$, $\sigma = (\alpha, \beta)$, are obtained by adding to the Hartree potential the spin-dependent exchange-correlation correction $v_{xc}^\sigma(\mathbf{r}) = \delta E_{xc}[\{\rho_\sigma\}]/\delta\rho_\sigma(\mathbf{r})$:

$$v_{KS}^\sigma(\mathbf{r}) = v_H(\mathbf{r}) + v_{xc}^\sigma(\mathbf{r})\,. \tag{411}$$

In the spin-polarized theory one also defines the local spin-FF (see Eq. 154),

$$f_\sigma(\mathbf{r}) = [\partial\rho_\sigma(\mathbf{r})/\partial N]_v\,, \tag{412}$$

with the spinless FF $f = f_\alpha + f_\beta$.

2. Electronic Softness Quantities in Kohn-Sham Theory

Following Parr and Yang [12, 39, 40] one defines separate local FF, $f^\pm(\mathbf{r})$ for the nucleophilic $(+,\ dN = 1)$ and electrophilic $(-,\ dN = -1)$ attacks, with the radical attack $(r,\ dN = 0)$ local reactivity trends being diagnosed from their arithmetic average: $f^r = (f^+ + f^-)/2$. Obviously, only when one neglects the orbital relaxation due to an addition or removal of an electron, these FF are defined solely by the relevant frontier orbital densities, $i = \{N^0 + 1\ (+,\ \mathrm{LUMO}),\ N^0\ (-,\ \mathrm{HOMO})\}$:

$$\rho_F^{\pm}(\mathbf{r}) = \sum_\sigma |\psi_{\binom{N^0+1}{N^0}}(\mathbf{x})|^2, \tag{413}$$

where N^0 denotes the starting (integer) number of electrons. In a general case, for the non-degenerate g.s.:

$$f^{\pm}(\mathbf{r}) = \rho_F^{\pm}(\mathbf{r}) + \sum_\sigma \sum_{i=1}^{N} n_i \, [\partial |\psi_i(\mathbf{x})|^2 / \partial N]_v^{\pm}, \tag{414}$$

where the upper + (-) index in the relaxational derivatives of the orbital densities denotes the appropriate change in the global number of electrons, i.e., that the derivative with respect to N is taken just above (below) an integer value N^0. The discontinuity at integer values of N arises at T = 0 grand canonical ensemble (see Section 7 of Chapter 2) [48]. It is also linked to the resulting discontinuity in the system chemical potential (Eq. 154). Clearly, this discontinuity also implies two measures of all derivatives involving the chemical potential, e.g., the softnesses (Eq. 186, 187),

$$s^{\pm}(\mathbf{r}) = [\partial \rho(\mathbf{r}) / \partial \mu]_v^{\pm}, \qquad S^{\pm} = \int s^{\pm}(\mathbf{r})\, d\mathbf{r} = (\partial N / \partial \mu)_v^{\pm}, \tag{415}$$

and $f^{\pm}(\mathbf{r}) = s^{\pm}(\mathbf{r}) / S^{\pm}$. The same distinction has to be made in the case of the relevant hardness quantities.

The explicit expressions for the local FF and softness quantities can be conveniently expressed in terms of the frontier density of Eq. 413 and the density of the KS states at the specified chemical potential (Fermi level) μ [40, 109]: $g(\mathbf{r}, E = \mu)$. The electron density $\rho(\mathbf{r})$ is linked to μ through the relation:

$$\rho(\mathbf{r}) = \int^{\mu} g(\mathbf{r}, E)\, dE, \tag{416}$$

which implies

$$s(\mathbf{r}) = g(\mathbf{r}, \mu) + \int^{\mu} [\partial g(\mathbf{r}, E) / \partial \mu]_v \, dE. \tag{417}$$

In the ensemble formulation of the KS theory [12, 16, 40, 142] one introduces the KS Green's function:

$$G(\mathbf{x}, \mathbf{x}'; E^-) \equiv \langle \mathbf{x} | [E^- - H_{KS}]^{-1} | \mathbf{x}' \rangle, \tag{418}$$

where $E^- = E - i\,\delta$, $\delta \to 0^+$. It determines the local density of KS states at the chemical potential μ:

$$g(\mathbf{r}, \mu) = \sum_\sigma \frac{1}{\pi} \operatorname{Im} G(\mathbf{x}, \mathbf{x}; \mu^-). \tag{419}$$

The resulting KS expressions for the local FF and softness quantities from the relevant functional chain rule transformations are:

$$f^{\pm}(\mathbf{r}) = \int \mathrm{K}^{-1}(\mathbf{r}, \mathbf{r}')\, \rho_F^{\pm}(\mathbf{r}')\, d\mathbf{r}' \,, \tag{420}$$

$$s^{\pm}(\mathbf{r}) = \int \mathrm{K}^{-1}(\mathbf{r}, \mathbf{r}')\, g^{\pm}(\mathbf{r}', \mu)\, d\mathbf{r}' \,; \tag{421}$$

the kernel $\mathrm{K}^{-1}(\mathbf{r}, \mathbf{r}')$ entering these expressions is the inverse of

$$\mathrm{K}(\mathbf{r}, \mathbf{r}') = \delta(\mathbf{r} - \mathbf{r}') + \int \mathrm{P}(\mathbf{r}, \mathbf{r}'')\, \mathrm{W}(\mathbf{r}'', \mathbf{r}')\, d\mathbf{r}'' \,, \tag{422}$$

where the effective KS interaction,

$$\mathrm{W}(\mathbf{r}, \mathbf{r}') = \left[\delta v_{KS}(\mathbf{r}) / \delta \rho(\mathbf{r}')\right]_v = |\mathbf{r} - \mathbf{r}'|^{-1} + \partial v_{xc}(\mathbf{r}) / \partial \rho(\mathbf{r}')$$
$$\equiv \mathrm{W}_H(\mathbf{r}, \mathbf{r}') + \mathrm{W}_{xc}(\mathbf{r}, \mathbf{r}') \,, \tag{423}$$

and the KS static polarization propagator is defined as:

$$\mathrm{P}(\mathbf{r}, \mathbf{r}') = -[\delta \rho(\mathbf{r}) / \delta v_{KS}(\mathbf{r}')]_{\mu} = \mathrm{P}(\mathbf{r}', \mathbf{r})$$
$$= -\sum_{\sigma} \sum_{\sigma'} \int^{\mu} \frac{1}{\pi} \operatorname{Im} [\mathrm{G}(\mathbf{x}, \mathbf{x}'; E^-)\, \mathrm{G}(\mathbf{x}', \mathbf{x}; E^-)\, dE] \,. \tag{424}$$

As the second functional derivative $\mathrm{W}_{xc}(\mathbf{r}, \mathbf{r}')$ is symmetric in $\mathbf{r}$ and $\mathbf{r}'$ and of short range, so that asymptotically $\mathrm{W}(\mathbf{r}, \mathbf{r}') \to \mathrm{W}_H(\mathbf{r}, \mathbf{r}')$ (the bare Coulomb interaction).

The kernel $\mathrm{K}(\mathbf{r}, \mathbf{r}')$ is the transpose of $\mathscr{K}(\mathbf{r}, \mathbf{r}')$, the KS potential-response function, $\mathrm{K}(\mathbf{r}, \mathbf{r}') = \mathscr{K}(\mathbf{r}, \mathbf{r}')^T$, where

$$\mathscr{K}(\mathbf{r}, \mathbf{r}') = [\delta v(\mathbf{r}) / \delta v_{KS}(\mathbf{r}')]_{\mu} = \delta(\mathbf{r} - \mathbf{r}') + \int \mathrm{W}(\mathbf{r}, \mathbf{r}'')\, \mathrm{P}(\mathbf{r}'', \mathbf{r}')\, d\mathbf{r}''$$
$$\sim \delta(\mathbf{r} - \mathbf{r}') + \int |\mathbf{r} - \mathbf{r}''|^{-1}\, \mathrm{P}(\mathbf{r}'', \mathbf{r}')\, d\mathbf{r}'' \,. \tag{425}$$

The last asymptotic relation states that $\mathscr{K}(\mathbf{r}, \mathbf{r}')$ is equal to the Hartree-KS static dielectric function.

Thus, the local FF and local softness of Eqs. 420 and 421 are the short-range linear mappings of the frontier orbital density and the local density of states, respectively, with the latter providing the "drivers of local reactivity" [109]. In other words, $f^{\pm}(\mathbf{r})$ is just a screened $\rho_F^{\pm}(\mathbf{r})$ and $s^{\pm}(\mathbf{r})$ is similarly a screened $g^{\pm}(\mathbf{r}', \mu)$.

3. Nuclear Reactivity Indices

As we have already argued in Section 5 of Chapter 4 the electronic and geometric structure properties are closely related by the mapping relations, since any charge displacement creates the corresponding change in forces acting on nuclei, which can be expressed using the familiar Hellmann- Feynman theorem as:

$$\boldsymbol{F}_\alpha = -\partial E[N,\ v(\mathcal{X})]/\partial \boldsymbol{R}_\alpha$$

$$= -\sum_{\beta\neq\alpha}^{m} Z_\alpha Z_\beta \partial(|\boldsymbol{R}_\alpha - \boldsymbol{R}_\beta|^{-1})/\partial \boldsymbol{R}_\alpha + \int \rho(\mathbf{r})\,\partial(Z_\alpha|\boldsymbol{R}_\alpha - \mathbf{r}|^{-1})/\partial \boldsymbol{R}_\alpha\, d\mathbf{r}\,,$$

$$\alpha = 1, 2, \ldots, m\,. \tag{426}$$

These forces represent the "intensive" state variables conjugate to the nuclear position ("extensive") variables of the system Born-Oppenheimer hamiltonian and the g. s. energy of Eq. 406. In what follows we drop the ± superscripts to simplify notation . By analogy to the electronic local FF one can define the nuclear FF [109],

$$\vec{\mathcal{F}}_\alpha = (\partial \boldsymbol{F}_\alpha/\partial N)_v\,, \tag{427}$$

and the nuclear softness,

$$\vec{\Sigma}_\alpha = (\partial \boldsymbol{F}_\alpha/\partial \mu)_v\,, \tag{428}$$

to quantify the g. s. interaction between the electronic (N or μ) and geometric (external potential or $\{\boldsymbol{R}_\alpha\}$) degrees of freedom. These nuclear reactivity indices can be expressed in terms of their electronic analogs [109]:

$$\vec{\mathcal{F}}_\alpha = \int f(\mathbf{r})\,\partial(Z_\alpha|\boldsymbol{R}_\alpha - \mathbf{r}|^{-1})/\partial \boldsymbol{R}_\alpha\, d\mathbf{r}\,, \tag{429}$$

$$\vec{\Sigma}_\alpha = \int s(\mathbf{r})\,\partial(Z_\alpha|\boldsymbol{R}_\alpha - \mathbf{r}|^{-1})/\partial \boldsymbol{R}_\alpha\, d\mathbf{r}\,. \tag{430}$$

These equations can be further expressed in terms of the screened Coulomb force:

$$\boldsymbol{F}^s_\alpha(\mathbf{r}) \equiv \int \mathcal{K}^{-1}(\mathbf{r},\mathbf{r}')\,\partial(Z_\alpha|\boldsymbol{R}_\alpha - \mathbf{r}'|^{-1})/\partial \boldsymbol{R}_\alpha\, d\mathbf{r}'\,, \tag{431}$$

$$\vec{\mathcal{F}}_\alpha = \int \boldsymbol{F}^s_\alpha(\mathbf{r})\,\tilde{\rho}_F(\mathbf{r})\, d\mathbf{r}\,,\quad \vec{\Sigma}_\alpha = \int \boldsymbol{F}^s_\alpha(\mathbf{r})\, g(\mathbf{r},\mu)\, d\mathbf{r}\,, \tag{432}$$

$$\tilde{\rho}_F(\mathbf{r}) = \rho_F(\mathbf{r})\Big/\!\int \mathrm{K}^{-1}(\mathbf{r},\mathbf{r}')\,\rho_F(\mathbf{r})\, d\mathbf{r}\, d\mathbf{r}'\,; \tag{433}$$

here $\tilde{\rho}_F(\mathbf{r})$ denotes the rescaled frontier-orbital density. Therefore, the frontier electron density and the local density of states at the Fermi level also provide the "drivers" for the above nuclear reactivity indices.

4. Softness and Hardness Kernels and Their Spectral Analysis

Similar expressions can be obtained for the nonlocal response properties such as the hardness and softness kernels of Eqs. 183-185, respectively. For example, the softness kernel is represented by the screened KS static polarization propagator $\mathrm{P}(\mathbf{r}, \mathbf{r}')$:

$$\sigma(\mathbf{r}, \mathbf{r}') = \int \mathrm{K}^{-1}(\mathbf{r}, \mathbf{r}'')\, \mathrm{P}(\mathbf{r}'', \mathbf{r}')\, d\mathbf{r}'' = \eta^{-1}(\mathbf{r}, \mathbf{r}') \,. \tag{434}$$

It has been shown [143] that it is useful to examine the eigenvalue problem of these kernels,

$$\int \eta(\mathbf{r}, \mathbf{r}')\, \xi_\nu(\mathbf{r}')\, d\mathbf{r}' = h_\nu\, \xi_\nu(\mathbf{r}) \quad \text{and} \quad \int \sigma(\mathbf{r}, \mathbf{r}')\, \xi_\nu(\mathbf{r}')\, d\mathbf{r}' = s_\nu\, \xi_\nu(\mathbf{r}) \,, \tag{435}$$

where the eigenvalues (principal hardnesses and softnesses) $\{ h_\nu = s_\nu^{-1} \}$ and the complete set of the ortho-normal eigenfunctions $\{ \xi_\nu(\mathbf{r}) \}$ determine the spectral resolution of both these kernels:

$$\eta(\mathbf{r}, \mathbf{r}') = \sum_\nu \xi_\nu^*(\mathbf{r}')\, h_\nu\, \xi_\nu(\mathbf{r}) \quad \text{and} \quad \sigma(\mathbf{r}, \mathbf{r}') = \sum_\nu \xi_\nu^*(\mathbf{r}')\, s_\nu\, \xi_\nu(\mathbf{r}) \,. \tag{436}$$

For stable systems the Le Châtelier principle,

$$\eta(\mathbf{r}, \mathbf{r}) = [\delta^2 E_v[\rho] / \delta\rho(\mathbf{r})\, \delta\rho(\mathbf{r})]_\mu = -[\delta u(\mathbf{r}) / \delta\rho(\mathbf{r})]_\mu > 0 \,, \tag{437}$$

all principal hardnesses must be positive, $\{ h_\nu > 0 \}$, so that the softness spectrum is limited by the eigenvalue s_M of the softest (most reactive) mode $\xi_M(\mathbf{r})$:

$$0 < s_\nu < s_M \,. \tag{438}$$

The inequality of Eq. 437 reads that an increase in density at a given point in the equilibrium state preserving the system chemical potential, imposed by the macroscopic external reservoir, has to be associated with a lowering of the external potential at this point, since $u(\mathbf{r}) = v(\mathbf{r}) - \mu$.

Using Eqs. 194 and 195 and the standard perturbational expansion for the linear response kernel gives:

$$\sigma(\mathbf{r}, \mathbf{r}') = \sum_{n>0} \frac{<0|\hat{\rho}(\mathbf{r})|n><n|\hat{\rho}(\mathbf{r}')|0> + <0|\hat{\rho}(\mathbf{r}')|n><n|\hat{\rho}(\mathbf{r})|0>}{E_n - E_0} + \frac{s(\mathbf{r})\, s(\mathbf{r}')}{S} \,, \tag{439}$$

where the summation is over all excited states of the system, $\Psi_n \equiv |n\rangle$, and $<0|\hat{\rho}(\mathbf{r})|n>$ denotes the matrix element of the density operator involving the g.s., $\Psi_0 = |0\rangle$, and the excited state Ψ_n.

The softness kernel can also be expressed in terms of the inverse of the static dielectric function kernel [109],

$$\epsilon^{-1}(\mathbf{r}, \mathbf{r}') \equiv \delta(\mathbf{r} - \mathbf{r}') - \int W_H(\mathbf{r}, \mathbf{r}'')\, \sigma(\mathbf{r}'', \mathbf{r}')\, d\mathbf{r}'' , \tag{440}$$

or in short notation: $\epsilon^{-1} = 1 - W_H \sigma$. Solving it for σ gives:

$$\sigma = W_H^{-1} [1 - \epsilon^{-1}] , \tag{441}$$

where $W_H^{-1}(\mathbf{r}, \mathbf{r}') = -(1/4\pi)\, \nabla^2 \delta(\mathbf{r} - \mathbf{r}')$.

Following Cohen *et al.* [109] one can also define the nuclear softness kernel:

$$\phi_\alpha(\mathbf{r}) = -\delta F_\alpha / \delta u(\mathbf{r}) = \partial [\int Z_\alpha |\boldsymbol{R}_\alpha - \mathbf{r}'|^{-1} \sigma(\mathbf{r}', \mathbf{r})\, d\mathbf{r}'] / \partial \boldsymbol{R}_\alpha$$

$$= \int \boldsymbol{F}_\alpha^{s}(\mathbf{r}')\, \mathrm{P}(\mathbf{r}', \mathbf{r})\, d\mathbf{r}' . \tag{442}$$

Therefore, as we have already observed in Section 5 of Chapter 4, the linear response kernel provides also the driving force for nuclear reactivity. Finally, expressing $\phi_\alpha(\mathbf{r})$ in terms of $\epsilon^{-1}(\mathbf{r}, \mathbf{r}')$ gives:

$$\phi_\alpha(\mathbf{r}) = Z_\alpha\, \partial [\delta(\boldsymbol{R}_\alpha - \mathbf{r}) - \epsilon^{-1}(\boldsymbol{R}_\alpha, \mathbf{r})] / \partial \boldsymbol{R}_\alpha . \tag{443}$$

5. Isoelectronic Kernels

In the closed systems, for which the fixed N constraint applies, the isoelectronic (internal, i) softness kernel $\sigma^{i}(\mathbf{r}, \mathbf{r}')$ is defined by the negative linear response kernel $\beta(\mathbf{r}, \mathbf{r}')$ of Eqs. 193, $\sigma^{i}(\mathbf{r}, \mathbf{r}') = -\beta(\mathbf{r}, \mathbf{r}')$, and 194 (see also Eq. 215). Its eigenvalue problem,

$$-\int \beta(\mathbf{r}, \mathbf{r}')\, \xi^{i}_{\nu}(\mathbf{r}')\, d\mathbf{r}' = \sigma^{i}_{\nu}\, \xi^{i}_{\nu}(\mathbf{r}) , \quad \int \xi^{i}_{\nu}(\mathbf{r})\, d\mathbf{r} = 0 , \quad \forall \nu , \tag{444}$$

determines the local internal normal modes and the associated eigenvalues (see Eq. 323). The above property of $\xi^{i}_{\nu}(\mathbf{r})$ eigenfunctions immediately follows from integrating the preceding eigenvalue equation and using the closed-system property of $\beta(\mathbf{r}, \mathbf{r}')$: $\int \beta(\mathbf{r}, \mathbf{r}')\, d\mathbf{r} = \int \beta(\mathbf{r}, \mathbf{r}')\, d\mathbf{r}' = 0$. Clearly, the internal and external softness kernels do not commute (see Eq. 194) and therefore have different sets of eigensolutions.

In this internal spectral resolution the isoelectronic softness and hardness kernels are defined by the following diagonal quadratic forms:

$$\sigma^{i}(\mathbf{r}, \mathbf{r}') = \sum_{\nu} \xi^{i*}_{\nu}(\mathbf{r}')\, \sigma^{i}_{\nu}\, \xi^{i}_{\nu}(\mathbf{r}) , \tag{445a}$$

$$\eta^{i}(\mathbf{r}, \mathbf{r}') = \sum_{\nu} \xi_{\nu}^{i*}(\mathbf{r}') (\mathscr{s}_{\nu}^{i})^{-1} \xi_{\nu}^{i}(\mathbf{r}) \ . \tag{445b}$$

Again, the internal stability criterion requires [143]:

$$\eta^{i}(\mathbf{r}, \mathbf{r}) = [\delta^{2}E_{v}[\rho]/\delta\rho(\mathbf{r})\,\delta\rho(\mathbf{r})]_{N} = -[\delta v(\mathbf{r})/\delta\rho(\mathbf{r})]_{N} > 0 \ . \tag{446}$$

Hence, a local increase in density for constant N must be accompanied by a lowering of the external potential in this point. This internal Le Châtelier principle requires that in a stable system all internal principal softnesses must be positive and limited from above by the eigenvalue $\mathscr{s}_{M}^{i}$ of the softest internal normal mode $\xi_{M}^{i}(\mathbf{r})$:

$$0 < \mathscr{s}_{\nu}^{i} < \mathscr{s}_{M}^{i} \ . \tag{447}$$

It can be demonstrated using the standard orbital perturbation theory that indeed all these principal internal softnesses are positive [143].

It should be observed that, due to the N = const. constraint, both the local and global internal "softnesses", obtained by the kernel partial and total integrations, respectively, must vanish identically: $s^{i}(\mathbf{r}) = \int \sigma^{i}(\mathbf{r}, \mathbf{r}')\,d\mathbf{r}' = 0$, $S^{i} = \int s^{i}(\mathbf{r})\,d\mathbf{r} = 0$. Thus, the closed systems can also be considered as infinitely hard open systems.

Some of the previous chain-rule transformations can be easily generalized to the isoelectronic case. For example, replacing the external softness kernel in Eq. 440 by its internal analog gives:

$$\epsilon_{i}^{-1} = 1 - W_{H}\,\sigma^{i} \ , \qquad [\sigma^{i}]^{-1} = W_{H}^{-1}[1 - \epsilon_{i}^{-1}] \ . \tag{448}$$

Also, changing the constraint μ into N in the definitions of the KS static polarization propagator and the KS potential response kernel of Eqs. 422, 424, and 425, respectively defines their internal analogs:

$$P_{i}(\mathbf{r}, \mathbf{r}') = [\delta\rho(\mathbf{r})/\delta v_{KS}(\mathbf{r}')]_{N} \ , \tag{449}$$

$$\mathscr{K}_{i}(\mathbf{r}, \mathbf{r}') = [\delta v(\mathbf{r})/\delta v_{KS}(\mathbf{r}')]_{N} = K_{i}^{T}(\mathbf{r}, \mathbf{r}') \ . \tag{450}$$

In terms of these quantities (see Eq. 434):

$$\sigma^{i}(\mathbf{r}, \mathbf{r}') = \int K_{i}^{-1}(\mathbf{r}, \mathbf{r}'')\,P_{i}(\mathbf{r}'', \mathbf{r}')\,d\mathbf{r}'' \ . \tag{451}$$

6. Charge Sensitivities in the Constrained Equilibrium State

6.1. Electronegativity/Chemical Potential Equalization Equations

Let us again consider the reactive system $M^+ = (A^+|B^+)$ consisting of the acidic reactant A and the basic reactant B, before the B → A CT. The equilibrium state in this reactant partitioning represents a particular case of the constrained equilibrium of Fig. 1b. Such reactant perspective is of great importance for the theory of chemical reactivity, the main objective of which is to diagnose reactivity preferences from the isolated reactant properties in $M^0 = A^0 + B^0$, and a simplified representation of their interaction in M^+ representing the mutually closed, polarized reactants in the presence of each other.

This partitioning implies that we are dealing with the two component system, defined by two distinct densities (or the AIM populational vectors): $\rho_A(\mathbf{r})$, $\rho_B(\mathbf{r})$ (or $\boldsymbol{N}_A$, $\boldsymbol{N}_B$, in the AIM resolution). Therefore, in such a two-component description [55, 144], one can similarly separate the external potential and universal parts of the energy density functional (see Eq. 4):

$$E^{L}_{v_A, v_B}[\rho_A|\rho_B] = \int [\rho_A(\mathbf{r})\, v_A(\mathbf{r}) + \rho_B(\mathbf{r})\, v_B(\mathbf{r})]\, d\mathbf{r} + F[\rho_A|\rho_B] \,, \tag{452a}$$

where $F[\rho_A|\rho_B]$ generates the following part of the electronic energy (see Eq. 5):

$$F[\rho_A|\rho_B] = \min_{\Psi_A \to \rho_A;\, \Psi_B \to \rho_B} \langle \Psi | T_e + V_{ee} | \Psi \rangle \,; \tag{453}$$

the above formal constrained-search is over all wavefunctions of reactants, represented by the antisymmetrized product of the reactant wavefunctions [145], $\Psi = \mathcal{A}(\Psi_A \Psi_B)$, which integrate to their prescribed densities. In the constrained equilibrium case the energy function of Eq. 452a can also be written as a functional of the reactant external potentials and their global electron populations, since they uniquely define the corresponding densities, $\overset{+}{\rho}_X = \overset{+}{\rho}_X[N_A, v_A|N_B, v_B]$:

$$E^{L}_{v_A, v_B}[\overset{+}{\rho}_A|\overset{+}{\rho}_B] \equiv E[N_A, v_A|N_B, v_B] \,. \tag{452b}$$

Supplementing $E^{L}_{v_A, v_B}[\rho_A|\rho_B]$ with the relevant constraints on the numbers of electrons in each subsystems defines the auxiliary energy density functional,

$$\mathscr{E}^{L}_{v_A, v_B}[\rho_A|\rho_B] = E^{L}_{v_A, v_B}[\rho_A|\rho_B] - \overset{+}{\mu}_A\left(N_A[\rho_A] - N_A^0\right) - \overset{+}{\mu}_B\left(N_B[\rho_B] - N_B^0\right), \tag{454}$$

leading to the system of coupled Euler-Lagrange equations of the intra-reactant chemical potential equalizations at the constrained equilibrium densities $\{\overset{+}{\rho}_X(\mathbf{r})\}$, X = A, B:

$$\overset{+}{\mu}_A = v_A(\mathbf{r}) + \delta F[\rho_A|\rho_B]/\delta \rho_A(\mathbf{r})\big|_{\{\overset{+}{\rho}_X\}} \equiv \overset{+}{\mu}_A(\mathbf{r}) \,, \tag{455a}$$

$$\overset{+}{\mu}_B = v_B(\mathbf{r}) + \delta F[\rho_A|\rho_B]/\delta \rho_B(\mathbf{r})\big|_{\{\overset{+}{\rho}_X\}} \equiv \overset{+}{\mu}_B(\mathbf{r}) \,. \tag{455b}$$

They replace the global chemical potential/electronegativity equalization Eq. 182, valid for the unconstrained equilibrium case of Fig. 1a. It follows from these equations that the vector $\boldsymbol{u}^{(A|B)} \equiv \{u_X(\mathbf{r}) = v_X(\mathbf{r}) - \mu_X^+\} \equiv \boldsymbol{v}^{(A,|B)} - \boldsymbol{\mu}_{(A|B)}$ of the relative external potentials is uniquelly defined by the equilibrium densities of the polarized reactants $\boldsymbol{\rho}^{(A|B)} \equiv \{\rho_X^+\}$. It should be observed, that in the limit of separated reactants $M^0(\infty) = A^0 + B^0$: $v_X^0(\mathbf{r}) = -\sum_x^X Z_x / |\boldsymbol{R}_x - \mathbf{r}|$, while at the finite separations $v_A(\mathbf{r}) = v_B(\mathbf{r}) = \sum_X v_X^0(\mathbf{r}) = v_{AB}(\mathbf{r})$.

6.2. *Hardness (Interaction) and Softness (Response) Quantities*

Equations 455 define the first differentials of the relative external potentials:

$$du_A(\mathbf{r}) = dv_A(\mathbf{r}) - d\mu_A^+ = -\int \left[d\rho_A(\mathbf{r}')\,\eta^{A,A}(\mathbf{r}',\mathbf{r}) + d\rho_B(\mathbf{r}')\,\eta^{B,A}(\mathbf{r}',\mathbf{r})\right] d\mathbf{r}', \tag{456a}$$

$$du_B(\mathbf{r}) = dv_B(\mathbf{r}) - d\mu_B^+ = -\int \left[d\rho_A(\mathbf{r}')\,\eta^{A,B}(\mathbf{r}',\mathbf{r}) + d\rho_B(\mathbf{r}')\,\eta^{B,B}(\mathbf{r}',\mathbf{r})\right] d\mathbf{r}', \tag{456b}$$

where the reactant-resolved hardness kernel matrix,

$$\boldsymbol{\eta}^{(A|B)}(\mathbf{r}',\mathbf{r}) = \left\{ \eta^{X,Y}(\mathbf{r}',\mathbf{r}) \equiv \frac{\delta^2 F[\rho_A|\rho_B]}{\delta\rho_X(\mathbf{r}')\,\delta\rho_Y(\mathbf{r})} = -\frac{\delta u_Y(\mathbf{r})}{\delta\rho_X(\mathbf{r}')}, \quad (X,Y) = (A,B) \right\}. \tag{457}$$

Equations 456 can be combined into the single matrix equation:

$$d\boldsymbol{u}^{(A|B)}(\mathbf{r}) = d\boldsymbol{v}^{(A|B)}(\mathbf{r}) - d\boldsymbol{\mu}_{(A|B)} = -\int d\boldsymbol{\rho}^{(A|B)}(\mathbf{r}')\,\boldsymbol{\eta}^{(A|B)}(\mathbf{r}',\mathbf{r})\,d\mathbf{r}'. \tag{456}$$

The inverse of the hardness kernel matrix defines the softness kernel in M^+,

$$\boldsymbol{\sigma}^{(A|B)}(\mathbf{r}',\mathbf{r}) = \{\sigma^{Y,X}(\mathbf{r}',\mathbf{r}) = -\delta\rho_X(\mathbf{r})/\delta u_Y(\mathbf{r}')\} = [\boldsymbol{\eta}^{(A|B)}(\mathbf{r}',\mathbf{r})]^{-1}, \tag{458}$$

satisfying the generalized reciprocity relation (see Eq. 184):

$$\int \boldsymbol{\eta}^{(A|B)}(\mathbf{r}',\mathbf{r})\,\boldsymbol{\sigma}^{(A|B)}(\mathbf{r},\mathbf{r}'')\,d\mathbf{r} = \sum_Y \int \eta^{X,Y}(\mathbf{r}',\mathbf{r})\,\sigma^{Y,Z}(\mathbf{r},\mathbf{r}'')\,d\mathbf{r}$$
$$= \{\delta_{X,Z}\,\delta(\mathbf{r}'-\mathbf{r}'')\}, \quad (X,Y,Z) = (A,B). \tag{459}$$

One similarly defines the FF matrix and the linear response kernel matrix for M^+:

$$\mathbf{f}^{(A|B)}(\mathbf{r}) = \left\{ f^{X,Y}(\mathbf{r}) = \frac{\partial}{\partial N_X}\left(\frac{\delta E[N_A, v_A | N_B, v_B]}{\delta v_Y(\mathbf{r})}\right)\right.$$

$$\left. = \left(\frac{\partial \overset{+}{\rho}_Y(\mathbf{r})}{\partial N_X}\right)_{N_{Y(\neq X)},\, v_A,\, v_B} = \left(\frac{\delta \overset{+}{\mu}_X}{\delta v_Y(\mathbf{r})}\right)_{N_A,\, N_B,\, v_{X(\neq Y)}} \right\}, \tag{460}$$

$$\boldsymbol{\beta}^{(A|B)}(\mathbf{r}',\mathbf{r}) = \left\{ \beta^{X,Y}(\mathbf{r}',\mathbf{r}) = \left(\frac{\delta^2 E[N_A, v_A | N_B, v_B]}{\delta v_X(\mathbf{r}')\,\delta v_Y(\mathbf{r})}\right)_{N_A,\, N_B}\right.$$

$$\left. = \left(\frac{\delta \overset{+}{\rho}_Y(\mathbf{r})}{\delta v_X(\mathbf{r}')}\right)_{N_A,\, N_B,\, v_{X(\neq Y)}} \right\}, \quad (X, Y) = (A, B)\,. \tag{461}$$

In this section we have slightly modified the notation of Section 8 in Chapter 2, to clearly indicate the assumed reactant division of the system under consideration. More specifically, the previous vectors $\boldsymbol{N}$ and $\boldsymbol{\mu}^+$ of the global quantities of reactants are now denoted as $\boldsymbol{N}_{(A|B)}$ and $\boldsymbol{\mu}_{(A|B)}$, respectively. Similarly, the vectors of local functions, $\boldsymbol{\rho}^+(\mathbf{r})$ and $\boldsymbol{v}(\mathbf{r})$ defined in Chapter 2 are now represented by the function vectors $\boldsymbol{\rho}^{(A|B)}(\mathbf{r})$ and $\boldsymbol{v}^{(A|B)}(\mathbf{r})$, accordingly. Finally, the non-local matrix kernels of Chapter 2, $\boldsymbol{\beta}(\mathbf{r}',\mathbf{r}) = \{\beta_{X,Y}(\mathbf{r}',\mathbf{r})\}$ and $\mathbf{f}(\mathbf{r}) = \{f_{X,Y}(\mathbf{r})\}$ are now replaced by $\boldsymbol{\beta}^{(A|B)}(\mathbf{r}',\mathbf{r}) = \{\beta^{X,Y}(\mathbf{r}',\mathbf{r})\}$ and $\mathbf{f}^{(A|B)}(\mathbf{r}) = \{f^{X,Y}(\mathbf{r})\}$.

It is of interest to find how these response quantities are related to the softness kernel matrix of Eq. 458. In a search for these generalizations of Eqs. 194 and 195 we follow the same strategy as that used in Chapter 3 for the global (unconstrained) equilibrium state. Consider the differentials of the reactant densities, unique functionals of u_A and u_B (see Eqs. 455 and 458):

$$d\,\boldsymbol{\rho}^{(A|B)}(\mathbf{r}) \equiv [d\,\overset{+}{\rho}_A(\mathbf{r}), d\,\overset{+}{\rho}_B(\mathbf{r})] = -\int d\,\boldsymbol{u}^{(A|B)}(\mathbf{r}')\,\boldsymbol{\sigma}^{(A|B)}(\mathbf{r}',\mathbf{r})\,d\mathbf{r}'$$

$$= -\int d\,\boldsymbol{v}^{(A|B)}(\mathbf{r}')\,\boldsymbol{\sigma}^{(A|B)}(\mathbf{r}',\mathbf{r})\,d\mathbf{r}' + d\boldsymbol{\mu}_{(A|B)}\,\mathbf{s}^{(A|B)}(\mathbf{r})\,, \tag{462}$$

where the local softness matrix (see Eq. 186):

$$\mathbf{s}^{(A|B)}(\mathbf{r}) = \left\{ s^{X,Y}(\mathbf{r}) = [\partial \rho_Y(\mathbf{r}) / \partial \overset{+}{\mu}_X]_{v_A,\, v_B} = \int \sigma^{X,Y}(\mathbf{r}',\mathbf{r})\,d\mathbf{r}' \right\}. \tag{463}$$

Writing $d\overset{+}{\mu}_X$ (of du_X) as $d\overset{+}{\mu}_X[N_A, v_A | N_B, v_B]$ gives:

$$d\boldsymbol{\mu}_{(A|B)} \equiv [d\mu_A^{+}, d\mu_B^{+}] = d\boldsymbol{N}_{(A|B)}\, \boldsymbol{\eta}_{(A|B)} + \int d\boldsymbol{v}^{(A|B)}(\mathbf{r})\, [\mathbf{f}^{(A|B)}(\mathbf{r})]^{T}\, d\mathbf{r}\,, \quad (464)$$

where $d\boldsymbol{N}_{(A|B)} \equiv [dN_A, dN_B] = \int d\boldsymbol{\rho}^{(A|B)}(\mathbf{r})\, d\mathbf{r}$ and the condensed hardness matrix is defined in the usual way (see Eq. 172):

$$\boldsymbol{\eta}_{(A|B)} = \left(\frac{\partial \boldsymbol{\mu}_{(A|B)}}{\partial \boldsymbol{N}_{(A|B)}}\right)_{\boldsymbol{v}^{(A|B)}}$$

$$= \left\{ \eta^{M}_{X,Y} = \left(\frac{\partial^2 E[N_A, v_A \mid N_B, v_B]}{\partial N_X\, \partial N_Y}\right)_{v_A,\, v_B} = \left(\frac{\partial \mu_Y^{+}}{\partial N_X}\right)_{N_Y,\, v_A,\, v_B} \right\}. \quad (465)$$

One can also express this matrix in terms of the hardness kernel matrix of Eq. 457 and the FF matrix of Eq. 460, using the following chain-rule transformation:

$$\eta^{M}_{X,Y} = \sum_{C}^{(A,B)} \sum_{D}^{(A,B)} \iint \left(\frac{\partial \rho_C(\mathbf{r}')}{\partial N_X}\right)\left(\frac{\delta^2 F[\rho_A \mid \rho_B]}{\delta \rho_C(\mathbf{r}')\, \delta \rho_D(\mathbf{r})}\right)\left(\frac{\partial \rho_D(\mathbf{r})}{\partial N_Y}\right) d\mathbf{r}'\, d\mathbf{r}$$

$$= \sum_{C}^{(A,B)} \sum_{D}^{(A,B)} \iint f^{X,C}(\mathbf{r}')\, \eta^{C,D}(\mathbf{r}', \mathbf{r})\, f^{Y,D}(\mathbf{r})\, d\mathbf{r}'\, d\mathbf{r}$$

$$\equiv \iint \{\mathbf{f}^{(A|B)}(\mathbf{r}')\, \boldsymbol{\eta}^{(A|B)}(\mathbf{r}', \mathbf{r})\, [\mathbf{f}^{(A|B)}(\mathbf{r})]^{T}\}^{X,Y}\, d\mathbf{r}'\, d\mathbf{r}\,. \quad (466)$$

Substituting Eq. 464 into Eq. 462 gives:

$$d\boldsymbol{\rho}^{(A|B)}(\mathbf{r}) = d\boldsymbol{N}_{(A|B)}\, \boldsymbol{\eta}_{(A|B)}\, \mathbf{s}^{(A|B)}(\mathbf{r})$$

$$+ \int [d\boldsymbol{v}^{(A|B)}(\mathbf{r}')]\, \{-\boldsymbol{\sigma}^{(A|B)}(\mathbf{r}', \mathbf{r}) + [\mathbf{f}^{(A|B)}(\mathbf{r}')]^{T}\, \mathbf{s}^{(A|B)}(\mathbf{r})\}\, d\mathbf{r}'\,. \quad (467)$$

We can independently write the differentials of $d\rho_X^{+}[N_A, v_A \mid N_B, v_B]$:

$$d\boldsymbol{\rho}^{(A|B)}(\mathbf{r}) = d\boldsymbol{N}_{(A|B)}\, \mathbf{f}^{(A|B)}(\mathbf{r}) + \int d\boldsymbol{v}^{(A|B)}(\mathbf{r}')\, \boldsymbol{\beta}^{(A|B)}(\mathbf{r}', \mathbf{r})\, d\mathbf{r}'\,. \quad (468)$$

A comparison between the last two equations gives the generalized relations of Eqs. 194 and 195:

$$\boldsymbol{\beta}^{(A|B)}(\mathbf{r}', \mathbf{r}) = -\boldsymbol{\sigma}^{(A|B)}(\mathbf{r}', \mathbf{r}) + [\mathbf{f}^{(A|B)}(\mathbf{r}')]^{T}\, \mathbf{s}^{(A|B)}(\mathbf{r})\,, \quad (469)$$

$$\mathbf{f}^{(A|B)}(\mathbf{r}) = \boldsymbol{\eta}_{(A|B)}\, \mathbf{s}^{(A|B)}(\mathbf{r}) \quad \text{or} \quad \mathbf{s}^{(A|B)}(\mathbf{r}) = \mathbf{S}_{(A|B)}\, \mathbf{f}^{(A|B)}(\mathbf{r})\,, \quad (470)$$

where $\mathbf{S}_{(A|B)} = \{ S^M_{X,Y} = \partial N_Y / \partial \mu^+_X \ , \ (X,Y) = (A,B) \} = \boldsymbol{\eta}^{-1}_{(A|B)}$. We would like to observe that the above quantities satisfy the following sum rules:

$$\int f^{X,Y}(\mathbf{r})\, d\mathbf{r} = \delta_{X,Y} \,, \qquad \int \beta^{X,Y}(\mathbf{r},\mathbf{r}')\, d\mathbf{r} = \int \beta^{X,Y}(\mathbf{r},\mathbf{r}')\, d\mathbf{r}' = 0 \,. \tag{471}$$

Finally, the explicit joint local transformation of the "perturbations" $\mathfrak{P}(A|B) = [dN_A, dN_B, d v_A, dv_B]$ into responses $\mathfrak{R}(A|B) = [d\mu^+_A, d\mu^+_B, d\rho^+_A(\mathbf{r}), d\rho^+_B(\mathbf{r})]$ is (see Eqs. 464 and 468):

$$[d\mu^+_A, d\mu^+_B, d\rho^+_A(\mathbf{r}), d\rho^+_B(\mathbf{r})] = [dN_A, dN_B, \int d\mathbf{r}'\, dv_A(\mathbf{r}'), \int d\mathbf{r}'\, dv_B(\mathbf{r}')]$$
$$\times \begin{pmatrix} \eta_{A,A} & \eta_{A,B} & f^{A,A}(\mathbf{r}) & f^{A,B}(\mathbf{r}) \\ \eta_{B,A} & \eta_{B,B} & f^{B,A}(\mathbf{r}) & f^{B,B}(\mathbf{r}) \\ f^{A,A}(\mathbf{r}') & f^{B,A}(\mathbf{r}') & \beta^{A,A}(\mathbf{r}',\mathbf{r}) & \beta^{A,B}(\mathbf{r}',\mathbf{r}) \\ f^{A,B}(\mathbf{r}') & f^{B,B}(\mathbf{r}') & \beta^{B,A}(\mathbf{r}',\mathbf{r}) & \beta^{B,B}(\mathbf{r}',\mathbf{r}) \end{pmatrix} \tag{472}$$

or in the matrix form:

$$[d\boldsymbol{\mu}_{(A|B)}, d\boldsymbol{\rho}^{(A|B)}(\mathbf{r})] = [d\mathbf{N}_{(A|B)}, \int d\mathbf{r}'\, d\boldsymbol{v}^{(A|B)}(\mathbf{r}')]$$
$$\times \begin{pmatrix} \boldsymbol{\eta}_{(A|B)} & \mathbf{f}^{(A|B)}(\mathbf{r}) \\ [\mathbf{f}^{(A|B)}(\mathbf{r}')]^T & \boldsymbol{\beta}^{(A|B)}(\mathbf{r}',\mathbf{r}) \end{pmatrix} ; \tag{473}$$

in short: $\mathfrak{R}(A|B) = \mathfrak{P}(A|B)\, \mathscr{H}^{-1}(A|B) \equiv \mathfrak{P}(A|B)\, \mathscr{S}(A|B)$, where $\mathscr{S}(A|B)$ is the generalized softness kernel matrix for M^+. This local-resolution equation is analogous to Eq. 225 of the AIM description. The inverse transformation (see Eq. 456) is:

$$[dN_A, dN_B, d v_A(\mathbf{r}), dv_B(\mathbf{r})] = [d\mu^+_A, d\mu^+_B, \int d\mathbf{r}'\, d\rho^+_A(\mathbf{r}'), \int d\mathbf{r}'\, d\rho^+_B(\mathbf{r}')]$$
$$\times \begin{pmatrix} 0 & 0 & 1 & 0 \\ 0 & 0 & 0 & 1 \\ 1 & 0 & -\eta^{A,A}(\mathbf{r}',\mathbf{r}) & -\eta^{A,B}(\mathbf{r}',\mathbf{r}) \\ 0 & 1 & -\eta^{B,A}(\mathbf{r}',\mathbf{r}) & -\eta^{B,B}(\mathbf{r}',\mathbf{r}) \end{pmatrix} \tag{474}$$

or in short: $\mathfrak{P}(A|B) = \mathfrak{R}(A|B)\, \mathscr{H}(A|B)$, where $\mathscr{H}(A|B)$ is the generalized hardness kernel matrix for M^+. This equation, which can be written in a more compact matrix form,

$$[d\boldsymbol{N}_{(A|B)}, d\boldsymbol{v}^{(A|B)}(\mathbf{r})] = [d\boldsymbol{\mu}_{(A|B)}, \int d\mathbf{r}'\, d\boldsymbol{\rho}^{(A|B)}(\mathbf{r}')] \times \begin{pmatrix} \mathbf{0}_{(A|B)} & \mathbf{I}_{(A|B)} \\ \mathbf{I}_{(A|B)} & -\boldsymbol{\eta}^{(A|B)}(\mathbf{r}',\mathbf{r}) \end{pmatrix}, \quad (475)$$

where $\mathbf{0}_{(A|B)}$ and $\mathbf{I}_{(A|B)}$ matrices are the 2×2 zero and identity matrices, respectively, constitutes the local generalization of Eq. 224.

It can be verified that these two transformations indeed satisfy the following reciprocity relation:

$$\begin{pmatrix} \boldsymbol{\eta}_{(A|B)} & \int d\mathbf{r}\, \mathbf{f}^{(A|B)}(\mathbf{r}) \\ [\mathbf{f}^{(A|B)}(\mathbf{r}')]^T & \int d\mathbf{r}\, \boldsymbol{\beta}^{(A|B)}(\mathbf{r}',\mathbf{r}) \end{pmatrix} \begin{pmatrix} \mathbf{0}_{(A|B)} & \mathbf{I}_{(A|B)} \\ \mathbf{I}_{(A|B)} & -\boldsymbol{\eta}^{(A|B)}(\mathbf{r},\mathbf{r}'') \end{pmatrix} = \begin{pmatrix} \mathbf{I}_{(A|B)} & \mathbf{0}_{(A|B)} \\ \mathbf{0}_{(A|B)} & \mathbf{I}_{(A|B)}\delta(\mathbf{r}''-\mathbf{r}) \end{pmatrix}. \quad (476)$$

For example, the first diagonal block in the product matrix kernel, given by the identity matrix, immediately follows from the normalization of the FF (Eq. 471). The second diagonal block of the Dirac delta functions follows from Eqs. 459 and 469:

$$\begin{aligned} &[\mathbf{f}^{(A|B)}(\mathbf{r}')]^T - \int \boldsymbol{\beta}^{(A|B)}(\mathbf{r}',\mathbf{r})\, \boldsymbol{\eta}^{(A|B)}(\mathbf{r},\mathbf{r}'')\, d\mathbf{r} \\ &= [\mathbf{f}^{(A|B)}(\mathbf{r}')]^T - \int \{-\boldsymbol{\sigma}^{(A|B)}(\mathbf{r}',\mathbf{r}) + [\mathbf{f}^{(A|B)}(\mathbf{r}')]^T\, \mathbf{s}^{(A|B)}(\mathbf{r})\}\, \boldsymbol{\eta}^{(A|B)}(\mathbf{r},\mathbf{r}'')\, d\mathbf{r} \\ &= [\mathbf{f}^{(A|B)}(\mathbf{r}')]^T + \delta(\mathbf{r}''-\mathbf{r}')\, \mathbf{I}_{(A|B)} - [\mathbf{f}^{(A|B)}(\mathbf{r}')]^T \iint \boldsymbol{\sigma}^{(A|B)}(\mathbf{r}',\mathbf{r})\, \boldsymbol{\eta}^{(A|B)}(\mathbf{r},\mathbf{r}'')\, d\mathbf{r}\, d\mathbf{r}' \\ &= \delta(\mathbf{r}''-\mathbf{r}')\, \mathbf{I}_{(A|B)}. \end{aligned}$$

Similarly, using Eqs. 459 and 470, one verifies that the (1, 2)-block in the product of Eq. 476 vanishes:

$$\boldsymbol{\eta}_{(A|B)}\, \mathbf{I}_{(A|B)} - \boldsymbol{\eta}_{(A|B)} \int \boldsymbol{\sigma}^{(A|B)}(\mathbf{r}',\mathbf{r})\, \boldsymbol{\eta}^{(A|B)}(\mathbf{r},\mathbf{r}'')\, d\mathbf{r}\, d\mathbf{r}' = \mathbf{0}_{(A|B)}.$$

Finally, the vanishing of the (2, 1)-block in the r.h.s. of Eq. 476 directly follows from the sum-rule of Eq. 471 for the linear response kernel matrix.

6.3. Equalization of the Local Hardnesses in the Constrained Equilibrium State of $M^+ = (A^+|B^+)$

It has been demonstrated in Eq. 198 (see also Appendix A) that the local hardness $\eta(\mathbf{r}) = -[\partial u(\mathbf{r})/\partial N]_v = \int f(\mathbf{r}')\,\eta(\mathbf{r}',\mathbf{r})\,d\mathbf{r}'$ is equalized throughout the space in the global equilibrium state (Fig. 1a). It is of interest for the theory of chemical reactivity to formulate an analogous principle valid for the constrained equilibrium state (Fig. 1b) in $M^+ = (A^+|B^+)$, defined by the intra-reactant Euler (electronegativity equalization) Eqs. 455.

It immediately follows from the above definition of the local hardness that the general local hardness matrix element $\eta^{X,Y}(\mathbf{r}) = \{\partial/\partial N_X[\delta E/\delta\rho_Y(\mathbf{r})]\}_{v_{AB}} = [\partial\mu_Y^+(\mathbf{r})/\partial N_X]_{v_{AB}} = \{\delta/\delta\rho_Y(\mathbf{r})[\partial E/\partial N_X]\}_{v_{AB}} = [\delta\mu_X^+/\delta\rho_Y(\mathbf{r})]_{v_{AB}}$ is indeed equalized throughout the space at the condensed hardness matrix element value:

$$\eta^{X,Y}(\mathbf{r}) = -[\partial u_Y(\mathbf{r})/\partial N_X]_{v_{AB}} = (\partial\mu_Y^+/\partial N_X)_{v_{AB}} = \eta^M_{X,Y}$$

$$= -\sum_C^{(A,B)} \int [\partial\rho_C(\mathbf{r}')/\partial N_X]_{v_{AB}}\,[\delta u_Y(\mathbf{r})/\delta\rho_C(\mathbf{r}')]_{v_{AB}}\,d\mathbf{r}'$$

$$= \sum_C^{(A,B)} \int f^{X,C}(\mathbf{r}')\,\eta^{C,Y}(\mathbf{r}',\mathbf{r})\,d\mathbf{r}' = \int [\mathbf{f}^{(A|B)}(\mathbf{r}')\,\boldsymbol{\eta}^{(A|B)}(\mathbf{r}',\mathbf{r})]^{X,Y}\,d\mathbf{r}'$$

$$(X,Y) = (A,B)\,. \tag{477}$$

Also, by the Maxwell relation:

$$\eta^{X,Y}(\mathbf{r}) = [\delta\mu_X^+/\delta\rho_Y(\mathbf{r})]_{v_{AB}} = \sum_C^{(A,B)} [\delta N_C/\delta\rho_Y(\mathbf{r})]\,[\partial\mu_X^+/\partial N_C]_{v_{AB}} = \eta^M_{Y,X}\,. \tag{478}$$

The symmetrical character of the condensed hardness tensor is also assured by the alternative expression of the local hardness:

$$\int [\boldsymbol{\eta}^{(A|B)}(\mathbf{r},\mathbf{r}')\,\mathbf{f}^{(A|B)}(\mathbf{r}')]^{X,Y}\,d\mathbf{r}' = \sum_C^{(A,B)} \int \eta^{X,C}(\mathbf{r},\mathbf{r}')\,f^{C,Y}(\mathbf{r}')\,d\mathbf{r}'$$

$$= -\sum_C^{(A,B)} \int [\delta u_C(\mathbf{r}')/\delta\rho_X(\mathbf{r})]_{v_{AB}}\,[\partial\rho_Y(\mathbf{r}')/\partial N_C]_{v_{AB}}\,d\mathbf{r}'$$

$$= \sum_C^{(A,B)} [\delta\mu_C^+/\delta\rho_X(\mathbf{r})]_{v_{AB}} \int [\partial\rho_Y(\mathbf{r}')/\partial N_C]_{v_{AB}}\,d\mathbf{r}'$$

$$= \eta^{Y,X}(\mathbf{r}) = \eta^M_{X,Y} = \eta^M_{Y,X}\,. \tag{479}$$

Both definitions of the condensed hardness matrix in M^+ (Eqs. 466 and 477) are consistent due to the above equalization rule and the FF normalization:

$$\sum_{C}^{(A,B)} \sum_{D}^{(A,B)} \int \left[\int f^{X,C}(\mathbf{r}')\, \eta^{C,D}(\mathbf{r}',\mathbf{r})\, d\mathbf{r}' \right] f^{Y,D}(\mathbf{r})\, d\mathbf{r}$$

$$= \sum_{D}^{(A,B)} \eta^{M}_{X,Y} \int f^{Y,D}(\mathbf{r})\, d\mathbf{r} = \sum_{C}^{(A,B)} \left[\int f^{X,C}(\mathbf{r}')\, d\mathbf{r}' \right] \eta^{M}_{C,Y} = \eta^{M}_{X,Y} . \qquad (480)$$

6.4. AIM Discretization

In the AIM-resolution kernels are replaced by the corresponding matrices (see Chapter 3). The energy function of Eq. 452b now becomes the function of the AIM population and external potential vectors of both reactants , $\boldsymbol{N}^{(A|B)} = (\boldsymbol{N}_A, \boldsymbol{N}_B)$ and $\boldsymbol{v}^{(A|B)} = (\boldsymbol{v}_A, \boldsymbol{v}_B)$ (see Eq. 216) :

$$E^{AIM}[\boldsymbol{N}_A, \boldsymbol{v}_A \,|\, \boldsymbol{N}_B, \boldsymbol{v}_B] = \sum_a^A N_a v_a + \sum_b^B N_b v_b + F^{AIM}[\boldsymbol{N}_A \,|\, \boldsymbol{N}_B] . \qquad (481)$$

When supplemented with the constraints on the number of electrons in each subsystem (see Eq. 454) it defines the corresponding auxiliary function giving rise to the following Euler (EE) equations,

$$\overset{+}{\mu}_A \boldsymbol{1}_A = \boldsymbol{v}_A + \left.\frac{\partial F[\boldsymbol{N}_A \,|\, \boldsymbol{N}_B]}{\partial \boldsymbol{N}_A}\right|_{\boldsymbol{N}_{(A|B)}}, \quad \overset{+}{\mu}_B \boldsymbol{1}_B = \boldsymbol{v}_B + \left.\frac{\partial F[\boldsymbol{N}_A \,|\, \boldsymbol{N}_B]}{\partial \boldsymbol{N}_B}\right|_{\boldsymbol{N}_{(A|B)}}, \qquad (482a)$$

or in short,

$$\partial F[\boldsymbol{N}_A \,|\, \boldsymbol{N}_B] / \partial \boldsymbol{N}_X \big|_{\boldsymbol{N}_{(A|B)}} = -\overset{+}{\boldsymbol{u}}_X \equiv \boldsymbol{v}^X - \overset{+}{\mu}_X \boldsymbol{1}_X , \quad X = A, B . \qquad (482b)$$

The FF matrix $\mathbf{f}^{(A|B)} \equiv \partial \boldsymbol{N}^{(A|B)} / \partial \boldsymbol{N}_{(A|B)} = \{ \boldsymbol{f}^{X,Y} = \partial \boldsymbol{N}_Y / \partial N_X = \partial \mu_X / \partial \boldsymbol{v}_Y$, $(X, Y) = (A, B) \}$ (see Eq. 154) replaces in the AIM-resolution the matrix function of Eq. 460. The rigid AIM hardness matrix $\boldsymbol{\eta}^{(A|B)} = \{ \boldsymbol{\eta}^{X,Y} = -\partial \boldsymbol{u}_Y / \partial \boldsymbol{N}_X, (X, Y) = (A, B) \}$ (Eq. 181) defines the softness matrix in this reactant resolution: $\boldsymbol{\sigma}^{(A|B)} = \{ \boldsymbol{\sigma}^{X,Y} = -\partial \boldsymbol{N}_Y / \partial \boldsymbol{u}_X$, $(X, Y) = (A, B) \} = [\boldsymbol{\eta}^{(A|B)}]^{-1}$, and the associated matrices of local, $\mathbf{s}^{(A|B)} = \{ \boldsymbol{s}^{X,Y} = \partial \boldsymbol{N}_Y / \partial \mu_X = \{ s_y^{X,Y} \equiv \sum_x^X \sigma_{x,y}^{X,Y} \}$, $(X, Y) = (A, B) \}$, and the condensed, $\mathbf{S}_{(A|B)} = \{ \sigma^M_{X,Y} = \partial N_Y / \partial \mu_X = \sum_y^Y s_y^{X,Y}$, $(X, Y) = (A, B) \}$, softness matrix in the reactant resolution. Similarly, the reactant resolved blocks of the linear response matrix, $\boldsymbol{\beta}^{(A|B)} = \{ \boldsymbol{\beta}^{X,Y} = (\partial \boldsymbol{N}_Y / \partial \boldsymbol{v}_X)_{N_X, N_Y}$, $(X, Y) = (A, B) \}$, with the off-diagonal elements defining responses in the AIM electron populations of one fragment to displacements in the external potential variables of another fragment, under the closed reactants constraints.

In terms of these AIM matrices the relations of Eqs. 469 and 470 become (see Eq. 229):

$$\boldsymbol{\beta}^{(A|B)} = -\boldsymbol{\sigma}^{(A|B)} + [\mathbf{f}^{(A|B)}]^T \mathbf{s}^{(A|B)} , \qquad (483)$$

$$\mathbf{f}^{(A|B)} = \boldsymbol{\eta}_{(A|B)}\, \mathbf{s}^{(A|B)}\,, \qquad \text{or} \qquad \mathbf{s}^{(A|B)} = \mathbf{S}_{(A|B)}\, \mathbf{f}^{(A|B)}\,, \tag{484}$$

where $\boldsymbol{\eta}_{(A|B)} = (\mathbf{S}_{(A|B)})^{-1}$ has the same meaning as in Eqs. 466 and 228:

$$\eta^{M}_{X,Y} = \sum_{C}^{(A,B)} \sum_{D}^{(A,B)} \sum_{c}^{C} \sum_{d}^{D} \left(\frac{\partial N_c}{\partial N_X}\right)\left(\frac{\partial^2 F[\boldsymbol{N}_A|\boldsymbol{N}_B]}{\partial N_c\, \partial N_d}\right)\left(\frac{\partial N_d}{\partial N_Y}\right)$$

$$= \sum_{C}^{(A,B)} \sum_{D}^{(A,B)} \sum_{c}^{C} \sum_{d}^{D} f_c^{X,C}\, \eta_{c,d}^{C,D}\, f_d^{Y,D} \equiv \{\mathbf{f}^{(A|B)}\, \boldsymbol{\eta}^{(A|B)}\, [\mathbf{f}^{(A|B)}]^T\}^{X,Y}\,, \tag{485}$$

and the sum rules of Eq. 471 now read:

$$\sum_{y}^{Y} f_y^{X,Y} = \delta_{X,Y}\,, \qquad \sum_{y}^{Y} \beta_{x,y}^{X,Y} = \sum_{x}^{X} \beta_{x,y}^{X,Y} = 0\,. \tag{486}$$

It should be observed that Eqs. 483 and 484 give Eq. 213:

$$\boldsymbol{\beta}^{(A|B)} = -\boldsymbol{\sigma}^{(A|B)} + [\mathbf{f}^{(A|B)}]^T\, \mathbf{S}_{(A|B)}\, \mathbf{f}^{(A|B)}\,. \tag{483a}$$

The local hardness equalization rule of the preceding section also holds in the AIM discretization:

$$\eta_x^{X,Y} = \frac{\partial}{\partial N_x}\left(\frac{\partial E}{\partial N_Y}\right) = \left(\frac{\partial \mu_Y^{+}}{\partial N_x}\right)_{\boldsymbol{v}^{(A|B)}} = \sum_{C}^{(A,B)} \left(\frac{\partial N_C}{\partial N_x}\right)_{\boldsymbol{v}^{(A|B)}} \left(\frac{\partial \mu_Y^{+}}{\partial N_C}\right)_{\boldsymbol{v}^{(A|B)}} = \eta^{M}_{X,Y}$$

$$= \frac{\partial}{\partial N_Y}\left(\frac{\partial E}{\partial N_x}\right) = \left(\frac{\partial \mu_x^{+}}{\partial N_Y}\right)_{\boldsymbol{v}^{(A|B)}} = \left(\frac{\partial \mu_X^{+}}{\partial N_Y}\right)_{\boldsymbol{v}^{(A|B)}} = \eta^{M}_{Y,X}\,, \qquad (X,Y) = (A,B)\,. \tag{487}$$

Clearly, these local hardnesses are simple functions of the FF indices:

$$\eta_X^{X,Y} = (\partial\mu_Y^{+}/\partial \boldsymbol{N}_X) = \sum_{C}^{(A,B)} (\partial \boldsymbol{N}_C/\partial N_X)\,(\partial\mu_Y^{+}/\partial \boldsymbol{N}_C) = \sum_{C}^{(A,B)} \boldsymbol{f}^{X,C}\, \boldsymbol{\eta}^{C,Y}$$

$$= \mathbf{1}_X\, \eta^{M}_{X,Y}\,, \qquad \text{or} \qquad [(\mathbf{f}^{(A|B)})\, \boldsymbol{\eta}^{(A|B)}]^{X,Y} = \mathbf{1}_X\, \eta^{M}_{X,Y}\,, \tag{488a}$$

$$\eta_X^{Y,X} = (\partial\boldsymbol{\mu}_X/\partial N_Y) = \sum_{C}^{(A,B)} (\partial \boldsymbol{N}_C/\partial N_Y)\,(\partial\boldsymbol{\mu}_X/\partial \boldsymbol{N}_C) = \eta^{M}_{Y,X}\, \mathbf{1}_X$$

$$\text{or} \qquad [\mathbf{f}^{(A|B)}\, \boldsymbol{\eta}^{(A|B)}]^{Y,X} = \eta^{M}_{Y,X}\, \mathbf{1}_X\,. \tag{488b}$$

In the AIM description the explicit transformations between the perturbation and response vectors are:

$$[dN_A, dN_B, d\boldsymbol{v}_A, d\boldsymbol{v}_B] = [d\mu_A^+, d\mu_B^+, d\boldsymbol{N}_A^+, d\boldsymbol{N}_B^+]\begin{pmatrix} 0 & 0 & \boldsymbol{1}_A & \boldsymbol{0}_B \\ 0 & 0 & \boldsymbol{0}_A & \boldsymbol{1}_B \\ \boldsymbol{1}_A^T & \boldsymbol{0}_A^T & -\boldsymbol{\eta}^{A,A} & -\boldsymbol{\eta}^{A,B} \\ \boldsymbol{0}_B^T & \boldsymbol{1}_B^T & -\boldsymbol{\eta}^{B,A} & -\boldsymbol{\eta}^{B,B} \end{pmatrix}, \quad (489)$$

$$[d\mu_A^+, d\mu_B^+, d\boldsymbol{N}_A^+, d\boldsymbol{N}_B^+] = [dN_A, dN_B, d\boldsymbol{v}_A, d\boldsymbol{v}_B]\begin{pmatrix} \eta_{A,A} & \eta_{A,B} & \boldsymbol{f}^{A,A} & \boldsymbol{f}^{A,B} \\ \eta_{B,A} & \eta_{B,B} & \boldsymbol{f}^{B,A} & \boldsymbol{f}^{B,B} \\ (\boldsymbol{f}^{A,A})^T & (\boldsymbol{f}^{B,A})^T & \boldsymbol{\beta}^{A,A} & \boldsymbol{\beta}^{A,B} \\ (\boldsymbol{f}^{A,B})^T & (\boldsymbol{f}^{B,B})^T & \boldsymbol{\beta}^{B,A} & \boldsymbol{\beta}^{B,B} \end{pmatrix} \cdot (490)$$

It can be easily verified that these transformations do indeed satisfy the reciprocity relation:

$$\begin{pmatrix} 0 & 0 & \boldsymbol{1}_A & \boldsymbol{0}_B \\ 0 & 0 & \boldsymbol{0}_A & \boldsymbol{1}_B \\ \boldsymbol{1}_A^T & \boldsymbol{0}_A^T & -\boldsymbol{\eta}^{A,A} & -\boldsymbol{\eta}^{A,B} \\ \boldsymbol{0}_B^T & \boldsymbol{1}_B^T & -\boldsymbol{\eta}^{B,A} & -\boldsymbol{\eta}^{B,B} \end{pmatrix}\begin{pmatrix} \eta_{A,A} & \eta_{A,B} & \boldsymbol{f}^{A,A} & \boldsymbol{f}^{A,B} \\ \eta_{B,A} & \eta_{B,B} & \boldsymbol{f}^{B,A} & \boldsymbol{f}^{B,B} \\ (\boldsymbol{f}^{A,A})^T & (\boldsymbol{f}^{B,A})^T & \boldsymbol{\beta}^{A,A} & \boldsymbol{\beta}^{A,B} \\ (\boldsymbol{f}^{A,B})^T & \boldsymbol{f}^{B,B})^T & \boldsymbol{\beta}^{B,A} & \boldsymbol{\beta}^{B,B} \end{pmatrix}$$

$$= \begin{pmatrix} 1 & 0 & \boldsymbol{0}_A & \boldsymbol{0}_B \\ 0 & 1 & \boldsymbol{0}_A & \boldsymbol{0}_B \\ \boldsymbol{0}_A^T & \boldsymbol{0}_A^T & \mathbf{I}_A & \mathbf{0}^{A,B} \\ \boldsymbol{0}_B^T & \boldsymbol{0}_B^T & \mathbf{0}^{B,A} & \mathbf{I}_B \end{pmatrix}, \quad (491)$$

where $\mathbf{I}_X$ denotes the identity matrix of reactant X, dimension of which determines the number of its constituent atoms. Again, the first two diagonal elements of the product matrix directly follow from the FF normalization. The next two diagonal (identity) blocks follow from Eqs. 483, 484, 487, and 488, e.g.:

$$\mathbf{1}_A^T \mathbf{f}^{A,A} - [\boldsymbol{\eta}^{(A|B)}\,\boldsymbol{\beta}^{(A|B)}]^{A,A}$$

$$= \mathbf{1}_A^T \mathbf{f}^{A,A} + [\boldsymbol{\eta}^{(A|B)}\,\boldsymbol{\sigma}^{(A|B)}]^{A,A} - \{\boldsymbol{\eta}^{(A|B)}\,[\mathbf{f}^{(A|B)}]^T\,\mathbf{S}_{(A|B)}\,\mathbf{f}^{(A|B)}\}^{A,A}$$

$$= \mathbf{1}_A^T \mathbf{f}^{A,A} + \mathbf{I}_A - \left\{\begin{pmatrix} \mathbf{1}_A^T & \mathbf{0}_A^T \\ \mathbf{0}_B^T & \mathbf{1}_B^T \end{pmatrix} \boldsymbol{\eta}_{(A|B)}\,\mathbf{S}_{(A|B)}\,\mathbf{f}^{(A|B)}\right\}^{A,A} = \mathbf{I}_A\,.$$

6.5. *Kohn-Sham-Type Scheme for Subsystems*

We would like to conclude this section with a comment on a possible Kohn-Sham-type algorithm for determining the optimum polarized reactant densities and the polarization energy in the reactive system, in the iterative, self-consistent procedure similar to the variation theory for subsystems of Klessinger and McWeeny [145].

Let us first partition the energy function of Eq. 452,

$$\begin{aligned} E_{v_{AB}}[\rho_A|\rho_B] &= \int v_{AB}(\mathbf{r})\,\rho_{AB}(\mathbf{r})\,d\mathbf{r} + J[\rho_{AB}] + T_S[\rho_{AB}] + E_{xc}[\rho_{AB}] \\ &= E^A_{v_{AB}}[\rho_A] + E^B_{v_{AB}}[\rho_B] + J^{AB}[\rho_A|\rho_B] + E^{AB}_{xc}[\rho_A|\rho_B] \\ &= E^A_{\rm eff}[\rho_A|\rho_B] + E^B_{v_{AB}}[\rho_B] = E^A_{v_{AB}}[\rho_A] + E^B_{\rm eff}[\rho_A|\rho_B]\,, \end{aligned} \qquad (492)$$

where $\rho_{AB}(\mathbf{r}) = \rho_A(\mathbf{r}) + \rho_B(\mathbf{r})$, $T_S[\rho_{AB}] = T_S[\rho_A] + T_S[\rho_B]$ is the KS kinetic energy functional of non-interacting electrons moving in the effective one-body potential $v_{KS}(\mathbf{r}) = \delta E_{XC}[\rho_{AB}]/\delta\rho_{AB}(\mathbf{r})$, and:

$$J[\rho] = \frac{1}{2}\iint \frac{\rho(\mathbf{r})\,\rho(\mathbf{r}')}{|\mathbf{r}-\mathbf{r}'|}\,d\mathbf{r}\,d\mathbf{r}'\,,$$

$$J^{AB}[\rho_A|\rho_B] = J[\rho_{AB}] - J[\rho_A] - J[\rho_B] = \iint \frac{\rho_A(\mathbf{r})\,\rho_B(\mathbf{r}')}{|\mathbf{r}-\mathbf{r}'|}\,d\mathbf{r}\,d\mathbf{r}'\,,$$

$$E^{AB}_{xc}[\rho_A|\rho_B] = E_{xc}[\rho_{AB}] - E_{xc}[\rho_A] - E_{xc}[\rho_B]\,.$$

Here the reactant energy $E^X_{v_{AB}}[\rho_X]$ denotes the energy of the X - group electrons only, while $E^X_{\rm eff}[\rho_A|\rho_B]$ is the energy of the X - group electrons in the presence of those belonging to the other group, in the field of all nuclei. The reactant densities are generated by the respective spin-orbitals $\{\psi_i^X\}$, $X = (A, B)$, of the group wavefunctions

$\Phi^X = \det|\psi_1^X, \psi_2^X, \dots, \psi_{N_X}^X|$ defining the system wavefunction $\Psi^{(A|B)} = \mathscr{N}\mathscr{A}[\Phi^A\Phi^B]$; here the subsystem (reactant) wavefunctions are strong-orthogonal and normalized, so that the antisymmetrizer $\mathscr{A}$ and the corresponding normalization constant $\mathscr{N}$ are associated with transpositions that exchange electron variables between two groups A and B. The group wavefunctions are constructed from the disjoint subsets of orthogonal reactant spin-orbitals. The inter-reactant orthogonality will keep the reactant densities effectively localized in the group of atoms they describe. The optimum KS densities of polarized reactants result from the condition of the stationary value of the effective group energy functional $E_{eff}^X[\rho_A|\rho_B]$ subject to the normalization constraint $\langle \Phi^X|\Phi^X\rangle = 1$. The resulting system of coupled KS-equations defining the optimum spin-orbitals for subsystem X = (A, B) are:

$$\rho_X(\mathbf{r}) = \sum_i^X \sum_\sigma n_i^X |\psi_i^X(\mathbf{r},\sigma)|^2 \equiv \rho_\alpha^X(\mathbf{r}) + \rho_\beta^X(\mathbf{r}), \quad 0 \le n_i^X \le 1, \quad \sum_i n_i^X = N_X,$$

$$(-\frac{1}{2}\Delta + v_{KS})\,\psi_i^X \equiv \mathrm{H}_{KS}\,\psi_i^X = \varepsilon_i^X\,\psi_i^X,$$

$$v_{KS}(\mathbf{r}) = \phi(\mathbf{r}) + v_{xc}(\mathbf{r}), \qquad \phi(\mathbf{r}) = v_{AB}(\mathbf{r}) + \int d\mathbf{r}'\,\rho_{AB}(\mathbf{r}')/|\mathbf{r}-\mathbf{r}'|,$$

$$v_{xc}(\mathbf{r}) = \frac{\delta E_{xc}^X[\rho_X]}{\delta\rho_X(\mathbf{r})} + \frac{\delta E_{xc}^{AB}[\rho_A|\rho_B]}{\delta\rho_X(\mathbf{r})} = \frac{\delta E_{xc}[\rho_{AB}]}{\delta\rho_{AB}(\mathbf{r})}; \tag{493}$$

here $\phi(\mathbf{r})$ is the Hartree (electrostatic) potential due to nuclei and electrons of both reactants, $\rho_{\alpha(\beta)}^X(\mathbf{r})$ denotes the density of the spin up (down) electrons in reactant X, and n_i^X stands for the orbital occupation number. Therefore, the change in the external potential of reactant X, relative to the separated reactant value, i.e., due to the presence of the other reactant Y is:

$$dv_X(\mathbf{r}) = \left(v_X(\mathbf{r}) + \phi_Y(\mathbf{r}) + \frac{\delta E_{xc}[\rho_Y]}{\delta\rho_Y(\mathbf{r})}\right) - v_X(\mathbf{r}) = \phi_Y(\mathbf{r}) + \frac{\delta E_{xc}[\rho_Y]}{\delta\rho_Y(\mathbf{r})}.$$

It includes the electrostatic potential due to the other subsystem and the exchange-correlation correction introduced in the reactive system by the presence of the Y-group electrons. The variational determination of the optimum group densities (KS orbitals) may be solved iteratively in much the same way as the variation wavefunction theory for subsystems [145]. From the initial group densities the KS Hamiltonian is constructed and its elements in the assumed basis set are calculated. Each group KS determinant is then optimized, and after improving wavefunctions of both subsystems, the cycle is repeated. Iteration is continued until densities (energies) in two successive cycles agree to any prescribed accuracy.

Let us conclude this section with a few remarks on the local chemical potentials and the hardness kernel matrix of the mutually closed reactants. Taking the functional derivative of the energy function of Eq. 494 with respect to the reactant density gives:

$$\frac{\delta E_{v_{AB}}[\rho_A|\rho_B]}{\delta\rho_X(\mathbf{r})} = \frac{\delta E_{eff}^{X}[\rho_A|\rho_B]}{\delta\rho_X(\mathbf{r})} = \frac{\delta T_S[\rho_X]}{\delta\rho_X(\mathbf{r})} + v_{KS}(\mathbf{r}) .$$

Therefore, the reactant local chemical potentials differ due to the kinetic energy contributions. Similarly, the second functional differentiation with respect to the local reactant densities gives rise to the corresponding expressions for the diagonal and off-diagonal hardness kernels of Eq. 457:

$$\eta^{X,X}(\mathbf{r}',\mathbf{r}) = \frac{\delta^2 T_S[\rho_X]}{\delta\rho_X(\mathbf{r}')\,\delta\rho_X(\mathbf{r})} + \frac{\delta v_{KS}(\mathbf{r})}{\delta\rho_{AB}(\mathbf{r}')} , \qquad \eta^{X,Y}(\mathbf{r}',\mathbf{r}) = \frac{\delta v_{KS}(\mathbf{r})}{\delta\rho_{AB}(\mathbf{r}')} .$$

Thus, also the diagonal and off-diagonal kernels differ by the kinetic energy contributions. The latter can be expressed in terms of the corresponding derivatives with respect to the KS orbitals, or they can be directly calculated from the approximate, explicit density functionals for the kinetic energy.

7. Conclusion

The above expressions in principle provide a framework for calculating the local charge sensitivities and related reactivity indices from the single- and two-component KS calculations, provided the sufficiently accurate density functionals are known. Also the hierarchy of the basic "driving" quantities behind various reactivity criteria has been established.

The KS theory can also be used to obtain approximate expressions for the local quantities in the global equilibrium case, and it can serve as a basis for defining the corresponding orbitally resolved indices, which we shall briefly discuss in the next chapter.

For example, following Gázquez *et al.* [42] one can expand spin densities $\{\rho_\sigma\}$ around the charge density from the spin-non-polarized KS calculations:

$$\rho_\alpha(\mathbf{r}) = \frac{1}{2}\,\rho(\mathbf{r})\Big|_{N_\alpha=N_\beta} + \frac{1}{2}(N_\alpha - N_\beta)\left.\frac{\partial\rho(\mathbf{r})}{\partial N}\right|_{N_\alpha=N_\beta} + \frac{1}{8}(N_\alpha - N_\beta)^2\left.\frac{\partial^2\rho(\mathbf{r})}{\partial N^2}\right|_{N_\alpha=N_\beta} + \ldots , \qquad (494)$$

$$\rho_\beta(\mathbf{r}) = \frac{1}{2}\,\rho(\mathbf{r})\,|_{N_\alpha = N_\beta} - \frac{1}{2}\,(N_\alpha - N_\beta)\,\frac{\partial\rho(\mathbf{r})}{\partial N}\bigg|_{N_\alpha = N_\beta}$$

$$+ \frac{1}{8}\,(N_\alpha - N_\beta)^2\,\frac{\partial^2\rho(\mathbf{r})}{\partial N^2}\bigg|_{N_\alpha = N_\beta} + \ldots, \qquad (495)$$

where N_σ is the global number of electrons corresponding to the spin $\sigma = [\alpha$(spin-up), β(spin-down)]. It should be observed that the first-order terms in these second-order Taylor expansions are defined by the FF of the spin-nonpolarized theory. By subtracting these two equations one establishes the following simple relation (correct up to second-order) between the local FF and the spin-density difference:

$$f(\mathbf{r}) = [\rho_\alpha(\mathbf{r}) - \rho_\beta(\mathbf{r})]\,/\,(N_\alpha - N_\beta)\,. \qquad (496)$$

This equation allows one to calculate $f(\mathbf{r})$ from the single spin-polarized KS calculations. Such an approach has generated the atomic FF plots exhibiting small negative values in the inner (chemically non-reactive) region of the atom, very close to the nucleus, and the large positive values in intermediate and outer (chemically reactive) regions of the atom.

CHAPTER 7

ELEMENTS OF THE ORBITALLY-RESOLVED CSA

This chapter deals specifically with the molecular orbital discretization of the CSA [30, 31, 59], in which the orbital electron occupation/CBO quantities provide the populational variables. As in the AIM and local descriptions we first examine the relevant energy expressions as well as the occupation-unconstrained populational gradients and Hessians within both the spin-restricted and spin-unrestricted Hartree-Fock and Kohn-Sham approaches. We then briefly summarize the idempotency-constrained (internal) energy and its populational/CBO derivatives, which enforce the occupational limits of molecular (spin) orbitals, at the final (equilibrium) charge distribution.

As before in the AIM and local descriptions we separately examine the global and constrained equilibrium cases in the two-reactant systems $M^* = (A^* \mid B^*)$ and $M^+ = (A^+ | B^+)$, respectively. The latter case is particularly important for defining the polarized reactants in reactive systems and to use their electronic structure and sensitivity characteristics to predict the gross features of both the intra- and inter-reactant charge flows, when a hypothetical barrier for the B → A CT is lifted. This process of reaching the global (inter-reactant) equilibrium from the initial constrained (intra-reactant) equilibrium state involves both the relaxation of orbitals and displacements in the reactant orbital occupations. The algorithms for determining the approximate equilibrium distribution in the quadratic approximation is briefly summarized. A given loop of the iterative self-consistent process of reaching the equilibrium density may be organized in familiar two steps: first, determining the optimum electron occupations for the current shapes of orbitals, and second - determining new orbitals for the changed occupations. A use of the (S)MO-resolved charge sensitivities in such SCF procedure is proposed.

1. Populational Variables and Energy Gradients

The orbital resolution provides a chemically interesting level of the CSA approach, intermediate between the local and AIM descriptions. In this representation of the canonical SCF or KS-DFT molecular orbitals (MO-resolution) the orbital occupation numbers, $\boldsymbol{n}^{\varphi} = (n_1^{\varphi}, \ldots, n_s^{\varphi})$, are considered as the system populational variables. For the "frozen" shapes of the one-particle functions (orbitals) $\boldsymbol{\varphi} = \{\varphi_i\}$ of the simplest, one-determinantal wave-function, the ground-state energy $E[N, v] = E_v[\rho] = E[\boldsymbol{n}^{\varphi}[N, v], \boldsymbol{\varphi}[N, v]]$ can be treated as a function of orbital occupations $\boldsymbol{n}^{\varphi}$ [30]:

$$E[\boldsymbol{n}^{\varphi}[N, v], \boldsymbol{\varphi}[N, v]]\,|_{\boldsymbol{\varphi}} \equiv E_{\varphi}(\boldsymbol{n}^{\varphi}) \,, \tag{497}$$

since $\rho(\mathbf{r})$ and $\rho^{\sigma}(\mathbf{r})$ are diagonal quadratic forms in terms of the canonical MO and KS orbitals (see Eq. 410):

$$\rho(\mathbf{r}) = \sum_i n_i^{\varphi} \varphi_i^*(\mathbf{r})\, \varphi_i(\mathbf{r}) \equiv \sum_i n_i^{\varphi}\, \Omega_{i,i}^{\varphi}(\mathbf{r}) \equiv \sum_i \rho_i^{\varphi}(\mathbf{r}) \,, \tag{498a}$$

$$\rho^{\sigma}(\mathbf{r}) = \sum\nolimits_{i}^{\sigma} n_i^{\varphi,\sigma}\, \varphi_i^{\sigma *}(\mathbf{r})\, \varphi_i^{\sigma}(\mathbf{r}) \equiv \sum\nolimits_{i}^{\sigma} n_i^{\varphi,\sigma}\, \Omega_{i,i}^{\varphi,\sigma}(\mathbf{r}) \equiv \sum\nolimits_{i}^{\sigma} \rho_i^{\sigma}(\mathbf{r})\,, \tag{498b}$$

where the vector $\boldsymbol{n}^{\varphi,\sigma} = \{n_i^{\varphi,\sigma}\}$ denotes the occupations of the canonical $\boldsymbol{\varphi}^{\sigma} = \{\varphi_i^{\sigma}\}$ spin-orbitals (SO).

The spin-non-polarized energy function of Eq. 497 can be naturally extended into the spin-polarized energy function, e.g., in the unrestricted HF (UHF) and the spin-polarized KS (LSDA) theories (see Appendix E):

$$E[N, v] = E_v[\{\rho_\sigma\}] = E[\{\boldsymbol{n}^{\varphi,\sigma}[N, v]\}, \{\boldsymbol{\varphi}^{\sigma}[N, v]\}]\,, \tag{499a}$$

$$E[\{\boldsymbol{n}^{\varphi,\sigma}[N, v]\}, \{\boldsymbol{\varphi}^{\sigma}[N, v]\}]\,|_{\boldsymbol{\varphi}} \equiv E_{\varphi}(\{\boldsymbol{n}^{\varphi,\sigma}\}). \tag{499b}$$

In the open-shell cases *the average energy of configuration* expression [145] can be adopted to obtain these functional dependencies and the corresponding populational derivatives [30, 31, 146, 147].

This dependence quite naturally appears in the KS DFT in the context of the Janak theorem [148]:

$$\partial E_{KS}[\{n_i\}, \{\psi_i\}]/\partial n_i = \varepsilon_i = \langle \psi_i | \mathrm{H}_{KS} | \psi_i \rangle \quad \text{or} \quad \partial E_{\psi}^{KS}/\partial \boldsymbol{n}^{\psi} = \boldsymbol{e}^{\psi}\,, \tag{500}$$

where E_{KS} is the KS electronic energy expressed as the function of the occupations $\boldsymbol{n}^{\psi} \equiv \{n_i\}$ of the KS orbitals $\boldsymbol{\psi} \equiv \{\psi_i\}$ corresponding to the eigenvalues $\boldsymbol{e}^{\psi} = \{\varepsilon_i\}$ (see Section 1 of Chapter 6). Clearly, a qualitatively similar result is obtained in the Hartree-Fock (HF) theory for the fixed shapes of the canonical HF MO, $\boldsymbol{\phi} \equiv \{\phi_i\}$:

$$\partial E_{HF}[\{n_i^{\phi}\}, \{\phi_i\}]/\partial n_i^{\phi} = e_i = \langle \phi_i | \mathrm{F} | \phi_i \rangle \quad \text{or} \quad \partial E_{\phi}^{HF}/\partial \boldsymbol{n}^{\phi} = \boldsymbol{e}^{\phi}\,; \tag{501}$$

here the vectors $\boldsymbol{e}^{\phi} = (e_1, \ldots, e_s)$ and $\boldsymbol{n}^{\phi} = (n_1^{\phi}, \ldots, n_s^{\phi})$, group the HF canonical MO energies and occupations, respectively, and F denotes the familiar Fock operator. In what follows we shall often use the joint short notation:

$$\partial E_{\varphi}(\boldsymbol{n}^{\varphi})/\partial \boldsymbol{n}^{\varphi} = \boldsymbol{e}^{\varphi} \quad (\text{RHF, LDA}),$$

$$\partial E_{\varphi}(\{\boldsymbol{n}^{\varphi,\sigma}\})/\partial \boldsymbol{n}^{\varphi,\sigma} = \boldsymbol{e}^{\varphi,\sigma} \quad (\text{UHF, LSDA})\,, \tag{502}$$

to cover both the HF(SCF) (Eq. 501) and KS(DFT) (Eq. 500) cases; here the usual abbreviations of the familiar wavefunction and DFT methods: RHF (*Restricted HF*), UHF (*Unrestricted HF*, see Appendix E), LDA (*Local Density Approximation*), and LSDA (*Local Spin-Density Approximation*) have been used.

Clearly, in a general orbital representation, $\boldsymbol{\lambda} = \{\lambda_i\}$, e.g., atomic canonical and hybrid orbitals, etc., the density cannot be expressed as the diagonal form of Eq. (498), being instead represented by the non-diagonal form:

$$\rho(\mathbf{r}) = \sum_i \sum_j \mathrm{P}^{\lambda}_{i,j}\, \lambda_i^*(\mathbf{r})\, \lambda_j(\mathbf{r}) \equiv \sum_i \sum_j \mathrm{P}^{\lambda}_{i,j}\, \Omega^{\lambda}_{i,j}(\mathbf{r}) , \tag{503a}$$

$$\rho^{\sigma}(\mathbf{r}) = \sum_i^{\sigma} \sum_j^{\sigma} \mathrm{P}^{\lambda,\sigma}_{i,j}\, \lambda_i^{\sigma\,*}(\mathbf{r})\, \lambda_j^{\sigma}(\mathbf{r}) \equiv \sum_i^{\sigma} \sum_j^{\sigma} \mathrm{P}^{\lambda,\sigma}_{i,j}\, \Omega^{\lambda,\sigma}_{i,j}(\mathbf{r}) , \tag{503b}$$

where $\mathbf{P}^{\lambda} \equiv \{\mathrm{P}^{\lambda}_{i,j} = \Sigma_{\sigma} \mathrm{P}^{\lambda,\sigma}_{i,j}\} \equiv \mathbf{P}^{\lambda,\alpha} + \mathbf{P}^{\lambda,\beta}$ is the CBO matrix element (see Section 4 in Chapter 1) in the λ representation. Indeed, the canonical HF or KS MO-occupations of Eq. 498, define the diagonal CBO matrices in the canonical φ representation:

$$\mathbf{P}^{\varphi} \equiv \{\mathrm{P}^{\varphi}_{i,j} = n_i^{\varphi}\, \delta_{i,j}\} , \qquad \mathbf{P}^{\varphi,\sigma} \equiv \{\mathrm{P}^{\varphi,\sigma}_{i,j} = n_i^{\varphi,\sigma}\, \delta_{i,j}\} . \tag{504}$$

In general representations the energy can be expressed as a function of the CBO matrices (see, e.g., Eq. 19 in Chapter 1). Again, for the fixed orbital shapes:

$$E[\mathbf{P}^{\lambda}[N, v], \boldsymbol{\lambda}[N, v]]|_{\boldsymbol{\lambda}} \equiv E_{\lambda}(\mathbf{P}^{\lambda}) , \tag{505}$$

$$E[\{\mathbf{P}^{\lambda,\sigma}[N, v]\}, \{\boldsymbol{\lambda}^{\sigma}[N, v]\}]|_{\boldsymbol{\lambda}} \equiv E_{\lambda}(\{\mathbf{P}^{\lambda,\sigma}\}). \tag{506}$$

A reference to Chapter 1 also shows that the explicit expression for this dependence in the RHF theory is:

$$E_{\lambda}^{\mathrm{HF}}(\mathbf{P}^{\lambda}) = \mathrm{tr}\left\{[\mathbf{h}^{\lambda} + \frac{1}{2}\mathbf{G}^{\lambda}(\mathbf{P}^{\lambda})]\,\mathbf{P}^{\lambda}\right\} = \frac{1}{2}\,\mathrm{tr}\,\{[\mathbf{h}^{\lambda} + \mathbf{F}^{\lambda}(\mathbf{P}^{\lambda})]\,\mathbf{P}^{\lambda}\} . \tag{507}$$

This immediately gives the CBO energy gradient of Eq. 20:

$$\partial E_{\lambda}^{\mathrm{HF}}(\mathbf{P}^{\lambda})/\partial \mathrm{P}^{\lambda}_{i,j} = \mathrm{h}^{\lambda}_{j,i} + \frac{1}{2}\mathrm{G}^{\lambda}_{j,i} + \frac{1}{2}\mathrm{tr}\{[\partial \mathbf{G}^{\lambda}(\mathbf{P}^{\lambda})/\partial \mathrm{P}^{\lambda}_{i,j}]\mathbf{P}^{\lambda}\} \equiv \mathrm{F}^{\lambda}_{j,i} , \tag{508}$$

where $\mathbf{F}^{\lambda} = \{\mathrm{F}^{\lambda}_{j,i}\} = \langle \boldsymbol{\lambda}|\mathrm{F}|\boldsymbol{\lambda}\rangle$ denotes the Fock matrix in the λ representation. One can use this result in the CBO chain-rule transformations of populational derivatives in the HF theory. For example, the MO occupation gradient of Eq. 501 immediately follows from the transformation (see Eq. 504):

$$\frac{\partial E_{\phi}^{\mathrm{HF}}(\{n_j^{\phi}\})}{\partial n_i^{\phi}} = \sum_k \sum_l \left(\frac{\partial E_{\phi}^{\mathrm{HF}}}{\partial \mathrm{P}^{\phi}_{k,l}}\right)\left(\frac{\partial \mathrm{P}^{\phi}_{k,l}}{\partial n_i^{\phi}}\right) = \sum_k \sum_l \mathrm{F}^{\phi}_{l,k}\,\delta_{i,k}\,\delta_{i,l} = \mathrm{F}^{\phi}_{i,i} = e_i . \tag{509}$$

In the KS DFT the chain-rule transformations involve functional derivatives with respect to the electronic density. Writing the kinetic and electron-nucleus attraction energies in $E_v[\rho]$ (Eq. 409) in terms of the KS orbitals $\boldsymbol{\psi}$ and their occupations $\boldsymbol{n}^{\psi}$ gives:

$$E_{KS}[n^{\psi}, \boldsymbol{\psi}, \rho] = \sum_i n_i \langle \psi_i | -\frac{1}{2}\nabla^2 + v | \psi_i \rangle + \frac{1}{2}\iint \rho(\mathbf{r})\, W_H(\mathbf{r}, \mathbf{r}')\, \rho(\mathbf{r}')\, d\mathbf{r}\, d\mathbf{r}'$$

$$+ E_{xc}[\rho] \equiv h[n^{\psi}, \boldsymbol{\psi}] + J[\rho] + E_{xc}[\rho]\,, \qquad (510)$$

where $W_H(\mathbf{r}, \mathbf{r}')$ stands for the bare-Coulomb electron interaction (Eq. 423). The Janak occupation derivative of Eq. 500 (see also Eq. 410) then immediately follows from the following chain-rule transformation (see Eq. 498a):

$$\frac{\partial E_{\psi}^{KS}(\{n_j\})}{\partial n_i} = \frac{\partial E_{KS}}{\partial n_i} + \int \left(\frac{\delta J}{\delta \rho(\mathbf{r})}\right)\left(\frac{\partial \rho(\mathbf{r})}{\partial n_i}\right) d\mathbf{r} + \int \left(\frac{\delta E_{xc}}{\delta \rho(\mathbf{r})}\right)\left(\frac{\partial \rho(\mathbf{r})}{\partial n_i}\right) d\mathbf{r}$$

$$= \langle \psi_i | -\frac{1}{2}\nabla^2 | \psi_i \rangle + \int v_{KS}(\mathbf{r})\, \Omega_{i,i}(\mathbf{r})\, d\mathbf{r} \equiv \varepsilon_i\,. \qquad (511)$$

In the KS theory one can similarly define the single particle operator matrix in a general λ representation,

$$\mathbf{H}^{KS,\lambda}(\mathbf{P}^{\lambda}) = \langle \boldsymbol{\lambda} | H_{KS} | \boldsymbol{\lambda} \rangle = \mathbf{T}^{\lambda} + \mathbf{v}^{KS,\lambda}(\mathbf{P}^{\lambda}) = \mathbf{h}^{\lambda} + \mathbf{v}_{ee}^{KS,\lambda}(\mathbf{P}^{\lambda})\,, \qquad (512)$$

where $\mathbf{T}^{\lambda}$ and $\mathbf{v}^{KS,\lambda}$ stand for the respective matrix representations in the $\boldsymbol{\lambda}$ basis set of the kinetic energy operator and the effective KS potential, respectively, $\mathbf{h}^{\lambda}$ is the matrix corresponding to the one-electron operator, $h = -(1/2)\nabla^2 + v$, and $\mathbf{v}_{ee}^{KS,\lambda} = <\boldsymbol{\lambda} | v_{KS} - v | \boldsymbol{\lambda}> \equiv <\boldsymbol{\lambda} | v_{ee}^{KS} | \boldsymbol{\lambda}>$.

Let us define the KS energy expression in the λ representation using the HF analogy (Eq. 19 in Chapter 1). The obvious replacement for the Fock matrix $\mathbf{F}^{\lambda}$ in DFT is the $\mathbf{H}^{KS,\lambda}$ matrix, while $\mathbf{G}^{\lambda}$ should be replaced by $\mathbf{v}_{ee}^{KS,\lambda}$. The KS equivalent of Eq. 507 is:

$$E_{\lambda}^{KS}(\mathbf{P}^{\lambda}) = \mathrm{tr}\,\{[\mathbf{h}^{\lambda} + \frac{1}{2}\mathbf{v}_{ee}^{KS,\lambda}(\mathbf{P}^{\lambda})]\mathbf{P}^{\lambda}\} \equiv \frac{1}{2}\mathrm{tr}\,\{[\mathbf{h}^{\lambda} + \mathbf{H}^{KS,\lambda}(\mathbf{P}^{\lambda})]\,\mathbf{P}^{\lambda}\}\,. \qquad (513)$$

Its CBO gradient is defined by the following equations (see Eq. 503a):

$$\frac{\partial E_{\lambda}^{KS}(\mathbf{P}^{\lambda})}{\partial P_{i,j}^{\lambda}} = \left(\mathbf{h}^{\lambda} + \frac{1}{2}\mathbf{v}_{ee}^{KS,\lambda}\right)_{j,i} + \frac{1}{2}\mathrm{tr}\left\{\left(\frac{\partial \mathbf{v}_{ee}^{KS,\lambda}(\mathbf{P}^{\lambda})}{\partial P_{i,j}^{\lambda}}\right)\mathbf{P}^{\lambda}\right\}, \qquad (514)$$

$$\left(\frac{\partial (\mathbf{v}_{ee}^{KS,\lambda})_{k,l}}{\partial P_{i,j}^{\lambda}}\right) = \langle \lambda_k | \int \left(\frac{\delta v_{ee}^{KS}}{\delta \rho(\mathbf{r})}\right)\left(\frac{\partial \rho(\mathbf{r})}{\partial P_{i,j}^{\lambda}}\right) d\mathbf{r} | \lambda_l \rangle$$

$$= \iint \Omega_{k,l}^{\lambda}(\mathbf{r}')\, W(\mathbf{r}', \mathbf{r})\, \Omega_{i,j}^{\lambda}(\mathbf{r})\, d\mathbf{r}\, d\mathbf{r}' \equiv (kl \,\vdots\vdots\, ij)^{\lambda}\,. \qquad (515)$$

The latter derivative represents the effective KS interaction which replaces the Coulomb-exchange HF interaction $(\mathrm{kl}\,||\,\mathrm{ij})^\lambda$ defined in Chapter 1; the $\mathrm{W}(\mathbf{r},\mathbf{r}')$ kernel is defined by Eq. 423.

It follows from the last two equations that

$$\frac{\partial E_\lambda^{\mathrm{KS}}(\mathbf{P}^\lambda)}{\partial \mathrm{P}_{\mathrm{i,j}}^\lambda} = \left(\mathbf{h}^\lambda + \frac{1}{2}\mathbf{v}_{\mathrm{ee}}^{\mathrm{KS},\lambda}\right)_{\mathrm{j,i}} + \frac{1}{2}\sum\nolimits_{\mathrm{k}}\sum\nolimits_{\mathrm{l}} \mathrm{P}_{\mathrm{k,l}}^\lambda\,(\mathrm{kl}\,||\,\mathrm{ij})^\lambda$$

$$= (\mathbf{h}^\lambda + \mathbf{v}_{\mathrm{ee}}^{\mathrm{KS},\lambda})_{\mathrm{j,i}} = (\mathbf{H}^{\mathrm{KS},\lambda})_{\mathrm{j,i}}\,. \tag{516}$$

This equation represents the KS equivalent of Eq. 508, which may also be useful in the chain-rule transformations in general AO representation λ. For example, in the canonical KS representation (see Eq. 504), $\boldsymbol{\lambda} = \boldsymbol{\psi}$, one immediately obtains the Janak Theorem (Eq. 500):

$$\frac{\partial E_\psi^{\mathrm{KS}}(\{n_{\mathrm{j}}\})}{\partial n_{\mathrm{i}}} = \sum\nolimits_{\mathrm{k}}\sum\nolimits_{\mathrm{l}}\left(\frac{\partial E_\psi^{\mathrm{KS}}}{\partial \mathrm{P}_{\mathrm{k,l}}^\psi}\right)\left(\frac{\partial \mathrm{P}_{\mathrm{k,l}}^\psi}{\partial n_{\mathrm{i}}}\right) = \sum\nolimits_{\mathrm{k}}\sum\nolimits_{\mathrm{l}} \mathrm{H}_{\mathrm{l,k}}^{\mathrm{KS},\psi}\,\delta_{\mathrm{i,k}}\,\delta_{\mathrm{i,l}} = \mathrm{H}_{\mathrm{i,i}}^{\mathrm{KS},\psi} = \varepsilon_{\mathrm{i}}\,. \tag{517}$$

2. Hardness Derivatives

The second partials with respect to the MO occupations define the rigid, external hardness matrix in the MO resolution [30, 31]. For example, in the RHF theory using Eqs. 18, 504, and 508 gives via the CBO chain rule:

$$\eta_{\mathrm{i,j}}^{\mathrm{HF}} \equiv \partial^2 E_\phi^{\mathrm{HF}}/\partial n_{\mathrm{i}}^\phi\,\partial n_{\mathrm{j}}^\phi = \partial e_{\mathrm{i}}/\partial n_{\mathrm{j}}^\phi = \partial e_{\mathrm{j}}/\partial n_{\mathrm{i}}^\phi$$

$$= \sum\nolimits_{\mathrm{p}}\sum\nolimits_{\mathrm{q}}\sum\nolimits_{\mathrm{k}}\sum\nolimits_{\mathrm{l}} (\partial \mathrm{P}_{\mathrm{p,q}}^\phi/\partial n_{\mathrm{i}}^\phi)\,(\partial^2 E_\phi^{\mathrm{HF}}/\partial \mathrm{P}_{\mathrm{p,q}}^\phi\,\partial \mathrm{P}_{\mathrm{k,l}}^\phi)\,(\partial \mathrm{P}_{\mathrm{k,l}}^\phi/\partial n_{\mathrm{j}}^\phi)$$

$$= \sum\nolimits_{\mathrm{p}}\sum\nolimits_{\mathrm{q}}\sum\nolimits_{\mathrm{k}}\sum\nolimits_{\mathrm{l}} (\delta_{\mathrm{i,p}}\,\delta_{\mathrm{i,q}})\,(\phi_{\mathrm{p}}\phi_{\mathrm{q}}\,||\,\phi_{\mathrm{k}}\phi_{\mathrm{l}})\,(\delta_{\mathrm{j,k}}\,\delta_{\mathrm{j,l}})$$

$$= (\phi_{\mathrm{i}}\phi_{\mathrm{i}}\,|\,\phi_{\mathrm{j}}\phi_{\mathrm{j}}) - \frac{1}{2}(\phi_{\mathrm{i}}\phi_{\mathrm{j}}\,|\,\phi_{\mathrm{j}}\phi_{\mathrm{i}}) \equiv \mathrm{J}_{\mathrm{i,j}} - \frac{1}{2}\mathrm{K}_{\mathrm{i,j}}\,. \tag{518}$$

Therefore, the rigid external hardness matrix element, coupling a given pair of MO, represents the effective Coulomb-exchange interaction between two electrons occupying these orbitals. We have used a similar approach to model the hardness tensor in the AIM resolution (Eqs. 21-25). Equation 518 also identifies the electron repulsion quantity associated with the second partial with respect to a general, e.g., AO basis set $\boldsymbol{\lambda}$:

$$\mathscr{H}_{\mathrm{p,q;k,l}}^{\mathrm{HF},\lambda} = \partial^2 E_\lambda^{\mathrm{HF}}/\partial \mathrm{P}_{\mathrm{p,q}}^\lambda\,\partial \mathrm{P}_{\mathrm{k,l}}^\lambda) = (\mathrm{p\,q}\,||\,\mathrm{k\,l})^\lambda\,. \tag{519}$$

The corresponding orbitally resolved tensor in the KS theory similarly represents the sensitivity of the KS eigenvalues per unit displacement of the KS occupations:

$$\eta_{i,j}^{KS} \equiv \partial^2 E_{\psi}^{KS} / \partial n_i \partial n_j = \partial \varepsilon_i / \partial n_j = \partial \varepsilon_j / \partial n_i$$

$$= \sum_p \sum_q \sum_k \sum_l (\partial P_{p,q}^{\psi} / \partial n_i)(\partial^2 E_{\psi}^{KS} / \partial P_{p,q}^{\psi} \partial P_{k,l}^{\psi})(\partial P_{k,l}^{\psi} / \partial n_j)$$

$$= \sum_p \sum_q \sum_k \sum_l (\delta_{i,p}\, \delta_{i,q})(\psi_p \psi_q \,\vert\vert\, \psi_k \psi_l)(\delta_{j,k}\, \delta_{j,l})$$

$$= (\psi_i \psi_i \,\vert\vert\, \psi_j \psi_j) \,. \tag{520}$$

Therefore, this quantity has a similar electron repulsion interpretation as in the HF case; notice, however, that it effectively includes Coulomb-exchange-correlation interaction between two electrons occupying a given pair of the KS orbitals. It also follows from this chain-rule transformation that the second partials of the KS energy with respect to the CBO variables in a general basis set $\boldsymbol{\lambda}$ are:

$$\mathscr{H}_{p,q;k,l}^{KS,\lambda} = \partial^2 E_{\lambda}^{KS} / \partial P_{p,q}^{\lambda} \partial P_{k,l}^{\lambda} = (p\, q \,\vert\vert\, k\, l)^{\lambda}. \tag{521}$$

In Appendix E we have summarized the relevant UHF and Generalized UHF (GUHF) energy expressions and derivatives with respect to the one-electron density matrices. In the UHF approximation the spin off-diagonal density matrices and the associated blocks of the spin-resolved Fock matrix are neglected, while the full GUHF approach of McWeeny includes both the diagonal and off-diagonal couplings between the α- and β-SO [145].

The unconstrained first and second partials with respect to the orbital CBO/occupation variables have implicitly assumed that they are truly independent. They can therefore be considered as reflecting the system sensitivities with respect to an external CT, involving the external electron reservoirs, separate for each orbital, without any consideration for the orthogonality of orbitals and the occupational limits imposed by the Pauli principle. Thus, the above orbital occupation/CBO variables are in reality dependent. One should therefore additionally examine the corresponding derivative properties of the *internal* auxiliary energy function with all relevant constraints already built in, in which these occupational/CBO degrees of freedom can indeed be considered as truly independent.

3. Nature of the Orbital Discretization

3.1. Local and Global State Variables in the Orbital Description

The Janak-type theorems (Eqs. 500 and 501) can be interpreted as defining the *"global" (spin) orbital chemical potentials* in the "frozen" SO ($\boldsymbol{\varphi}$) approximation, $(\partial E_{g.s.}^{SO} / \partial \boldsymbol{n}^{SO})_{\boldsymbol{\varphi}} = \boldsymbol{e}^{\varphi}$, in terms of the *local orbital chemical potentials*, $\boldsymbol{\mu}^{SO}(\mathbf{r}) = [\delta E_v^{SO} / \delta \boldsymbol{\rho}^{SO}(\mathbf{r})]_{\boldsymbol{\varphi}} = [\boldsymbol{\mu}^{\alpha}(\mathbf{r}), \boldsymbol{\mu}^{\beta}(\mathbf{r})] = \{\mu_i^{\sigma}(\mathbf{r})\}$,

$$\left(\frac{\partial E_{g.s.}^{SO}}{\partial n_i^{\sigma}}\right)_{\varphi,\,v} = e_i^{\sigma} = \int \left(\frac{\partial \rho_i^{\sigma}(\mathbf{r})}{\partial n_i^{\sigma}}\right)_{\varphi} \left(\frac{\delta E_v^{SO}}{\delta \rho_i^{\sigma}(\mathbf{r})}\right)_{\varphi} d\mathbf{r} \equiv \int \Omega_{i,i}^{\sigma}(\mathbf{r})\, \mu_i^{\sigma}(\mathbf{r})\, d\mathbf{r}\,, \tag{522}$$

with the orbital local Fukui function, $[\partial \rho_i^{\sigma}(\mathbf{r})/\partial n_i^{\sigma}]_{\varphi} = \Omega_{i,i}^{\sigma}(\mathbf{r})$, providing the relevant weighting function; here $E_v^{SO}[\boldsymbol{\rho}^{SO}]$ and $E_{g.s.}^{SO}(\boldsymbol{n}^{SO}, v) = E_v^{SO}[\boldsymbol{\rho}_{opt}^{SO}]$ denote the system energies in the SO-resolution, for the fixed external potential $v(\mathbf{r})$, corresponding to the non-equilibrium and equilibrium orbital densities, respectively. The local SO chemical potentials, $\boldsymbol{\mu}^{SO}(\mathbf{r})$, are then the local SO populational gradients, while the orbital energies $\boldsymbol{e}^{SO}$ provide the global SO populational derivatives, i.e., the "intensive" conjugates (potentials) of the "extensive" (density and occupation) state variables.

Since in the orbital description each SO has its own level of the global "chemical potential", one can formally consider the orbital resolution as corresponding to the multicomponent representation (see Section 6 in Chapter 6) of the electron distribution; this constrained equilibrium state implies hypothetical barriers preventing flows of electrons between SO. Therefore, the spin-polarized energy can be regarded as the following functional of the orbital spin-densities:

$$E_v^{SO}[\rho_1^{\alpha}|\rho_2^{\alpha}|...|\rho_1^{\beta}|\rho_2^{\beta}|...] = \sum_i \sum_{\sigma} \int v(\mathbf{r})\, \rho_i^{\sigma}(\mathbf{r})\, d\mathbf{r} + F[\rho_1^{\alpha}|\rho_2^{\alpha}|...|\rho_1^{\beta}|\rho_2^{\beta}|...]$$

$$\equiv E_v[\boldsymbol{\rho}^{SO}] = \int \boldsymbol{\rho}^{SO}(\mathbf{r})\, [\boldsymbol{v}^{SO}(\mathbf{r})]^{\dagger}\, d\mathbf{r} + F[\boldsymbol{\rho}^{SO}]\,, \tag{523}$$

with the universal multicomponent functional $F[\boldsymbol{\rho}^{SO}]$ generating the sum of the electron kinetic and repulsion energies:

$$F[\boldsymbol{\rho}^{SO}] = \min_{\Psi \to \boldsymbol{\rho}^{SO}} \langle \Psi | \mathrm{T}_e + \mathrm{V}_{ee} | \Psi \rangle\,, \tag{524}$$

and the row function vector $\boldsymbol{v}^{SO}(\mathbf{r}) = v(\mathbf{r})\, \boldsymbol{1}^{SO}$.

The system energy for the fixed (optimum) orbitals can be expressed in terms of the global orbital quantities:

$$E_{g.s.}^{SO}(\boldsymbol{n}^{SO}, \boldsymbol{v}^{SO})_{\varphi} = \sum_i \sum_{\sigma} n_i^{\sigma}\, v_i^{\sigma} + F_{g.s.}^{SO}(\boldsymbol{n}^{SO})_{\varphi}\,, \tag{525}$$

where $\boldsymbol{n}^{SO} = (\boldsymbol{n}^{\alpha}, \boldsymbol{n}^{\beta}) = \{n_i^{SO}\}$ groups all SO occupations and $\boldsymbol{v}^{SO} = (\boldsymbol{v}^{\alpha}, \boldsymbol{v}^{\beta}) = \{v_i^{\sigma}\}$ stands for the vector of all weighted external potential quantities,

$$v_i^{\sigma} = \int \Omega_{i,i}^{\sigma}(\mathbf{r})\, v(\mathbf{r})\, d\mathbf{r}\,. \tag{526}$$

Equations 523 and 525 replace in the SO-resolution the phenomenological Eq. 216.

3.2. Equilibrium Criteria and Charge Sensitivities in the Local SO Description

Using the orbital energies as the Lagrange multipliers controlling the orbital occupations in the auxiliary functional for the closed system (N = const.):

$$\mathscr{E}_N^{SO}[...|\rho_i^\sigma|...] = E_v[\boldsymbol{\rho}^{SO}] - \sum_i \sum_\sigma e_i^\sigma (n_i^\sigma - n_i^{\sigma,0}) \equiv \mathscr{E}_N^{SO}[\boldsymbol{\rho}^{SO}], \quad (527)$$

then gives rise to the Euler-Lagrange equation:

$$\frac{\delta \mathscr{E}_N^{SO}}{\delta \rho_i^\sigma(\mathbf{r})} = \frac{\delta E_v^{SO}}{\delta \rho_i^\sigma(\mathbf{r})} - e_i^\sigma = \mu_i^\sigma(\mathbf{r}) - e_i^\sigma = 0, \quad \sigma = \alpha, \beta, \quad i = 1, 2, \ldots, \quad (528)$$

which can be interpreted as the SO *electronegativity equalization equation.*

Thus, the local orbital electronegativity is equalized throughout the space at the orbital energy level. Moreover, since $\delta E_v[\boldsymbol{\rho}^{SO}]/\delta\boldsymbol{\rho}^{SO}(\mathbf{r}) = v(\mathbf{r}) + \delta F[\boldsymbol{\rho}^{SO}]/\delta\boldsymbol{\rho}^{SO}(\mathbf{r})$, this matrix equation can be written in the equivalent form:

$$\boldsymbol{u}^{SO}(\mathbf{r}) \equiv [\boldsymbol{v}^{SO}(\mathbf{r}) - \boldsymbol{e}^{SO}] = -\delta F/\delta\boldsymbol{\rho}^{SO}(\mathbf{r}), \quad (529)$$

which implies that $\boldsymbol{\rho}^{SO} = \boldsymbol{\rho}^{SO}[\boldsymbol{u}^{SO}]$ and $\boldsymbol{u}^{SO} = \boldsymbol{u}^{SO}[\boldsymbol{\rho}^{SO}]$. Hence,

$$\mathrm{d}\boldsymbol{u}^{SO}(\mathbf{r}) = -\int \mathrm{d}\boldsymbol{\rho}^{SO}(\mathbf{r}')\,[\delta^2 F/\delta\boldsymbol{\rho}^{SO}(\mathbf{r}')\,\delta\boldsymbol{\rho}^{SO}(\mathbf{r})]\,\mathrm{d}\mathbf{r}'$$

$$= -\int \mathrm{d}\boldsymbol{\rho}^{SO}(\mathbf{r}')\,\boldsymbol{\eta}^{SO}(\mathbf{r}',\mathbf{r})\,\mathrm{d}\mathbf{r}', \quad (530)$$

$$\mathrm{d}\boldsymbol{\rho}^{SO}(\mathbf{r}) = -\int \mathrm{d}\boldsymbol{u}^{SO}(\mathbf{r}')\,\boldsymbol{\sigma}^{SO}(\mathbf{r}',\mathbf{r})\,\mathrm{d}\mathbf{r}', \quad (531)$$

where the SO softness kernel matrix, $\boldsymbol{\sigma}^{SO}(\mathbf{r}',\mathbf{r}) = -\delta\boldsymbol{\rho}^{SO}(\mathbf{r})/\delta\boldsymbol{u}^{SO}(\mathbf{r}')$, is the inverse of the SO hardness kernel matrix $\boldsymbol{\eta}^{SO}(\mathbf{r}',\mathbf{r}) = -\delta\boldsymbol{u}^{SO}(\mathbf{r})/\delta\boldsymbol{\rho}^{SO}(\mathbf{r}')$:

$$\int \boldsymbol{\eta}^{SO}(\mathbf{r},\mathbf{r}')\,\boldsymbol{\sigma}^{SO}(\mathbf{r}',\mathbf{r}'')\,\mathrm{d}\mathbf{r}' = \mathbf{I}^{SO}\,\delta(\mathbf{r}-\mathbf{r}''). \quad (532)$$

The matrix softness kernel defines the local softness function matrix in the SO resolution:

$$\mathbf{s}^{SO}(\mathbf{r}) = \int \boldsymbol{\sigma}^{SO}(\mathbf{r}',\mathbf{r})\,\mathrm{d}\mathbf{r}' = -\delta\boldsymbol{n}^{SO}/\delta\boldsymbol{u}^{SO}(\mathbf{r}), \quad (533)$$

and the matrix of global SO softnesses, defined as derivatives of the SO occupations with respect to the global external potential variables, $\boldsymbol{u}^{SO} \equiv \boldsymbol{v}^{SO} - \boldsymbol{e}^{SO} = \{u_i^\sigma\}$,

$$\boldsymbol{\sigma}^{SO} = \int \mathbf{s}^{SO}(\mathbf{r})\,\mathrm{d}\mathbf{r} = -\int [\delta\boldsymbol{n}^{SO}/\delta\boldsymbol{u}^{SO}(\mathbf{r})]\,\mathrm{d}\mathbf{r}$$

$$= -\{\int [\delta \boldsymbol{u}^{SO}/\delta \boldsymbol{u}^{SO}(\mathbf{r})]\, d\mathbf{r}\}\ (\partial \boldsymbol{n}^{SO}/\partial \boldsymbol{u}^{SO})$$

$$= -\partial \boldsymbol{n}^{SO}/\partial \boldsymbol{u}^{SO} = (\boldsymbol{\eta}^{SO})^{-1}\,, \qquad (534)$$

since $\delta u_i^{\sigma}/\delta u_j^{\sigma'}(\mathbf{r}) = \delta_{i,j}\,\delta_{\sigma,\sigma'}\,\Omega_{i,i}^{\sigma}(\mathbf{r})$. In Eq. 534 we have indicated that this global softness matrix represents the inverse matrix of global SO hardnesses,

$$\boldsymbol{\eta}^{SO} = (\partial^2 F_{g.s.}^{SO}(\boldsymbol{n}^{SO})/\partial \boldsymbol{n}^{SO}\,\partial \boldsymbol{n}^{SO})_{\boldsymbol{\varphi},v}\,. \qquad (535)$$

Let us examine its element coupling φ_i^{σ} and $\varphi_j^{\sigma'}$:

$$\eta_{i,j}^{\sigma,\sigma'} = (\partial^2 E_{g.s.}^{SO}/\partial n_i^{\sigma}\,\partial n_j^{\sigma'})_{\boldsymbol{\varphi},v} = (\partial^2 F_{g.s.}^{SO}/\partial n_i^{\sigma}\,\partial n_j^{\sigma'})_{\boldsymbol{\varphi}}$$

$$= -(\partial u_j^{\sigma'}/\partial n_i^{\sigma})_{\boldsymbol{\varphi},v} = (\partial e_j^{\sigma'}/\partial n_i^{\sigma})_{\boldsymbol{\varphi},v}$$

$$= \int\int [\partial \rho_i^{\sigma}(\mathbf{r})/\partial n_i^{\sigma}]_{\boldsymbol{\varphi}}\ [\delta^2 F^{SO}/\delta \rho_i^{\sigma}(\mathbf{r})\,\delta \rho_j^{\sigma'}(\mathbf{r}')]_{\boldsymbol{\varphi}}\ [\partial \rho_j^{\sigma'}(\mathbf{r}')/\partial n_i^{\sigma}]_{\boldsymbol{\varphi}}\, d\mathbf{r}\, d\mathbf{r}'$$

$$= \int\int \Omega_{i,i}^{\sigma}(\mathbf{r})\ \eta_{i,j}^{\sigma,\sigma'}(\mathbf{r},\mathbf{r}')\ \Omega_{j,j}^{\sigma'}(\mathbf{r}')\, d\mathbf{r}\, d\mathbf{r}'\,. \qquad (536)$$

Thus, as we have already observed in Eq. 522, the canonical orbital charge distributions $\{\Omega_{i,i}^{\sigma}(\mathbf{r})\}$ play the role of the orbital FF, which combine the orbital hardness kernel, $\boldsymbol{\eta}^{SO}(\mathbf{r}',\mathbf{r}) = \{\eta_{i,j}^{\sigma,\sigma'}(\mathbf{r},\mathbf{r}') = [\delta\mu_j^{\sigma'}(\mathbf{r}')/\delta\rho_i^{\sigma}(\mathbf{r})]_{\boldsymbol{\varphi},v}\}$, into the global hardness matrix elements in the SO-resolution. It should also be observed that, due to the orbital chemical potential equalization of Eq. 528, the local orbital hardnesses,

$$\eta_{i,j}^{\sigma,\sigma'}(\mathbf{r}') = \left(\partial/\partial n_i^{\sigma}[\delta E_v^{SO}/\delta \rho_j^{\sigma'}(\mathbf{r}')]\right)_{\boldsymbol{\varphi},v}$$

$$= [\partial\mu_j^{\sigma'}(\mathbf{r}')/\partial n_i^{\sigma})_{\boldsymbol{\varphi},v} = [\delta e_i^{\sigma}/\delta \rho_j^{\sigma'}(\mathbf{r}')]_{\boldsymbol{\varphi},v} = \int \Omega_{i,i}^{\sigma}(\mathbf{r})\ \eta_{i,j}^{\sigma,\sigma'}(\mathbf{r},\mathbf{r}')\, d\mathbf{r}'$$

$$= \eta_{i,j}^{\sigma,\sigma'}(\mathbf{r}) = \left(\delta/\delta \rho_i^{\sigma}(\mathbf{r})[\partial E_g^{SO}/\partial n_j^{\sigma'}]\right)_{\boldsymbol{\varphi},v}$$

$$= [\delta e_j^{\sigma'}/\delta \rho_i^{\sigma}(\mathbf{r})]_{\boldsymbol{\varphi},v} = [\partial\mu_i^{\sigma}(\mathbf{r})/\partial n_j^{\sigma'}]_{\boldsymbol{\varphi},v} = \int \eta_{i,j}^{\sigma,\sigma'}(\mathbf{r},\mathbf{r}')\ \Omega_{j,j}^{\sigma'}(\mathbf{r}')\, d\mathbf{r}'$$

$$= \eta_{i,j}^{\sigma,\sigma'}\,, \qquad (537)$$

are also equalized at the corresponding values of the global hardness matrix element.

For the inverse of $\boldsymbol{\eta}^{SO}$, i.e., the SO-resolved softness matrix $\boldsymbol{\sigma}^{SO}$ (see Eq. 534), one similarly finds the following, additive combinational formula for the global orbital softness matrix element in terms of the orbital softness kernel $\boldsymbol{\sigma}^{SO}(\mathbf{r}',\mathbf{r})$,

$$\sigma_{i,j}^{\sigma,\sigma'} \equiv -\left(\frac{\partial n_j^{\sigma'}}{\partial u_i^{\sigma}}\right)_{\varphi,\nu} = \left(\frac{\partial n_j^{\sigma'}}{\partial e_i^{\sigma}}\right)_{\varphi,\nu}$$

$$= \iint \left(\frac{\partial \mu_i^{\sigma}(\mathbf{r})}{\partial e_i^{\sigma}}\right)_{\varphi,\nu} \left(\frac{\partial \rho_j^{\sigma'}(\mathbf{r}')}{\partial \mu_i^{\sigma}(\mathbf{r})}\right)_{\varphi,\nu} d\mathbf{r}\, d\mathbf{r}' = \iint \sigma_{i,j}^{\sigma,\sigma'}(\mathbf{r},\mathbf{r}')\, d\mathbf{r}\, d\mathbf{r}' . \tag{538}$$

One can also write the vector differentials $d\boldsymbol{\rho}^{SO}(\mathbf{r}) = d\boldsymbol{\rho}^{SO}[\boldsymbol{n}^{SO}, \nu; \mathbf{r}]$ and $d\boldsymbol{e}^{SO}[\boldsymbol{n}^{SO}, \nu]$:

$$d\boldsymbol{\rho}^{SO}[\boldsymbol{n}^{SO}, \nu; \mathbf{r}] = d\boldsymbol{n}^{SO}\, \mathbf{f}^{SO}(\mathbf{r}) + \int d\boldsymbol{\nu}^{SO}(\mathbf{r}')\, \boldsymbol{\beta}^{SO}(\mathbf{r}',\mathbf{r})\, d\mathbf{r}', \tag{539}$$

$$d\boldsymbol{e}^{SO}[\boldsymbol{n}^{SO}, \nu] = d\boldsymbol{n}^{SO}\, \boldsymbol{\eta}^{SO} + \int d\boldsymbol{\nu}^{SO}(\mathbf{r})\, \mathbf{f}^{SO}(\mathbf{r})\, d\mathbf{r} , \tag{540}$$

where the SO Fukui Function matrix,

$$\mathbf{f}^{SO}(\mathbf{r}) = \frac{\partial}{\partial \boldsymbol{n}^{SO}} \left\{ \frac{\delta E_{g.s.}^{SO}[\boldsymbol{n}^{SO}, \boldsymbol{\nu}]}{\delta \boldsymbol{\nu}^{SO}(\mathbf{r})} \right\} = \left[\frac{\delta \boldsymbol{e}^{SO}}{\delta \boldsymbol{\nu}^{SO}(\mathbf{r})} \right]_{\boldsymbol{n}^{SO}} = \left[\frac{\partial \boldsymbol{\rho}^{SO}(\mathbf{r})}{\partial \boldsymbol{n}^{SO}} \right]_{\nu}, \tag{541}$$

and the SO linear response kernel matrix

$$\boldsymbol{\beta}^{SO}(\mathbf{r}',\mathbf{r}) = \left(\delta \boldsymbol{\rho}^{SO}(\mathbf{r}) / \delta \boldsymbol{\nu}^{SO}(\mathbf{r}')\right)_{\boldsymbol{n}^{SO}} . \tag{542}$$

Combining Eqs. 534, 539 and 540 gives the familiar, Berkowitz-Parr-type relations in the SO resolution (see Eqs. 469 and 470):

$$\mathbf{f}^{SO}(\mathbf{r}) = \boldsymbol{\eta}^{SO}\, \mathbf{s}^{SO}(\mathbf{r}) \quad \text{or} \quad \boldsymbol{\sigma}^{SO}\, \mathbf{f}^{SO}(\mathbf{r}) = \mathbf{s}^{SO}(\mathbf{r}) , \tag{543}$$

$$\boldsymbol{\beta}^{SO}(\mathbf{r}',\mathbf{r}) = -\boldsymbol{\sigma}^{SO}(\mathbf{r}',\mathbf{r}) + [\mathbf{f}^{SO}(\mathbf{r}')]^{\dagger}\, \boldsymbol{\sigma}^{SO}\, \mathbf{f}^{SO}(\mathbf{r}) . \tag{544}$$

This analogy can be pursued still further in terms of the integral matrix equations linking the displacements (perturbations) and responses in the SO resolution (see Eqs. 473 and 475):

$$[d\boldsymbol{e}^{SO}, d\boldsymbol{\rho}^{SO}(\mathbf{r})] = [d\boldsymbol{n}^{SO}, \int d\mathbf{r}'\, d\boldsymbol{\nu}^{SO}(\mathbf{r}')] \begin{bmatrix} \boldsymbol{\eta}^{SO} & \mathbf{f}^{SO}(\mathbf{r}) \\ \mathbf{f}^{SO,\dagger}(\mathbf{r}') & \boldsymbol{\beta}^{SO}(\mathbf{r}',\mathbf{r}) \end{bmatrix}, \tag{545}$$

$$[\mathrm{d}\boldsymbol{n}^{SO}, \mathrm{d}\boldsymbol{v}^{SO}(\mathbf{r})] = [\mathrm{d}\boldsymbol{e}^{SO}, \int \mathrm{d}\mathbf{r}'\ \mathrm{d}\boldsymbol{\rho}^{SO}(\mathbf{r}')] \begin{bmatrix} \mathbf{0}^{SO} & \mathbf{I}^{SO} \\ \mathbf{I}^{SO} & -\boldsymbol{\eta}^{SO}(\mathbf{r}', \mathbf{r}) \end{bmatrix} . \quad (546)$$

3.3. Equilibrium Criteria and Charge Sensitivities in the Global Description

The energy function of Eq. 525 corresponds to the optimum orbital densities satisfying the above constrained equilibrium equations. The orbital chemical potential/ electronegativity equalization Eqs. 528 and 529 are equivalent to the HF/KS equations since they may also be interpreted as equations marking the equalization of local orbital energies (see Eqs. 500 and 501):

$$(\phi_i^\sigma)^{-1}\,\mathrm{F}\,\phi_i^\sigma \equiv \mu_{i,\mathrm{HF}}^\sigma(\mathbf{r}) = e_i^\sigma \quad \text{and} \quad (\psi_i^\sigma)^{-1}\,\mathrm{H}_{\mathrm{KS}}\,\psi_i^\sigma \equiv \mu_{i,\mathrm{KS}}^\sigma(\mathbf{r}) = \varepsilon_i^\sigma . \quad (547)$$

The optimum SO occupations, $\boldsymbol{n}^{SO}(\boldsymbol{u}^{SO})$, minimize the auxiliary energy function of the SO global state parameters:

$$\mathscr{E}_{\mathrm{g.s.}}^{SO}(\boldsymbol{n}^{SO}, \boldsymbol{v}^{SO}) = E_{\mathrm{g.s.}}^{SO}(\boldsymbol{n}^{SO}, \boldsymbol{v}^{SO}) - \boldsymbol{e}^{SO}(\boldsymbol{n}^{SO} - \boldsymbol{n}^0)^\dagger , \quad (548)$$

and hence follow from the corresponding SO global Euler equations:

$$-(v_i^\sigma - e_i^\sigma) \equiv -u_i^\sigma = \int \Omega_{i,i}^\sigma(\mathbf{r}) \left(\frac{\delta F^{SO}}{\delta \rho_i^\sigma(\mathbf{r})} \right)_{\boldsymbol{\varphi}} \mathrm{d}\mathbf{r} \equiv \frac{\partial F_{\mathrm{g.s.}}^{SO}(\boldsymbol{n}^{SO})}{\partial n_i^\sigma} , \quad (549)$$

which can be written in the following matrix form:

$$-\boldsymbol{u}^{SO} = \partial F_{\mathrm{g.s.}}^{SO}(\boldsymbol{n}^{SO}) / \partial \boldsymbol{n}^{SO} . \quad (550)$$

We therefore conclude that the optimum orbital occupations for the "frozen" shapes of orbitals are uniquely determined in the closed system (N = const.) by the $\boldsymbol{u}^{SO} = (\boldsymbol{u}^\alpha, \boldsymbol{u}^\beta)$ quantities of Eq. 549, and so are the global SO occupation derivative properties.

In the open system in contact with the macroscopic electron reservoir characterized by the chemical potential μ_r the energy function has to be supplemented by the constraint term corresponding to the subsidiary condition of the fixed system average number of electrons $N(\boldsymbol{n}^{SO}) = \boldsymbol{n}^{SO}(\boldsymbol{1}^{SO})^\dagger = N^0$:

$$\mathscr{E}^{SO}(N, \boldsymbol{v}^{SO}) = E_{\mathrm{g.s.}}^{SO}[\boldsymbol{n}^{SO}(N, \boldsymbol{v}^{SO}), \boldsymbol{v}^{SO}] - \mu_r[N(\boldsymbol{n}^{SO}) - N^0] . \quad (551)$$

It gives rise to the SO external Euler-Lagrange equation for the optimum average number of electrons N:

$$\partial \mathscr{E}^{SO}(N, \boldsymbol{v}^{SO})/\partial N = \boldsymbol{\mathscr{F}}^{SO}(\boldsymbol{e}^{SO})^{\dagger} - \mu_r \equiv \mu - \mu_r = 0\,, \tag{552}$$

where the *external* global FF vector in the SO representation,

$$\boldsymbol{\mathscr{F}}^{SO} = \frac{\partial^2 E_{g.s.}^{SO}(N, \boldsymbol{v}^{SO})}{\partial N\, \partial \boldsymbol{v}^{SO}} = \left(\frac{\partial \mu}{\partial \boldsymbol{v}^{SO}}\right)_N = \left(\frac{\partial \boldsymbol{n}^{SO}}{\partial N}\right)_{\boldsymbol{v}^{SO}}. \tag{553}$$

Thus, at the external equilibrium, the reservoir chemical potential μ_r must be equal to the global chemical potential of the system in question, $\mu_r = \mu(N, \boldsymbol{v}^{SO})$, represented by the $\boldsymbol{\mathscr{F}}$-weighted sum of orbital energies. The same SO weighting factors determine the SO hardnesses,

$$\boldsymbol{\eta}^{SO} = \frac{\partial^2 E_{g.s.}^{SO}(N, \boldsymbol{v}^{SO})}{\partial N\, \partial \boldsymbol{n}^{SO}} = \frac{\partial \mu}{\partial \boldsymbol{n}^{SO}} = \frac{\partial \boldsymbol{e}^{SO}}{\partial N} = \boldsymbol{\mathscr{F}}^{SO}\, \boldsymbol{\eta}^{SO} = \{\eta_i^{\sigma}\}\,, \tag{554}$$

and the SO global hardness:

$$\eta^{SO} = \frac{\partial^2 E_{g.s.}^{SO}(N, \boldsymbol{v}^{SO})}{\partial N^2} = \boldsymbol{\eta}^{SO}(\boldsymbol{\mathscr{F}}^{SO})^{\dagger}\,. \tag{555}$$

In what follows we assume that the occupational limits imposed by the Pauli principle are build-into these external FF indices, which also depend upon the temperature of the reservoir (molecular environment).

The relevant expression for the global linear response matrix in the fixed SO approximation, $\boldsymbol{\beta}^{SO} \equiv (\partial \boldsymbol{n}^{SO}/\partial \boldsymbol{v}^{SO})_{\varphi, N}$, follows from examining the differential of the orbital energies $\boldsymbol{e}^{SO} = \boldsymbol{e}^{SO}(\boldsymbol{n}^{SO}(N, \boldsymbol{v}^{SO}), \boldsymbol{v}^{SO})$ for constant N:

$$d\boldsymbol{e}_N^{SO} = d\boldsymbol{v}^{SO} + d\boldsymbol{v}^{SO}(\partial \boldsymbol{n}^{SO}/\partial \boldsymbol{v}^{SO})_{\varphi, N}\,\boldsymbol{\eta}^{SO} = d\boldsymbol{v}^{SO}(\mathbf{I}^{SO} + \boldsymbol{\beta}^{SO}\,\boldsymbol{\eta}^{SO})\,. \tag{556}$$

Hence,

$$\boldsymbol{\beta}^{SO} = [(\partial \boldsymbol{e}^{SO}/\partial \boldsymbol{v}^{SO})_N - \mathbf{I}^{SO}]\,\boldsymbol{\sigma}^{SO}\,. \tag{557}$$

Therefore, one has to determine the sensitivities of orbital energies with respect to the global external potential quantities to generate the global linear response matrix in the SO resolution.

Through its inverse, $(\boldsymbol{\beta}^{SO})^{-1}$, the linear response matrix can be expressed in terms of the relevant local quantities of the preceding section. This can be demonstrated using the following chain-rule transformation of the defining derivative:

$$(\boldsymbol{\beta}^{SO})^{-1} \equiv \left(\frac{\partial \boldsymbol{v}^{SO}}{\partial \boldsymbol{n}^{SO}}\right)_{\boldsymbol{\varphi}, N} = \iint \left(\frac{\partial \boldsymbol{\rho}^{SO}(\mathbf{r}')}{\partial \boldsymbol{n}^{SO}}\right)_N \left(\frac{\partial \boldsymbol{v}^{SO}(\mathbf{r})}{\partial \boldsymbol{\rho}^{SO}(\mathbf{r}')}\right) \left(\frac{\partial \boldsymbol{v}^{SO}}{\partial \boldsymbol{v}^{SO}(\mathbf{r})}\right)_{\boldsymbol{\varphi}} \mathbf{dr\, dr'}$$

$$\equiv \iint \bar{\mathbf{f}}^{SO}(\mathbf{r}')\, [\boldsymbol{\beta}^{SO}(\mathbf{r}', \mathbf{r})]^{-1}\, \boldsymbol{\Omega}^{SO}(\mathbf{r})\ \ \mathbf{dr\, dr'}\,, \tag{558}$$

where the diagonal matrix function $\boldsymbol{\Omega}^{SO}(\mathbf{r})$ groups the SO densities, $\{\Omega^{SO}_{i,j}(\mathbf{r}) = \Omega^{\sigma}_{i,i}\, \delta_{i,j}\, \delta_{\sigma,\sigma'}\}$, and the *internal relaxation matrix* function,

$$[\partial \rho_i(\mathbf{r})/\partial n_j]_N \equiv \bar{f}_{j,i}(\mathbf{r}) = f_{j,i}(\mathbf{r}) - \sum_{k \neq j} [\partial n_k/\partial n_j]_N\, f_{k,i}(\mathbf{r})\,. \tag{559}$$

In the last equation the indices i, j, and k refer to the SO index in vectors $\boldsymbol{n}^{SO}$ and $\boldsymbol{\rho}^{SO}(\mathbf{r})$, which uniquely identifies the SO, i.e., i ($\equiv \varphi^{\sigma}_i$); the same simplified notation will be used in the next section. This relation takes into account the implicit dependencies between SO occupations due to the fixed N constraint:

$$dn_j = -\sum_{k \neq j} dn_k\,. \tag{560}$$

4. Frontier Orbital Considerations

The external FF indices, $\mathcal{F}^{SO} = (\mathcal{F}^{\alpha}, \mathcal{F}^{\beta}) = \{\mathcal{F}_i\}$, depend upon the nature of the dN displacement (see Section 2 in Chapter 6). The only non-vanishing values in the absolute zero temperature will correspond to the frontier SO, HOMO (H) or LUMO (L), involved in the change under consideration. For example, when electron is added (removed) to (from) the system, $\mathcal{F}_{L(H)} = 1$, in the $dN > 0$ ($dN < 0$) global process, while the remaining indices vanish identically. This frontier orbital (F) perspective allows one to formally define the global chemical potential and orbital/global charge sensitivities in the "frozen" SO approximation. Namely, using Eq. 552 gives

$$\mu^F = \sum_i e_i\, \mathcal{F}_i = \begin{cases} \mu_n \equiv e_L, & dN = dn_L \quad \text{(nucleophilic external CT)} \\ \mu_e \equiv e_H, & dN = dn_H \quad \text{(electrophilic external CT)} \end{cases}. \tag{561}$$

This predicts the inflow of electrons to the system ($dN > 0$), when $\mu_r > e_L$ (nucleophilic environment), an outflow of electrons from the system ($dN < 0$) for $\mu_r < e_H$ (electrophilic environment), and an isoelectronic system polarization ($dN = 0$) in the range $e_H < \mu_r < e_L$ (isoelectronic, CT-neutral environment).

The orbital hardnesses $\boldsymbol{\eta}^{SO}$ of Eq. 554 can be similarly related to the corresponding frontier row of the SO-resolved hardness matrix:

$$\eta_i^{F} = \sum_j \mathcal{F}_j\, \eta_{j,i} = \begin{pmatrix} \eta_i^{n} \equiv \eta_{L,i}, & dN = dn_L \ \text{(nucleophilic external CT)} \\ \eta_i^{e} \equiv \eta_{H,i}, & dN = dn_H \ \text{(electrophilic external CT)} \end{pmatrix}. \tag{562}$$

Similarly, for the global hardness of Eq. 555 one similarly obtains:

$$\eta^{F} = \mathcal{F}^{SO}\, \boldsymbol{\eta}^{SO}\, (\mathcal{F}^{SO})^{\dagger} = \begin{pmatrix} \eta_{L,L} \equiv \eta_n^{F}, & dN = dn_L \ \text{(nucleophilic CT)} \\ \eta_{H,H} \equiv \eta_e^{F}, & dN = dn_H \ \text{(electrophilic CT)} \end{pmatrix}. \tag{563}$$

In the isoelectronic (radical) chemical attack, when the number of electrons in the system remains unchanged, yet another finite-difference measure of the global hardness can be adopted, which involves an estimate from an electron addition to φ_L $[\mathcal{F}_+^{SO} = (\mathcal{F}_L = 1, \{\mathcal{F}_j = 0, j \neq L\})]$ and an electron removal from φ_H $[\mathcal{F}_-^{SO} = (\mathcal{F}_H = 1, \ \{\mathcal{F}_j = 0, j \neq H\})]$:

$$\eta_i^{F} = \mathcal{F}_+^{SO}\, \boldsymbol{\eta}^{SO}\, (\mathcal{F}_-^{SO})^{\dagger} = \eta_{L,H}, \qquad (dN = 0, \text{ radical attack}). \tag{564}$$

It should also be mentioned at this point that the familiar finite-difference estimate of the SO hardness, obtained by the quadratic interpolation of the orbital energies for the neutral species, M^0, anion M^{-1} (obtained by adding electron to φ_L), and cation M^{+1} (resulting from a removal of a single electron from φ_H) gives yet another [*unbiased* (u)] F-estimate of the SO hardness:

$$\eta_u^{F} = e_L - e_H, \tag{565}$$

measuring the H-L orbital energy gap. It follows from this quantity that the soft species are identified in the orbital description by small such energy separation, while hard species correspond to large energy difference between the frontier SO.

When both α and β orbitals are changing their electron occupations in an external CT the above $\mathcal{F}^{SO}$ averaging will also involve the spin off-diagonal hardness matrix elements. Moreover, in the case of degenerate frontier orbital energies, the external populational displacement will generally involve all orbitals of such L(H) levels, thus projecting the whole sets into the averages determining the orbital and global SO hardnesses.

Recently, it has been argued [30] that one can use the orbital occupation probabilities,

$$\boldsymbol{p}^{SO} = \boldsymbol{n}^{SO}/N = \{p_i\}, \tag{566}$$

as weights in the averaging of the SO orbital energies and hardness matrix elements into the corresponding global SO quantities:

$$\langle \mu^{SO} \rangle = \boldsymbol{p}^{SO}\, (\boldsymbol{e}^{SO})^{\dagger}, \tag{567}$$

$$\langle \boldsymbol{\eta}^{SO} \rangle = \boldsymbol{p}^{SO}\, \eta^{SO}\,, \qquad \langle \eta^{SO} \rangle = \langle \boldsymbol{\eta}^{SO} \rangle\, (\boldsymbol{p}^{SO})^{\dagger} = \boldsymbol{p}^{SO}\, \eta^{SO}\, (\boldsymbol{p}^{SO})^{\dagger}\,. \tag{568}$$

When limited only to the valence-shell orbitals the quantity of Eq. 567 is the negative electronegativity of Allen [149]. The orbital occupation probabilities of Eq. 566 project out the virtual orbital contributions to charge sensitivities. One can remedy this by adopting Slater's transition-state approach [150] of half-occupations of the frontier electrons. This also effectively accounts for the orbital relaxation. As recently demonstrated by Grigorov et al. [147] the KS estimate of $\langle \eta^{SO} \rangle$ from the Slater transition-state approximation correlates very well with the experimental finite difference estimate of $\eta \approx I - A$, where I and A stand for the system ionization potential and electron affinity, respectively.

It should be noted that the derivative properties discussed in this section refer to the equilibrium displacements along the "ground-state" energy surface $E^{SO}_{g.s.}(\boldsymbol{n}^{SO}, \boldsymbol{v}^{SO})$ for the fixed SO shapes. For example, the $\boldsymbol{\beta}^{SO}$ matrix carries the information about changes in the SO occupations per unit displacements in the SO external potential quantities, $\boldsymbol{v}^{SO}$, due to changes in the external potential $v(\mathbf{r})$. Obviously, the extrapolated linear responses $d\boldsymbol{n}^{SO}$, can violate the SO occupation limit $0 \le n_i \le 1$, unless this physical range is not safeguard by the additional constraints or the way in which the displaced occupations are generated. One way to incorporate the occupational limits is to express the orbital occupations through the angle variables [16]:

$$n_i = \cos^2 a_i \qquad \text{or} \qquad dn_i = -\sin(2\,a_i)\, da_i\,, \tag{569}$$

with the angles $\boldsymbol{a}^{SO} = \{a_i\} = (\boldsymbol{a}^{\alpha}, \boldsymbol{a}^{\beta})$ providing the occupation generating parameters preserving the Pauli limits. Obviously, adopting such an approach calls for the transformation of the hardness/softness quantities discussed above to these angle representation.

In the process of determining the optimum orbital occupations in the SCF iteration for the current shapes of orbitals, one searches for the minimum of the *constrained energy function*, involving the cut of $E^{SO}_{g.s.}(\boldsymbol{n}^{SO}, \boldsymbol{v}^{SO})$ for the fixed $\boldsymbol{v}^{SO}$, $E^{SO}_{v}(\boldsymbol{n}^{SO})$, supplemented by the relevant diagonal idempotency constraints, which assure the correct occupational limits of the optimum solutions [31]. We shall discuss such constrained energy functions later in Sections 6 and 7 of this chapter.

5. Global Frontier Orbital Properties of Open Reactive Systems

Let us qualitatively examine, using the frontier orbital approximation, the case of a global equilibrium in the system (A¦B¦r), consisting of reactants A and B in contact with a hypothetical (macroscopic) reservoir r characterized by its chemical potential, μ_r. We first observe that the acidic (acceptor, harder) reactant A will generally exhibit a lower value of e_H (higher value of e_L), in comparison to the corresponding levels of the basic (donor, softer) reactant B; this is due to a relatively higher electronegativity and hardness (larger H-L gap) of A. A straightforward analysis shows that there are five regions of μ_r, which define different actions of the reservoir relative to both these reactants in (A¦B¦r):

(i) $\mu_r < e_H^A$: outflow of electrons from both reactants to r : $dN_A < 0$ and $dN_B < 0$;
(ii) $e_H^A \le \mu_r < e_H^B$: outflow of electrons from B: $dN_A = 0$, $dN_B < 0$;
(iii) $e_H^B \le \mu_r \le e_L^B$: isoelectronic environment: $dN_A = dN_B = 0$;
(iv) $e_L^B < \mu_r \le e_L^A$: inflow of electrons to B: $dN_A = 0$, $dN_B > 0$;
(v) $e_L^A < \mu_r$: inflow of electrons to A and B: $dN_A > 0$ and $dN_B > 0$.

Thus, the reservoir chemical potential levels below e_H^B give rise to an outflow of electrons from M (electrophilic environment of M), while those above e_L^B generate an inflow of electrons to M (nucleophilic environment of M). In other words, the frontier orbital levels of the softer reactant (B) mark the iso-electronic range of the reservoir chemical potentials, and the upper (lower) limit of the chemical potential of the electrophilic (nucleophilic) environment. These five regions of the reservoir chemical potential values determine all qualitative possibilities of the external charge and spin transfer in (M¦r) in such a reactant resolved frontier orbital perspective.

The above analysis also shows that in the region (i) of the chemical potential of a strongly electrophilic reservoir the global chemical potential of M is given by the FF weighted average of the H orbital energies of both reactants (see Eqs. 108 and 114):

$$\mu_M^F(e,e) = \mathscr{F}_{A(e)}^{M(F)}\, e_H^A + \mathscr{F}_{B(e)}^{M(F)}\, e_H^B, \tag{570}$$

where:

$$\mathscr{F}_{A(e)}^{M(F)} = \eta_{B(e)}^{M(F)} / [\eta_{A(e)}^{M(F)} + \eta_{B(e)}^{M(F)}], \quad \mathscr{F}_{B(e)}^{M(F)} = \eta_{A(e)}^{M(F)} / [\eta_{A(e)}^{M(F)} + \eta_{B(e)}^{M(F)}], \tag{571}$$

and the corresponding electrophilic (e) frontier SO measures of the condensed hardness matrix elements in the reactant resolution are:

$$\eta_{A(e)}^{M(F)} = \eta_{H,H}^{A,A} - \eta_{H,H}^{A,B}, \quad \eta_{B(e)}^{M(F)} = \eta_{H,H}^{B,B} - \eta_{H,H}^{A,B}. \tag{572}$$

Both FF indices of Eq. 571 must be positive, to satisfy the SO occupation limits, i.e. predicting the fractional electron removal from the H levels of both reactants in the global $dN < 0$ electron displacement; it should be emphasized that they will be satisfied for an internally stable system only when both $\eta_A^{M(F)}$ and $\eta_B^{M(F)}$ are positive.

The vector of these external FF indices, $\mathscr{F}^F(e,e) \equiv [\mathscr{F}_{A(e)}^{M(F)}, \mathscr{F}_{B(e)}^{M(F)}]$, also determines the global hardness of M at these reservoir chemical potential levels [region (i)]:

$$\eta_M^F(e,e) = \boldsymbol{\mathscr{F}}^F(e,e)\, \boldsymbol{\eta}^F(e,e)\, [\boldsymbol{\mathscr{F}}^F(e,e)]^\dagger, \tag{573}$$

where $\boldsymbol{\eta}^F(e,e)$ is the (2 × 2) F-hardness matrix including the HOMO hardness matrix elements in the (A|B) resolution:

$$\boldsymbol{\eta}^{F}(e, e) = \begin{pmatrix} \eta_{H,H}^{A,A} & \eta_{H,H}^{A,B} \\ \eta_{H,H}^{B,A} & \eta_{H,H}^{B,B} \end{pmatrix}. \tag{574}$$

In the region (ii) of the chemical potential of an electrophilic reservoir only the HOMO of B looses electrons, so that $\mathscr{F}^{F}(i, e) = (0, 1)$ and hence

$$\mu_{M}^{F}(i, e) = e_{H}^{B} \quad \text{and} \quad \eta_{M}^{F}(i, e) = \eta_{H,H}^{B,B}. \tag{575}$$

The isoelectronic (i) environment of the region (iii), where the frontier orbitals of reactants in M do not change their electron occupations, can be formally described as corresponding to the zero global softness (infinite hardness) of M.

A similar analysis for a nucleophilic (n) environment, represented by the region (iv) of the reservoir chemical potentials, gives $\mathscr{F}^{F}(i, n) = (0, 1)$ and

$$\mu_{M}^{F}(i, n) = e_{L}^{B} \quad \text{and} \quad \eta_{M}^{F}(i, n) = \eta_{L,L}^{B,B}. \tag{576}$$

Finally, for a strongly nucleophilic environment [region (v) of μ_r] one finds

$$\mu_{M}^{F}(n, n) = \mathscr{F}_{A(n)}^{M(F)}\, e_{L}^{A} + \mathscr{F}_{B(n)}^{M(F)}\, e_{L}^{B}, \tag{577}$$

where:

$$\mathscr{F}_{A(n)}^{M(F)} = \eta_{B(n)}^{M(F)} / [\eta_{A(n)}^{M(F)} + \eta_{B(n)}^{M(F)}], \quad \mathscr{F}_{B(n)}^{M(F)} = \eta_{A(n)}^{M(F)} / [\eta_{A(n)}^{M(F)} + \eta_{B(n)}^{M(F)}], \tag{578}$$

$$\eta_{A(n)}^{M(F)} = \eta_{L,L}^{A,A} - \eta_{L,L}^{A,B}, \quad \eta_{B(n)}^{M(F)} = \eta_{L,L}^{B,B} - \eta_{L,L}^{A,B}. \tag{579}$$

In this case the global hardness follows from the LUMO part of the SO hardness tensor in the reactant resolution,

$$\boldsymbol{\eta}^{F}(n, n) = \begin{pmatrix} \eta_{L,L}^{A,A} & \eta_{L,L}^{A,B} \\ \eta_{L,L}^{B,A} & \eta_{L,L}^{B,B} \end{pmatrix}, \tag{580}$$

$$\eta_{M}^{F}(n, n) = \mathscr{F}^{F}(n, n)\, \boldsymbol{\eta}^{F}(n, n)\, [\mathscr{F}^{F}(n, n)]^{\dagger}, \tag{581}$$

where $\mathscr{F}^{F}(n, n) \equiv (\mathscr{F}_{A(n)}^{M(F)}, \mathscr{F}_{B(n)}^{M(F)})$.

In the above qualitative considerations we have implicitly incorporated the orbital occupation limits by excluding flows violating the Pauli exclusion principle, e.g., the electron transfer from the HOMO of B to the HOMO of A.

We would like to conclude this section with a comment on the SCF MO-type algorithm for determining the interreactant charge flows leading to the global equilibrium state in $M^* = (A^* \mid B^*)$. This can be done routinely using the independent SCF MO (or DFT) calculations on the isolated reactants $M^0 = (A^0 | B^0)$ or the polarized reactants in $M^+ = (A^+ | B^+)$, using, e.g., the variation theory for subsystems [145], and the SCF MO calculations for a supermolecule $M^* = (A^* \mid B^*)$. Clearly, in a more qualitative analysis a perturbational approach to the interaction between the reactant orbitals can also be envisaged at the second stage [100, 101]. The closed reactant calculations define the initial (reference) global number of electrons, $N_X^0 = \boldsymbol{n}_X^0 \boldsymbol{1}_X^\dagger$, and occupations $\boldsymbol{n}_\varphi^0 = (\boldsymbol{n}_A^0, \boldsymbol{n}_B^0)$ of the optimum SO in this state, $\boldsymbol{\varphi}_X^0$, X = A, B; to avoid the basis set superposition error these SCF calculations should use the whole basis set of all AIM in M. Due to the orthogonality of $\boldsymbol{\varphi}_X^0$ these (gross) SO occupations are equal to the corresponding diagonal elements of the overall (spinless) CBO matrices of reactants: $n_{A,a}^0 = P_{a,a}^{0,A}$ and $n_{B,b}^0 = P_{b,b}^{0,B}$, where a and b denote the SO of reactants A and B, respectively.

One then expands the SO of M as a whole, $\boldsymbol{\varphi}_M$, in the supermolecule calculations on $M^* = (A^* \mid B^*)$ in terms of the initial SO of reactants $\boldsymbol{\varphi}^0 = (\boldsymbol{\varphi}_A^0, \boldsymbol{\varphi}_B^0)$, $\boldsymbol{\varphi}_M = \boldsymbol{\varphi}^0 \mathbf{C}_\varphi$, and solves the corresponding secular equations for the optimum *Linear Combinations of (reactant) MO* (LCMO) expansion coefficients $\mathbf{C}_\varphi$:

$$\mathbf{H}_\varphi \mathbf{C}_\varphi = \mathbf{S}_\varphi \mathbf{C}_\varphi \mathbf{e}_\varphi, \qquad \mathbf{C}_\varphi^\dagger \mathbf{S}_\varphi \mathbf{C}_\varphi = \mathbf{I}_M, \tag{582}$$

where $\mathbf{H}_\varphi = \langle \boldsymbol{\varphi}^0 | H_M | \boldsymbol{\varphi}^0 \rangle$, $\mathbf{S}_\varphi = \langle \boldsymbol{\varphi}^0 | \boldsymbol{\varphi}^0 \rangle$, $\mathbf{e}_\varphi$ denotes the matrix of the canonical SO energies in M^*, and H_M stands for the molecular electronic hamiltonian. The resulting matrix $\mathbf{C}_\varphi$ and the optimum SO occupations in M, $\mathbf{n}_M = \{n_i \delta_{i,j}\}$, determine the CBO matrix in the $\boldsymbol{\varphi}^0$ representation,

$$\mathbf{P}_\varphi = \mathbf{C}_\varphi \mathbf{n}_M \mathbf{C}_\varphi^\dagger, \tag{583}$$

which can be used to define the effective occupations of $\boldsymbol{\varphi}^0$ in M^*, using the familiar populational analysis approach. For example, in the Mulliken scheme one attributes to the initial SO of reactants the following gross electron occupations:

$$n_a^M = (\mathbf{P}_\varphi)_{a,a} + \sum_b^B (\mathbf{P}_\varphi)_{a,b} (\mathbf{S}_\varphi)_{b,a},$$

$$n_b^M = (\mathbf{P}_\varphi)_{b,b} + \sum_a^A (\mathbf{P}_\varphi)_{b,a} (\mathbf{S}_\varphi)_{a,b}. \tag{584}$$

In the last equation we have recognized that only the inter-reactant overlap integrals in $\mathbf{S}_\varphi$ do not vanish identically among the off-diagonal elements of the overlap matrix in the $\boldsymbol{\varphi}^0$ representation. These effective SO populations, $\boldsymbol{n}_\varphi^M = [\{n_a^M\}, \{n_b^M\}] \equiv (\boldsymbol{n}_A^M, \boldsymbol{n}_B^M)$, define the occupational displacements due to the B → A CT, $\Delta \boldsymbol{n}_\varphi^* = \boldsymbol{n}_\varphi^M - \boldsymbol{n}_\varphi^0 \equiv (\Delta \boldsymbol{n}_A^*, \Delta \boldsymbol{n}_B^*)$, and the corresponding shifts in the average numbers of electrons of reactants in M: $\Delta N_X^* = \Delta \boldsymbol{n}_X^* \boldsymbol{1}_X^\dagger$. These finite-differences in the reactant SO resolution can then be used to estimate the CT FF of reactants, effectively combining the respective diagonal and off- diagonal FF contributions,

$$\mathscr{F}_X^{CT} = \Delta n_X^* / \Delta N_X^*, \quad X = A, B, \tag{585}$$

which define the overall CT FF vector in the SO-resolution: $\mathscr{F}^{CT} = (\mathscr{F}_A^{CT}, \mathscr{F}_B^{CT})$. We would also like to mention at this point the recently proposed, elegant approach by Ciosłowski and Stefanov [151], based upon the physical space partitioning of molecular electronic charge distribution into subsystems, in accordance with the topological theory of atoms in molecules [7]. It can also be used to generate the subsystem *in situ* chemical potential (electronegativity) differences and charge sensitivities of molecular subsystems from the reactant and supermolecule SCF CI calculations.

6. Constrained Energy Functions and Lagrange Multipliers

In variational principles formulated in terms of the CBO matrix elements, which involve variations of the molecular orbital shapes, the system energy has to be supplemented by the relevant CBO (density matrix) idempotency constraints reflecting the subsidiary conditions of the orbital orthogonality and normalization [145]. Moreover, in open systems the corresponding constraints associated with the subsidiary conditions on the global number of electrons and orbital occupations, should also be incorporated into the relevant auxiliary energy functional [31]. For example, one has to ensure that the orbital occupation limits implied by the Fermi statistics have to be satisfied in the ground-state, unless this is not automatically guaranteed by the specific procedure used to modify the occupation numbers, e.g., through the angle variables mentioned in Section 4 of this chapter.

In the CSA one focuses mainly on the SO occupational degrees of freedom, for the fixed orbitals, e.g., the optimum SO for the specific orbital occupations, $\boldsymbol{n}^0$ (here we drop the SO superscript to simplify notation). This defines the energy function of the SO occupations (see Eq. 499b) parametrically dependent upon the assumed orbitals and their occupations: $E_\varphi^0(\boldsymbol{n}) = E_\varphi[\boldsymbol{n}; \boldsymbol{\varphi}^0(\boldsymbol{n}^0)]$. When one is interested in the specific excited electron orbital configuration, generally involving also the empty (virtual) orbitals in the g.s. configuration, one has to ensure in the $\boldsymbol{n}$-variational principle that the optimum occupations will reproduce the very configuration in question. This can be effected via the linear occupational constraints added to the energy function $E_\varphi^0(\boldsymbol{n})$ (see also Eq. 548):

$$\mathscr{E}_\varphi^0(\boldsymbol{n}; \boldsymbol{n}^0) \equiv E_\varphi^0(\boldsymbol{n}) - \boldsymbol{e}\,(\boldsymbol{n} - \boldsymbol{n}^0)^\dagger . \tag{586}$$

However, the prior knowledge of orbital occupations is not available in the g.s. SCF calculations. Therefore, one has to replace the above linear occupational constraints with the corresponding non-linear (quadratic) auxiliary conditions, which do not *a priori* predetermine any specific electron configuration. In particular, for the molecular system they have to guarantee the right (extreme) occupations of the SO in the equilibrium (ground) state [31]. Preferably, such constraints should not fix the overall number of electrons, by not specifying the number of occupied SO, to also cover the open system case. Therefore, the relevant global constraint, involving the chemical potential of the external reservoir, equal at equilibrium to that of the molecule as a whole, will also enter the corresponding auxiliary energy function in the open system case (see Eq. 551).

Let us consider the quadratic occupational constraints in the UHF/LSDA approximation. Should one want to assure only the integral occupations of spin-orbitals in the optimum configuration, it suffices to associate with each (canonical) SO, φ_i, the quadratic occupation constraint term, $n_i(n_i - 1) = 0$, since this equation is satisfied for both extreme SO occupations: $n_{i,opt} = 1$ or $n_{i,opt} = 0$. When fractional occupations are assumed or some g.s. virtual SO are to be occupied in the excited electron configuration under consideration, $\boldsymbol{n}^0$, the same type of the orbital occupation constraint $n_i(n_i - n_i^0) = 0$ can be used instead of the linear constraints of Eq. 586, since it also assures the desired solutions: $n_{i,opt} = n_i^0 > 0$ for occupied orbitals and $n_{i,opt} = n_i^0 = 0$ for virtual orbitals in the assumed configuration. Notice, that no global constraint is required when the electron configuration is fixed, since then $N_{opt} = \sum_i n_i^0 \equiv N(\boldsymbol{n}^0) = N^0$.

These orbital occupation constraints are sufficient when the initial (orthogonal) orbitals $\boldsymbol{\varphi}^0(\boldsymbol{n}^0)$ are held "frozen" during populational displacements from $\boldsymbol{n}^0$. However, when orbital relaxation is allowed, one has to include additional idempotency constraints,

$$(\boldsymbol{\gamma}^\sigma)^2 - \boldsymbol{\gamma}^\sigma = \mathbf{0}\,, \qquad \sigma = \alpha, \beta\,. \tag{587}$$

The one-electron density matrix for spin σ, $\boldsymbol{\gamma}^\sigma$, in a given representation $|\boldsymbol{\lambda}\rangle$ can be defined in terms of the corresponding SO expansion coefficients:

$$\boldsymbol{\gamma}^\sigma = \langle \boldsymbol{\lambda} | \boldsymbol{\varphi}^\sigma \rangle \langle \boldsymbol{\varphi}^\sigma | \boldsymbol{\lambda} \rangle = \mathbf{C}^{\lambda,\sigma} (\mathbf{C}^{\lambda,\sigma})^\dagger\,; \tag{588}$$

here the LCAO MO matrix $\mathbf{C}^{\lambda,\sigma} = \langle \boldsymbol{\lambda} | \boldsymbol{\varphi}^\sigma \rangle$ defines the σ-SO, $|\boldsymbol{\varphi}^\sigma\rangle$, in terms of the basis functions $|\boldsymbol{\lambda}\rangle$: $|\boldsymbol{\varphi}^\sigma\rangle = |\boldsymbol{\lambda}\rangle\, \mathbf{C}^{\lambda,\sigma}$. Clearly, the direct orbital normalization and orthogonality constraints,

$$\langle \varphi_l^\sigma | \varphi_k^\sigma \rangle - \delta_{k,l} = 0\,, \tag{589}$$

can replace the idempotency relations of Eq. 587 as the alternative constraints in the relevant Euler-Lagrange variational principle.

Therefore, the auxiliary energy functional of the SO and their occupations for the most general case involving the constraints of the fixed number of electrons, integral electron occupations of SO [$n_i^{\sigma,0}$ = (0 or 1)], and the SO orthonormality is:

$$\mathscr{E}_\varphi[\boldsymbol{n}, \boldsymbol{\varphi}; N^0] = E[\boldsymbol{n}, \boldsymbol{\varphi}] - \mu\,[N(\boldsymbol{n}) - N^0]$$

$$- \sum_i \omega_i [n_i(n_i - 1)] - \sum_{i \le j}\sum \vartheta_{i,j}^\sigma \left(\langle \varphi_i | \varphi_j \rangle - \delta_{i,j}\right), \tag{590a}$$

where $\boldsymbol{n} = (\boldsymbol{n}^\alpha, \boldsymbol{n}^\beta)$ groups the SO occupations, while $\boldsymbol{\omega} = (\boldsymbol{\omega}^\alpha, \boldsymbol{\omega}^\beta) = \{\omega_i^\sigma\}$ and $\boldsymbol{\vartheta} = (\boldsymbol{\vartheta}^\alpha, \boldsymbol{\vartheta}^\beta) = \{\vartheta_{i,j}^\sigma\}$ are the Lagrange multipliers associated with the orbital occupation and orthonormality constraints, respectively. In the canonical representation, in which $\boldsymbol{\vartheta}^{\sigma}$ is diagonal for $\sigma = \alpha, \beta$, the corresponding diagonal ortho-normality constraints are the orbital energies $\{\vartheta_{i,j}^{\sigma,\mathrm{can}} = e_i^{\sigma}\,\delta_{i,j}\}$. Alternatively, one could use the auxiliary function of the orbital occupations and spin-density matrices:

$$\mathscr{E}_{\gamma}[\boldsymbol{n}, \{\boldsymbol{\gamma}^{\sigma}\}; N^0] = E_v[\boldsymbol{n}, \{\boldsymbol{\gamma}^{\sigma}\}] - \mu\,[N(\boldsymbol{n}) - N^0]$$
$$- \sum_i \omega_i [n_i(n_i - 1)] - \sum_{\sigma}\sum_{i \le j} \theta_{i,j}^{\sigma}\left(\sum_k \gamma_{i,k}^{\sigma}\,\gamma_{k,j}^{\sigma} - \gamma_{i,j}^{\sigma}\right), \tag{590b}$$

with the Lagrange multipliers $\boldsymbol{\theta} = (\boldsymbol{\theta}^{\alpha}, \boldsymbol{\theta}^{\beta}) = \{\theta_{i,j}^{\sigma}\}$ being associated with the density matrix idempotency constraints.

In the simpler case, when orbital shapes are held fixed, the corresponding auxiliary functional involves only the first group of populational constraints, valid when only orbital occupations are changed; similarly, when occupations are "frozen", only the second group of orthonormality/idempotency constraints is needed for optimization of orbitals or, equivalently, density matrices in the fixed basis set. In the canonical representation, the diagonal Lagrange multipliers associated with the orbital orthonormality constraints become the SO energies, and the relevant Euler equations for the orbitals are the familiar HF/KS equations. The same result follows from the optimization of the density matrix function of Eq. 590b [145]. One can therefore envisage solving the general problem of simultaneously optimizing both orbitals and their occupations in two stages: in one the orbitals are optimized for the current fixed occupations, and in the other the optimum occupations are determined for the current shapes of orbitals. Such two stages are indeed involved in the familiar SCF calculations, with the *minimum orbital energy* (MOE) criterion being usually adopted when fixing orbital occupations in the current loop of the SCF iterative scheme.

Consider the occupation determining step, of main interest in the CSA, with the auxiliary energy functional for the fixed (canonical) orbitals $\boldsymbol{\varphi}^0$:

$$\mathscr{E}_{\varphi}^{0}(\boldsymbol{n}; N^0) = E(\boldsymbol{n}; \boldsymbol{\varphi}^0) - \mu\,[N(\boldsymbol{n}) - N^0] - \sum_i \omega_i\, n_i\,(n_i - 1)\,. \tag{591}$$

The physical meaning of $\boldsymbol{\omega}$ follows from the associated Euler-Lagrange equations determining the extremum of $\mathscr{E}_{\varphi}^{0}(\boldsymbol{n}; N^0)$ for the occupations satisfying the global constraint:

$$\left.\frac{\partial E}{\partial n_i}\right|_{n^{\mathrm{opt}},\,N^0} = e_i[N^0, v] = \omega_i\,(2\,n_i^{\mathrm{opt}}[N^0, v] - 1)\,, \tag{592}$$

where $n_i^{\mathrm{opt}}[N^0, v]$ and $e_i[N^0, v]$ denote the optimum SO occupation and the associated orbital energy (at this electron configuration) for the specified global number of electrons, external potential, and the assumed shapes of orbitals. Hence, for the g.s. occupied SO ($n_i^{\mathrm{opt}}[N^0, v] = 1$),

$$\omega_i = e_i[N^0, v] \qquad i \in \text{occupied SO}\ , \tag{593a}$$

and for the g.s. virtual orbitals ($n_i^{opt}[N^0, v] = 0$):

$$\omega_i = -e_i[N^0, v] \qquad i \in \text{virtual SO}\ . \tag{593b}$$

Determining the minimum of the energy function, which fixes the assumed electron configuration $\boldsymbol{n}^0$ through the quadratic SO occupational constraints,

$$\mathscr{E}_\varphi^0(\boldsymbol{n}; \boldsymbol{n}^0) = E[\boldsymbol{n}; \boldsymbol{\varphi}^0] - \sum_i \omega_i^0\, n_i\, (n_i - n_i^0)\ , \tag{594}$$

and thus the prescribed global number of electrons $N(\boldsymbol{n}^0) = N^0$, gives the following Euler-Lagrange equations:

$$\left.\frac{\partial E}{\partial n_i}\right|_{\boldsymbol{n}^0} = \omega_i^0\, n_i^0 = e_i[\boldsymbol{n}^0, v]\ . \tag{595}$$

Hence, for the occupied SO ($n_i^0 > 0$)

$$\omega_i^0 = e_i[\boldsymbol{n}^0, v]/n_i^0\ , \qquad i \in \text{occupied SO}\ . \tag{596a}$$

It should be observed that the above Lagrange multiplier becomes infinite for virtual orbitals ($n_i^0 = 0$, $e_i[\boldsymbol{n}^0, v] > 0$):

$$\omega_i^0 = \infty\ , \qquad i \in \text{virtual SO}\ . \tag{596b}$$

7. Occupational Derivatives of the Constrained Energy Functions

Let us examine the electron occupation gradient (chemical potential) and Hessian (hardness tensor) of the auxiliary energy functions of the preceding section. They both determine the equilibrium distribution of electrons in the quadratic approximation. Within the CSA the most interesting case for determining the reactant interaction is the constrained energy function of Eq. 591, which does not predetermine the electron configuration. For simplicity we consider the occupation determining step for the "frozen" canonical SO; in this part of the SCF iteration step the orthonormality/ idempotency constraints do not intervene.

We first examine the occupational gradient of the constrained energy function for the closed molecular system, i.e., when the constraint $N(\boldsymbol{n}) = N^0$ is automatically satisfied; in such a case the constrained energy function becomes:

$$\mathscr{E}_N^0(\boldsymbol{n}) = E[\boldsymbol{n}; \boldsymbol{\varphi}^0] - \sum_i \omega_i\, n_i\, (n_i - 1)$$

$$= E[\boldsymbol{n}; \boldsymbol{\varphi}^0] - \sum_i [n_i(n_i - 1)/(2n_i^{opt} - 1)]\, e_i \equiv E[\boldsymbol{n}; \boldsymbol{\varphi}^0] - \sum_i p_i\, e_i\ . \tag{597}$$

The i-th component of its occupational gradient, $\bar{\boldsymbol{u}} = (\bar{\boldsymbol{u}}^{\alpha}, \bar{\boldsymbol{u}}^{\beta}) = \partial \mathscr{E}_N^0 / \partial \boldsymbol{n} = \{\bar{u}_i\}$, is given by the following expression:[1]

$$\frac{\partial \mathscr{E}_N^0}{\partial n_i} \equiv \bar{u}_i = \frac{2(n_i^{opt} - n_i)}{2n_i^{opt} - 1} e_i - \sum_j p_j \eta_{j,i} \equiv q_i e_i - \sum_j p_j \eta_{j,i}, \tag{598}$$

where the unconstrained SO hardness matrix elements (see Eqs. 518, 520, and 535-537) group the second partials of the energy with respect to the SO occupations: $\boldsymbol{\eta} = \partial^2 E / \partial \boldsymbol{n}\, \partial \boldsymbol{n} = \partial \boldsymbol{e} / \partial \boldsymbol{n}$. We would like to observe that for the integral (optimum) SO occupations, i.e., when $\{ n_i^{opt} = (0 \text{ or } 1)\}$,

$$\bar{\boldsymbol{u}}(\boldsymbol{n}^{opt}) = \boldsymbol{0}, \tag{599}$$

since then $\{ q_i = p_i = 0\}$.

In the open system case, when N is not preserved, this electron occupation gradient is modified by the external potential of the reservoir μ_r (see Eq. 591):

$$\bar{\boldsymbol{u}}_{ext} = \frac{\partial \mathscr{E}_{\varphi}^0(\boldsymbol{n}; N^0)}{\partial \boldsymbol{n}} = \bar{\boldsymbol{u}} - \mu_r \boldsymbol{1}. \tag{600}$$

One similarly derives the constrained hardness matrix, $\bar{\mathbf{H}} = \partial^2 \mathscr{E}_N^0 / \partial \boldsymbol{n}\, \partial \boldsymbol{n} = \partial \bar{\boldsymbol{u}} / \partial \boldsymbol{n} = \{\bar{\mathrm{H}}_{i,j}\}$,

$$\bar{\mathrm{H}}_{i,j} = \frac{\partial^2 \mathscr{E}_N^0}{\partial n_i\, \partial n_j} = \frac{\partial \bar{u}_i}{\partial n_j} = \frac{\partial \bar{u}_j}{\partial n_i} = \frac{-2}{2n_i^{opt} - 1} \delta_{i,j}\, e_i + (q_i + q_j - 1)\, \eta_{j,i}. \tag{601}$$

For the final, integral SO occupations, $\{ n_i^{opt} = (0 \text{ or } 1)\}$, this expression gives:

$$\bar{\mathrm{H}}_{i,i}(\boldsymbol{n}^{opt}) = \begin{cases} -2e_i - \eta_{i,i} & (\text{i} \in \text{occupied SO}) \\ 2e_i - \eta_{i,i} & (\text{i} \in \text{virtual SO}) \end{cases},$$

$$\bar{\mathrm{H}}_{i,j}(\boldsymbol{n}^{opt}) = -\eta_{i,j} \qquad (\text{i} \neq \text{j}). \tag{602}$$

Thus, in the g.s. equilibrium (i.e., for the minimum of $\mathscr{E}_N^0$), $\bar{\mathbf{H}}(\boldsymbol{n}^{opt}) = -\mathbf{D} - \boldsymbol{\eta}$, where the diagonal ("level shifting") contribution, $-\mathbf{D}$, is positive definite since $e_i < 0$ (occupied SO) and $e_i > 0$ (virtual SO). Moreover, by the familiar wavefunction variational principle, $\bar{\mathbf{H}}(\boldsymbol{n}^{opt})$ must also be positive definite, despite the negative SO

[1] The constrained energy derivatives reported in Ref. [31] include an error due to an erroneous occupation dependence of the Lagrange multipliers.

hardness matrix contribution.

The quadratic Taylor expansion of $\mathscr{E}_N^0$ at the optimum (integral) occupations is:

$$\Delta\mathscr{E}_N^0(\boldsymbol{n}) = \overline{\boldsymbol{u}}(\boldsymbol{n}^{opt})\, \mathrm{d}\boldsymbol{n}^\dagger + \frac{1}{2}\mathrm{d}\boldsymbol{n}\, \overline{\mathbf{H}}(\boldsymbol{n}^{opt})\, \mathrm{d}\boldsymbol{n}^\dagger + \ldots$$

$$= \frac{1}{2}\mathrm{d}\boldsymbol{n}\, \overline{\mathbf{H}}(\boldsymbol{n}^{opt})\, \mathrm{d}\boldsymbol{n}^\dagger + \ldots > 0\,. \tag{603}$$

Thus $\mathscr{E}_N^0$ reaches minimum at the optimum (integral) configuration for the assumed shapes of orbitals.

For a general starting $\boldsymbol{n}^0$, for which $\overline{\boldsymbol{u}}(\boldsymbol{n}^0) \neq \boldsymbol{0}$, the optimum shifts of the SO occupations, $\mathrm{d}\boldsymbol{n}^{opt}(\boldsymbol{n}^0) = \boldsymbol{n}^{opt} - \boldsymbol{n}^0$, are determined in the above quadratic approximation from the condition of the vanishing gradient:

$$\overline{\boldsymbol{u}}(\boldsymbol{n}^{opt}) = \partial\mathscr{E}_N^0/\partial\boldsymbol{n}|_{\boldsymbol{n}^{opt}} \cong \overline{\boldsymbol{u}}(\boldsymbol{n}^0) + \mathrm{d}\boldsymbol{n}^{opt}\, \overline{\mathbf{H}}(\boldsymbol{n}^0) = \boldsymbol{0}\,. \tag{604}$$

These *internal electronegativity equalization* (IEE) equations in the SO resolution are solved by inverting $\overline{\mathbf{H}}(\boldsymbol{n}^0)$ to the corresponding softness matrix, $\overline{\mathbf{S}}(\boldsymbol{n}^0) = [\overline{\mathbf{H}}(\boldsymbol{n}^0)]^{-1}$:

$$\mathrm{d}\boldsymbol{n}^{opt}(\boldsymbol{n}^0) = -\,\overline{\boldsymbol{u}}(\boldsymbol{n}^0)\, \overline{\mathbf{S}}(\boldsymbol{n}^0)\,. \tag{605}$$

The above IEE procedure may provide an alternative to the MOE principle, which is commonly used to determine the SO occupations in the current SCF iteration. It can be run self-consistently, by using the optimum occupations,

$$\boldsymbol{n}_i^{opt} = \boldsymbol{n}_i^0 + \mathrm{d}\boldsymbol{n}_i^{opt}(\boldsymbol{n}_i^0)\,, \tag{606}$$

of the i-th estimate as the starting configuration $\boldsymbol{n}_{i+1}^0$ for the next estimate, until the subsequent iterations give selfconsistent SO occupations. Although the current occupation numbers in this iterative procedure may be fractional and exceeding the SO occupational limits, the optimum selfconsistent values will be integral, $\{\,\boldsymbol{n}_i^{opt} = (0 \text{ or } 1)\}$, as required by the Pauli principle in the one-determinant approximation.

APPENDIX A

VARIATIONAL PRINCIPLE FOR THE FUKUI FUNCTION AND THE MINIMUM HARDNESS PRINCIPLE

Consider the second-order Taylor expansion of the system electronic energy $E[\rho, v]$ (Eq. 13), for the constant external potential and the initial equilibrium density, $\rho = \rho_{g.s.}$, marking the chemical potential equalization (Eq. 6):

$$\Delta^{1+2}E_v[\rho] = \mu\, dN + \frac{1}{2}\int\int d\rho(\mathbf{r})\,\eta(\mathbf{r},\mathbf{r}')\,d\rho(\mathbf{r}')\,d\mathbf{r}\,d\mathbf{r}' . \tag{A1}$$

Let us further assume a general variation in the system electron density,

$$d\rho(\mathbf{r}) = dN\, g(\mathbf{r}) , \tag{A2}$$

with the unity-normalized shape-function $g(\mathbf{r})$:

$$\int g(\mathbf{r})\,d\mathbf{r} = 1 . \tag{A3}$$

Since for the starting equilibrium (g.s.) density the first-order energy does not involve g, only the second-order contribution, determined by the hardness kernel $\eta(\mathbf{r},\mathbf{r}')$, will determine the g-dependent part of $\Delta^{1+2}E_v[\rho]$,

$$\Delta^2 E_v[g] \equiv \frac{1}{2}(dN)^2\int\int g(\mathbf{r})\,\eta(\mathbf{r},\mathbf{r}')\,g(\mathbf{r}')\,d\mathbf{r}\,d\mathbf{r}' . \tag{A4}$$

Treating now $g(\mathbf{r})$ as the variational function, one finds that the optimum density displacement function, for which the system energy reaches the extremum value, must satisfy the following equation:

$$\int\left[\int g(\mathbf{r})\,\eta(\mathbf{r},\mathbf{r}')\,d\mathbf{r}\right]dg(\mathbf{r}')\,d\mathbf{r}' \equiv \int \eta[g;\mathbf{r}']\,dg(\mathbf{r}')\,d\mathbf{r}' = 0 . \tag{A5}$$

This variational principle is satisfied only when local quantity $\eta[g_{opt};\mathbf{r}']$ is equalized throughout the space, for the optimum $g_{opt}(\mathbf{r}) = f(\mathbf{r})$, due to the normalization constraint of Eq. A3, $\int dg(\mathbf{r})\,d\mathbf{r} = 0$. One recognizes this as the local hardness equalization principle (Eq. 198),

$$\eta[g_{opt};\mathbf{r}'] = \eta[g_{opt}] = \eta[f] = \eta , \tag{A6}$$

at the global hardness level

$$\eta = (\partial\mu/\partial N)_v = \eta[f, \mathbf{r}] \equiv \int f(\mathbf{r}')\,\eta(\mathbf{r}', \mathbf{r})\,d\mathbf{r}\,. \tag{A7}$$

This equalization takes place only when g becomes identical with the system local FF [152]:

$$g_{\text{opt}}(\mathbf{r}) = f(\mathbf{r}) = (\partial\rho(\mathbf{r})/\partial N)_v\,. \tag{A8}$$

One can therefore alternatively define the local FF through the EE equations, i.e. the inversion of the hardness kernel, or by the variational principle:

$$\delta_g(\Delta^2 E_v[g]) = 0\,. \tag{A9}$$

Let us denote the integral in Eq. A4 as the hardness functional for a general case for a nonequilibrium $g \neq f$,

$$\eta[g] \equiv \int\int g(\mathbf{r})\,\eta(\mathbf{r}, \mathbf{r}')\,g(\mathbf{r}')\,d\mathbf{r}\,d\mathbf{r}'\,. \tag{A10}$$

It follows from Eq. A9 that this functional reaches extremum, when the starting nonequilibrium "distribution" $g(\mathbf{r})$ becomes the true equilibrium (global) FF of the system under consideration. That this extremum is in fact a minimum for stable systems immediately follows from the spectral resolution of the hardness kernel (see Eqs. 435-437) and the general stability criterion, which requires that for stable systems the hardness kernel defines the positive definite quadratic form:

$$\begin{aligned}\eta[g] &= \int\int g(\mathbf{r})\left[\sum_\nu \xi_\nu(\mathbf{r})\,h_\nu\,\xi_\nu(\mathbf{r}')\right]g(\mathbf{r}')\,d\mathbf{r}\,d\mathbf{r}' \\ &= \sum_\nu h_\nu\left(\int g(\mathbf{r})\,\xi_\nu(\mathbf{r})\,d\mathbf{r}\right)^2 > 0\,.\end{aligned} \tag{A11}$$

Therefore, among all trial functions $g(\mathbf{r})$ the FF $f(\mathbf{r}) = g_{\text{opt}}(\mathbf{r})$, which minimizes $\eta[g]$ for stable systems. This principle complements the maximum global hardness principle [29], which examines changes of the system global hardness at constant global chemical potential. It should be observed that the above minimization of $\eta[g]$ has distinctly different content, since this quantity for a general, non-equilibrium g, cannot be interpreted as the system global hardness (Eq. A7).

Clearly, the same is true in any discrete resolution. For example, in the AIM description $\eta(g) = \boldsymbol{g}\,\boldsymbol{\eta}\,\boldsymbol{g}^\dagger$, $\boldsymbol{g}\,\boldsymbol{1}^\dagger = 1$ (or $d\boldsymbol{g}\,\boldsymbol{1}^\dagger = 0$), where $\boldsymbol{g}$ stands for the trial FF vector. The variational principle, $\boldsymbol{g}\,\boldsymbol{\eta}\,d\boldsymbol{g}^\dagger = 0$, requires equalization of the local hardnesses for the optimum $\boldsymbol{g} = \boldsymbol{f}$: $\boldsymbol{f}\boldsymbol{\eta} = \eta\,\boldsymbol{1}$ [20]. Again, the spectral (PNM) resolution of $\boldsymbol{\eta}$ shows that in a stable system $\eta(g)$ reaches a minimum for $\boldsymbol{g} = \boldsymbol{f}$.

Consider now a simple illustrative case of a division of a molecular system M into two complementary subsystems $M = (\mathcal{S}_1|\mathcal{S}_2) = (A|B)$. In this case the equilibrium FF indices are given by Eq. 114. Let us define the trial FF vector: $\boldsymbol{g} = (a, 1-a)$, for which the resulting hardness function becomes the following simple function of the variational parameter a:

$$\eta(g) = \eta^{M}_{A,A}\, a^2 + 2\eta^{M}_{A,B}\, a\,(1 - a) + \eta^{M}_{B,B}\,(1 - a)^2 \equiv \eta(a)\,. \qquad (A12)$$

The extremum of $\eta(a)$ immediately gives

$$a_{opt} = (\eta^{M}_{B,B} - \eta^{M}_{A,B})/[\eta^{M}_{A,A} - 2\eta^{M}_{A,B} + \eta^{M}_{B,B}] = \eta^{M}_{B}/(\eta^{M}_{A} + \eta^{M}_{B}) = f^{M}_{A}\,, \qquad (A13)$$

in accordance with Eq. 114. The second derivative with respect to a:

$$\partial^2\eta(a)/\partial a^2 = 2\,(\eta^{M}_{A} + \eta^{M}_{B})\,, \qquad (A14)$$

must be positive for the internally stable M (Eq. 361).

This principle of the minimum value of $\eta[g]$ at the global equilibrium distribution of the inflowing charge agrees with intuitive expectations. Namely, this quantity represents the average "repulsion" of this extra charge with the remaining electrons in the molecular system under consideration. One would therefore naturally expect it to be as low as possible, when the electrons reach the equilibrium distribution, since the delocalized electrons in a molecule should avoid each other to minimize their mutual repulsion.

APPENDIX B

RELAXED FRAGMENT TRANSFORMATION OF THE HARDNESS MATRIX IN M = (A|B) SYSTEMS

As we have already remarked in Section 6.3 of Chapter 3, the relaxed hardness matrix of M=(A|B) (Eq. 180),

$$\boldsymbol{\eta}^{rel} \equiv \boldsymbol{\eta} + \delta\boldsymbol{\eta}(\mathrm{M}) = \begin{pmatrix} \boldsymbol{\eta}^{\mathrm{M}}_{\mathrm{A,A}} & \boldsymbol{\eta}^{\mathrm{M}}_{\mathrm{A,B}} \\ \boldsymbol{\eta}^{\mathrm{M}}_{\mathrm{B,A}} & \boldsymbol{\eta}^{\mathrm{M}}_{\mathrm{B,B}} \end{pmatrix}, \tag{B1}$$

represents the transformed (rigid) AIM hardness tensor, $\boldsymbol{\eta}$, since it corresponds to the representation of the collective *relaxational modes* (REM), $\boldsymbol{e}^{rel}_{\mathrm{M}} = \boldsymbol{e}_{\mathrm{M}}\,\mathbf{t}$, defined by the transformation

$$\mathbf{t} = \begin{pmatrix} \mathbf{I}_{\mathrm{A}} & \mathbf{T}^{(\mathrm{A|B})} \\ \mathbf{T}^{(\mathrm{B|A})} & \mathbf{I}_{\mathrm{B}} \end{pmatrix} = (\boldsymbol{t}_1 | \boldsymbol{t}_2 | \ldots | \boldsymbol{t}_{\mathrm{m}}), \tag{B2}$$

where $\mathbf{T}^{(\mathrm{A|B})}$ and $\mathbf{T}^{(\mathrm{B|A})}$ are the relaxational matrices of Eqs. 175 and 178. The corresponding mode population variables and chemical potentials are:

$$\mathrm{d}\boldsymbol{N}^{rel} = \mathrm{d}\boldsymbol{N}\,\mathbf{t} \quad \text{and} \quad \boldsymbol{\mu}^{rel} = \boldsymbol{\mu}\,\mathbf{t}. \tag{B3}$$

Hence, by transforming the AIM EE equations, $\mathrm{d}\boldsymbol{\mu} = \mathrm{d}\boldsymbol{N}\,\boldsymbol{\eta}$, one obtains their REM representation,

$$\mathrm{d}\boldsymbol{\mu}^{rel} = (\mathrm{d}\boldsymbol{N}\,\mathbf{t})\,(\mathbf{t}^{-1}\boldsymbol{\eta}\,\mathbf{t}) \equiv \mathrm{d}\boldsymbol{N}^{rel}\,\boldsymbol{\eta}^{rel}, \tag{B4}$$

which identifies $\boldsymbol{\eta}^{rel}$ as the similarity transformed $\boldsymbol{\eta}$.

We would like to emphasize that REM basis vectors, $\boldsymbol{e}^{rel}_{\mathrm{M}} = \{\boldsymbol{e}^{rel}_{\mathrm{k}} = \boldsymbol{e}_{\mathrm{M}}\,\boldsymbol{t}_{\mathrm{k}}\}$, are not orthonormal, since

$$(\boldsymbol{e}^{rel}_{\mathrm{M}})^{\dagger}\,\boldsymbol{e}^{rel}_{\mathrm{M}} = \mathbf{t}^{\dagger}\,\boldsymbol{e}^{\dagger}_{\mathrm{M}}\,\boldsymbol{e}_{\mathrm{M}}\,\mathbf{t} = \mathbf{t}^{\dagger}\,\mathbf{t} \neq \mathbf{I}. \tag{B5}$$

One can express the corresponding blocks of the inverse transformation $\mathbf{t}^{-1} \equiv \{\mathbf{t}^{-1}_{\mathrm{X,Y}},\ (\mathrm{X,Y}) = (\mathrm{A,B})\}$ in terms of the corresponding blocks of $\mathbf{t}$ (Eq. B2):

$$\mathbf{t}^{-1}_{\mathrm{A,A}} = [\mathbf{I}_{\mathrm{A}} - \mathbf{T}^{(\mathrm{A|B})}\mathbf{T}^{(\mathrm{B|A})}]^{-1}, \qquad \mathbf{t}^{-1}_{\mathrm{B,B}} = [\mathbf{I}_{\mathrm{B}} - \mathbf{T}^{(\mathrm{B|A})}\mathbf{T}^{(\mathrm{A|B})}]^{-1},$$

$$\mathbf{t}^{-1}_{\mathrm{A,B}} = -\,[\mathbf{I}_{\mathrm{A}} - \mathbf{T}^{(\mathrm{A|B})}\mathbf{T}^{(\mathrm{B|A})}]^{-1}\,\mathbf{T}^{(\mathrm{A|B})}, \quad \mathbf{t}^{-1}_{\mathrm{B,A}} = -\,[\mathbf{I}_{\mathrm{B}} - \mathbf{T}^{(\mathrm{B|A})}\mathbf{T}^{(\mathrm{A|B})}]^{-1}\,\mathbf{T}^{(\mathrm{B|A})}. \tag{B6}$$

The matrix power expansion of $(\mathbf{I} - \mathbf{x})^{-1}$,

$$(\mathbf{I} - \mathbf{x})^{-1} = \sum_{i=0}^{\infty} \mathbf{x}^i , \tag{B7}$$

then allows one to explicitly express the relaxational corrections $\delta\boldsymbol{\eta}_{X,X}(M)$ and $\delta\boldsymbol{\eta}_{X,Y}(M)$, in powers of the relevant blocks in $\boldsymbol{\eta}$ and $\mathbf{t}$.

The transformation defined by Eq. B4 gives:

$$\boldsymbol{\eta}^{M}_{A,A} = \mathbf{t}^{-1}_{A,A}[\boldsymbol{\eta}^{A,A} + \boldsymbol{\eta}^{A,B}\mathbf{T}^{(B|A)}] + \mathbf{t}^{-1}_{A,B}[\boldsymbol{\eta}^{B,A} + \boldsymbol{\eta}^{B,B}\mathbf{T}^{(B|A)}] ,$$

$$\boldsymbol{\eta}^{M}_{A,B} = \mathbf{t}^{-1}_{A,A}[\boldsymbol{\eta}^{A,A}\mathbf{T}^{(A|B)} + \boldsymbol{\eta}^{A,B}] + \mathbf{t}^{-1}_{A,B}[\boldsymbol{\eta}^{B,A}\mathbf{T}^{(A|B)} + \boldsymbol{\eta}^{B,B}] , \tag{B8}$$

and the remaining expressions immediately follow from exchanging fragment symbols in Eq. B8. The power expansion (Eq. B7) of $(\mathbf{I}_A - \mathbf{T}^{(A|B)}\mathbf{T}^{(B|A)})^{-1} \equiv (\mathbf{I}_A - \mathbf{x}_A)^{-1}$ gives:

$$\mathbf{t}^{-1}_{A,A} \equiv \sum_{i=0}^{\infty} \mathbf{x}^i_A \quad \text{and} \quad \mathbf{t}^{-1}_{A,B} \equiv -\left[\sum_{i=0}^{\infty} \mathbf{x}^i_A\right]\mathbf{T}^{(A|B)} , \tag{B9}$$

and hence, to the second-order in $\mathbf{T}^{(X|Y)}$,

$$\boldsymbol{\eta}^{M}_{A,A} = \boldsymbol{\eta}^{A,A} + [\boldsymbol{\eta}^{A,B}\mathbf{T}^{(B|A)} - \mathbf{T}^{(A|B)}\boldsymbol{\eta}^{B,A} + \mathbf{x}_A\boldsymbol{\eta}^{A,A} - \mathbf{T}^{(A|B)}\boldsymbol{\eta}^{B,B}\mathbf{T}^{(B|A)} + \dots]$$

$$\equiv \boldsymbol{\eta}^{A,A} + \delta\boldsymbol{\eta}_{A,A}(M) , \tag{B10}$$

$$\boldsymbol{\eta}^{M}_{A,B} = \boldsymbol{\eta}^{A,B} + [\boldsymbol{\eta}^{A,A}\mathbf{T}^{(A|B)} - \mathbf{T}^{(A|B)}\boldsymbol{\eta}^{B,B} + \mathbf{x}_A\boldsymbol{\eta}^{A,B} - \mathbf{T}^{(A|B)}\boldsymbol{\eta}^{B,A}\mathbf{T}^{(A|B)} + \dots]$$

$$\equiv \boldsymbol{\eta}^{A,B} + \delta\boldsymbol{\eta}_{A,B}(M) . \tag{B11}$$

We would like to emphasize that the relaxed hardness data can also be obtained using the single matrix inversion procedure of Section 4 in Chapter 3. Namely, to obtain the diagonal block $\boldsymbol{\eta}^{M}_{A,A}$ one assumes the partitioning $M^{rel}_A = (1|2|\dots|n|B)$, in which all n AIM in A are mutually closed, while the remaining m-n constituent atoms belonging to B can freely exchange electrons. Equation 225 then gives $\{\boldsymbol{\eta}_{a,a'}\} = \boldsymbol{\eta}^{M}_{A,A}$, and the relaxed off-diagonal parameters $\boldsymbol{\eta}^{M}_{a,B}$ measuring the effective coupling hardness between $a \in A$ and the relaxed B as a whole.

APPENDIX C

THE AIM/PNM-RESOLVED INTERSECTING-STATE MODEL

In order to generalize the ISM of Section 9 in Chapter 4 we express the quadratic energy of the A(acid)-B(base) reactive system, for the constant external potential due to both reactants, in terms of the AIM populational displacements of A, $\boldsymbol{x} = \boldsymbol{N}_A - \boldsymbol{N}_A^0$, $\boldsymbol{y} = \boldsymbol{N}_B - \boldsymbol{N}_B^0$, grouped together in the overall AIM displacement vector $\boldsymbol{z} = (\boldsymbol{x}, \boldsymbol{y})$:

$$\Delta E_{AB}(\boldsymbol{z}) = \boldsymbol{\mu}\, \boldsymbol{z}^{\dagger} + \frac{1}{2}\boldsymbol{z}\,\boldsymbol{\eta}\,\boldsymbol{z}^{\dagger} = \left[\boldsymbol{\mu}_A\, \boldsymbol{x}^{\dagger} + \frac{1}{2}(\boldsymbol{x}\,\boldsymbol{\eta}^{A,A}\,\boldsymbol{x}^{\dagger} + \boldsymbol{y}\,\boldsymbol{\eta}^{B,A}\,\boldsymbol{x}^{\dagger})\right]$$

$$+\left[\boldsymbol{\mu}_B\, \boldsymbol{y}^{\dagger} + \frac{1}{2}(\boldsymbol{y}\,\boldsymbol{\eta}^{B,B}\,\boldsymbol{y}^{\dagger} + \boldsymbol{x}\,\boldsymbol{\eta}^{A,B}\,\boldsymbol{y}^{\dagger})\right] \equiv [\bar{E}_A(\boldsymbol{z}) + E_B(\boldsymbol{z})]\,, \tag{C1}$$

where the vector of initial AIM chemical potential of both reactants in their corresponding internal equilibria is defined by the vector $\boldsymbol{\mu} = (\boldsymbol{\mu}_A, \boldsymbol{\mu}_B) = (\mu_A^{+}\boldsymbol{1}_A, \mu_B^{+}\boldsymbol{1}_B)$. In what follows we shall use the CT label to identify quantities associated with the N-constrained CT. The full equilibrium CT along such a trajectory,

$$N_{CT} = \sum_i^M z_i^{CT} = dN_A^{CT} \equiv \sum_a^A x_a^{CT} = -dN_B^{CT} \equiv \sum_b^B y_b^{CT}\,, \tag{C2}$$

is given by the corresponding equilibrium AIM population displacements:

$$\boldsymbol{z}^{CT} = N_{CT}\,\boldsymbol{f}^{CT} \equiv N_{CT}\,(\boldsymbol{f}_A^{CT}, \boldsymbol{f}_B^{CT}) \equiv (\boldsymbol{x}^{CT}, \boldsymbol{y}^{CT})\,, \tag{C3}$$

where $\boldsymbol{f}_A^{CT} = \boldsymbol{f}^{A,A} - \boldsymbol{f}^{B,A}$ and $\boldsymbol{f}_B^{CT} = -(\boldsymbol{f}^{B,B} - \boldsymbol{f}^{A,B})$. The N-constrained trajectory $\boldsymbol{z}^{CT}(x)$ can be defined as the uniform scaling of the final displacement vector $\boldsymbol{z}^{CT}$, using the CT progress variable x:

$$\boldsymbol{z}^{CT}(x) = x\,\boldsymbol{z}^{CT} \equiv [\boldsymbol{x}^{CT}(x), \boldsymbol{y}^{CT}(x)]\,, \quad x \in [0, 1]\,. \tag{C4}$$

Again, one calibrates the zero energy of the acidic reactant by shifting $\bar{E}_A(\boldsymbol{z})$ by the corresponding activation energy of the basic reactant for the full amount of CT, $E_B(N_{CT}) \equiv E_B(\boldsymbol{z}^{CT})$,

$$E_A(\boldsymbol{z}) = \bar{E}_A(\boldsymbol{z}) + E_B(N_{CT})\,, \tag{C5}$$

in order to obtain the acid-stabilization energy of the transferred N_{CT} electrons equal the the overall $E_{CT} = \Delta E_{AB}(\boldsymbol{z}^{CT})$:

$$E_A(\boldsymbol{z}^{CT}) = E_{CT}\,. \tag{C6}$$

The matrix equation defining the intersection points of two such quadratic energy surfaces of reactants (see Fig. 40c), $E_A(\boldsymbol{z}^*) = E_B(\boldsymbol{z}^*)$, is:

$$E_B(\boldsymbol{z}^*) - \bar{E}_A(\boldsymbol{z}^*) = (-\boldsymbol{\mu}_A, \boldsymbol{\mu}_B)\,\boldsymbol{z}^{*\dagger} + \frac{1}{2}\boldsymbol{z}^*\begin{pmatrix} -\boldsymbol{\eta}^{A,A} & \boldsymbol{\eta}^{A,B} \\ -\boldsymbol{\eta}^{B,A} & \boldsymbol{\eta}^{B,B} \end{pmatrix}\boldsymbol{z}^{*\dagger} \tag{C7}$$

$$\equiv \tilde{\boldsymbol{\mu}}\, z^{*\dagger} + \frac{1}{2} z^{*}\, \tilde{\boldsymbol{\eta}}\, z^{*\dagger} = E_{\mathrm{B}}(N_{\mathrm{CT}}) \ . \tag{C8}$$

The last equation implies the dependence of a single populational displacement, say z^{*}_{m} , on the remaining ones, $\mathbf{z}^{*}_{m}$ [in $\mathbf{z}^{*} = (\mathbf{z}^{*}_{m},\ z^{*}_{m})$]:

$$z^{*}_{m} = z^{*}_{m}(\mathbf{z}^{*}_{m}) \ . \tag{C9}$$

The explicit form of this equation follows from the corresponding partitioning of the chemical potential vector $\tilde{\boldsymbol{\mu}} = (\tilde{\boldsymbol{\mu}}_{m}, \tilde{\mu}_{m})$ and the hardness matrix $\tilde{\boldsymbol{\eta}}$ of Eq. C8,

$$\tilde{\boldsymbol{\eta}} = \begin{pmatrix} \tilde{\boldsymbol{\eta}}_{m,m} & \tilde{\boldsymbol{\eta}}^{\dagger}_{m} \\ \tilde{\boldsymbol{\eta}}_{m} & \tilde{\eta}_{m,m} \end{pmatrix} , \tag{C10}$$

where $\tilde{\boldsymbol{\eta}}_{m,m}$ is the square block associated with the independent variables $\mathbf{z}^{*}_{m}$, $\tilde{\eta}_{m,m}$ is the diagonal element corresponding to z^{*}_{m} , and the row vector $\tilde{\boldsymbol{\eta}}_{m}$ groups the elements coupling the two subsets. In terms of these quantities Eq. C8 reads:

$$\tilde{\eta}_{m,m}(z^{*}_{m})^{2} + 2\,(\tilde{\mu}_{m} + 2\mathbf{z}^{*}_{m}\tilde{\boldsymbol{\eta}}^{\dagger}_{m})\, z^{*}_{m} + 2\tilde{\boldsymbol{\mu}}_{m}\, \mathbf{z}^{*\dagger}_{m} + \mathbf{z}^{*}_{m}\,\tilde{\boldsymbol{\eta}}_{m,m}\mathbf{z}^{*\dagger}_{m} - 2\,E_{\mathrm{B}}(N_{\mathrm{CT}})$$

$$\equiv a\,(z^{*}_{m})^{2} + b\,(\mathbf{z}^{*}_{m})\, z^{*}_{m} + c\,(\mathbf{z}^{*}_{m}) = 0 \ . \tag{C11}$$

Differentiating this equation with respect to $\mathbf{z}^{*}_{m}$ gives the intersection derivatives:

$$(\partial z_{m} / \partial \mathbf{z}_{m})\,|_{*} = -[(\partial c / \partial \mathbf{z}_{m})\,|_{*} + z^{*}_{m}\,(\partial b / \partial \mathbf{z}_{m})\,|_{*}] \,/\, [2a\, z^{*}_{m} + b\,(\mathbf{z}^{*}_{m})] \ . \tag{C12}$$

Finally, the minimum energy pass between the two intersecting paraboloids, which determines the N-unconstrained "transition state" $z^{\ddagger}$, is obtained by finding

$$\min\, E_{\mathrm{A}}[\mathbf{z}^{*}_{m},\ z^{*}_{m}(\mathbf{z}^{*}_{m})] = \min\, E_{\mathrm{B}}[\mathbf{z}^{*}_{m},\ z^{*}_{m}(\mathbf{z}^{*}_{m})] \ . \tag{C13}$$

This can be done by solving the corresponding zero-gradient equations, e.g.,

$$\left(\frac{\partial E_{\mathrm{A}}}{\partial \mathbf{z}_{m}}\right)_{*} = \left[\left(\frac{\partial E_{\mathrm{A}}}{\partial \mathbf{z}_{m}}\right)_{z_{m}} + \left(\frac{\partial E_{\mathrm{A}}}{\partial z_{m}}\right)\left(\frac{\partial z_{m}}{\partial \mathbf{z}_{m}}\right)\right|_{*}\Bigg] = \mathbf{0}_{m} \ . \tag{C14}$$

Obviously, one could also use the reactant PNM populational displacements, e.g., IDM, instead of the AIM reference frame, to define the intersecting reactant energy paraboloids and properties of the representative N-restricted and N-unrestricted transition-states. In the IDM case one would then have to transform the chemical potential/hardness quantities, which define the reactant quadratic surfaces and the transition-state equations, in accordance with the $\mathbf{U}_{\mathrm{int}}$ transformation defined by Eq. 323.

APPENDIX D

BOND-MULTIPLICITIES FROM ONE-DETERMINANTAL, DIFFERENCE APPROACH

In a series of recent articles [6] the new quantum mechanical, finite-difference approach to classical problems of the chemical valence and bond-multiplicity [4, 5] has been developed within the one-determinantal (Hartree-Fock or Kohn-Sham) descriptions of molecular systems. These indices are formulated in terms of the atomic and diatomic contributions (ionic and covalent) to the molecular (M) expectation value of the displacements in the CBO ($\hat{P}$) or density ($\hat{\rho}$) operators in a molecule, relative to the reference state of the *separated atoms limit*, identified by the superscript $^{(0)}$. The overall absolute valence in the spin non-polarized approximation [*restricted Hartree-Fock* (RHF) or the *local density approximation* (LDA) theories] and its partitioning into the total atomic and diatomic contributions are defined in the following way:

$$V = -\frac{1}{2}\langle \Delta\hat{P} \rangle \equiv \mathrm{Tr}\,\mathbf{P}\,\Delta\mathbf{P} \equiv \sum_{a}^{AIM} V_a + \sum_{a<b}^{AIM}\sum V_{ab}\,, \tag{D1}$$

where $\Delta\hat{P} = \hat{P}(\mathrm{M}) - \hat{P}^0(\mathrm{SAL}) = N(\hat{\rho} - \hat{\rho}^0) = N\Delta\hat{\rho}$, the CBO matrix in the orthogonal atomic orbital basis set $|\boldsymbol{\lambda}\rangle$ is given by the familiar expression $\mathbf{P} = \mathbf{C}\mathbf{o}\mathbf{C}^{\dagger} = \langle \boldsymbol{\lambda}\,|\,\hat{P}\,|\,\boldsymbol{\lambda}\rangle$, with the CBO operator $\hat{P} \equiv |\boldsymbol{\psi}\rangle\mathbf{o}\langle\boldsymbol{\psi}| \equiv N\hat{\rho}$ being defined by the MO $|\boldsymbol{\psi}\rangle = |\boldsymbol{\lambda}\rangle\mathbf{C}$ and their occupations $\mathbf{o} = \{n_i\,\delta_{i,j}\}$.

The atomic valences in the *unrestricted Hartree-Fock* (UHF) or the *local spin density* (LSDA) approximations are similarly defined by the spin-resolved CBO data: $\mathbf{P} = \sum_\sigma \mathbf{P}^\sigma$, $\sigma = \alpha, \beta$: $V_a = V_a(\{(\Delta P^\sigma_{\mu,\nu})^2\})$ where $(\mu, \nu) \in a$; they are diagonal quadratic forms of the intra-atomic CBO displacements, with the ionic contributions depending on the corresponding shifts in the MO occupations, and the covalent contributions being determined by the off-diagonal, intra-atomic CBO changes: $V_a = V_a^{ion} + V_a^{cov}$, $V_a^{ion} = V_a^{ion}(\{(\Delta P^\sigma_{\mu,\mu})^2\})$ and $V_a^{cov} = V_a^{cov}(\{(\Delta P^\sigma_{\mu,\nu})^2, \mu \neq \nu\})$. The diatomic valence, $V_{ab} = V_{ab}(\{(\Delta P^\sigma_{\mu,\tau})^2\})$, where $\mu \in a$ and $\tau \in b$, is of purely covalent origin, since it represents the diagonal quadratic form of the inter-atomic CBO displacements. In this set of valence indices the negative values correspond to an increase in the overall bonding, while the positive values stand for the antibonding contributions.

It has also been demonstrated previously [6] that these valence indices measure differences in the two-electron probabilities (in terms of electron pairs), being related to the second-order displacements in the pair-diagonal part of the two-electron density matrix in the $|\boldsymbol{\lambda}\rangle$ representation (Löwdin, electron-pair normalization), $\boldsymbol{\Gamma} = \{\Gamma(\mu, \rho) \equiv \Gamma(\mu, \rho\,|\,\mu, \rho)\}$:

$$\Delta\Gamma^{(2)} = \frac{1}{2}\sum_{\lambda,\mu}^{OAO}\sum_{\nu,\tau}\Delta P_{\lambda,\mu}\left(\frac{\partial^2\Gamma}{\partial P_{\lambda,\mu}\,\partial P_{\nu,\tau}}\right)^{o}\Delta P_{\nu,\tau}\,. \tag{D2}$$

The bond-multiplicities are obtained as the sum of diatomic and the weighted average of the relevant atomic valences, $\mathcal{N}_{ab} = V_{ab} + w_a(ab)\,V_a + w_b(ab)\,V_b$, where w_a denotes the unbiased weighting factor determined by the ratio of the diatomic valence of the a−b bond, relative to the sum of all diatomic valences including atom a. Such overall bond-order measures have been shown to provide a good representation of the chemical bond-multiplicities [6].

The explicit expressions for the atomic and diatomic valence indices in the UHF/LSDA theories, in terms of the spin-resolved CBO displacements are :

$$V_a^{ion} = -\frac{1}{2}\sum_{\mu}^{a}\,[\,(\Delta P^{\alpha}_{\mu,\mu})^2 + (\Delta P^{\beta}_{\mu,\mu})^2\,]\,, \tag{D3}$$

$$V_a^{cov} = -\sum\sum_{\mu<\nu}^{a}\,[\,(\Delta P^{\alpha}_{\mu,\nu})^2 + (\Delta P^{\beta}_{\mu,\nu})^2\,]\,, \tag{D4}$$

$$V_{ab}^{cov} = -\sum_{\mu}^{a}\sum_{\tau}^{b}\,[\,(\Delta P^{\alpha}_{\mu,\tau})^2 + (\Delta P^{\beta}_{\mu,\tau})^2\,]\,. \tag{D5}$$

The valence/bond-multiplicity changes corresponding to a transition between structures I and II are obtained as differences between the associated absolute values, $\Delta V(\mathrm{I}\rightarrow\mathrm{II}) = V_{\mathrm{II}} - V_{\mathrm{I}}$. For example, in catalytic systems the two structures of interest are the chemisorption system M (II) and the *separated reactants limit* (SRL, I). Let us denote for definiteness by A and B the surface cluster and allyl adsorbate, respectively, with (a, a', ...) and (b, b', ...) denoting their constituent atoms. In addition to the specific bond-order changes due to the chemisorption, $\Delta\mathcal{N}_{ij} = \mathcal{N}_{ij}(\mathrm{M}) - \mathcal{N}_{ij}(\mathrm{SRL})$, between atoms i and j, of interest also are the total intra-reactant bond valence changes in A and B, respectively,

$$\Delta V^{X} = \sum\sum_{x<x'}^{X}\Delta\mathcal{N}_{xx'}\,, \qquad X = A,\,B\,, \tag{D6}$$

as well as the total interaction valence between A and B:

$$V^{A\text{-}B} = \sum_{a}^{A}\sum_{b}^{B}\Delta\mathcal{N}_{ab}\,. \tag{D7}$$

These quantities define the overall change in the system bond-multiplicity and its resolution into the corresponding covalent and ionic components:

$$\Delta V = \Delta V^{A} + \Delta V^{B} + V^{A\text{-}B} \equiv \Delta V^{cov} + \Delta V^{ion} . \tag{D8}$$

When examining implications of the CSA determined AIM charge displacements due to the chemisorption for the substrate/adsorbate bond activations (Chapter 5) we have often conjectured that an increase (decrease) in the initial polarization of the coordination bond marks a strengthening (weakening) of the bond in question. The quantum-mechanical bond multiplicities summarized in this Appendix do indeed justify this conclusion [6, 124]. Namely, the overall bond multiplicity of the two-electron model of the coordination bond is equal to the amount of the charge donated (accepted) by the basic (acidic) atom, although the proportions between its ionic and covalent components do change with the bond polarity.

APPENDIX E

GUHF AND UHF ENERGY EXPRESSIONS AND THEIR DERIVATIVES WITH RESPECT TO THE SPIN-RESOLVED DENSITY MATRICES

The RHF development of Section 1 and 2 in Chapter 7 can be extended to the UHF type approaches. Following McWeeny [145] let us define the atomic spin-orbital basis set, $|\boldsymbol{\lambda}^{\mathrm{UHF}}\rangle = |\lambda_1\alpha, \lambda_2\alpha, \ldots; \lambda_1\beta, \lambda_2\beta, \ldots\rangle = |\boldsymbol{\lambda}^{\alpha}; \boldsymbol{\lambda}^{\beta}\rangle$, in terms of which the molecular spin-orbitals are defined by the spin-resolved LCAO MO matrix $\mathbf{C}^{\mathrm{UHF}}$:

$$|\boldsymbol{\psi}^{\mathrm{UHF}}\rangle = |\boldsymbol{\psi}^{\alpha}; \boldsymbol{\psi}^{\beta}\rangle = |\boldsymbol{\lambda}^{\alpha}; \boldsymbol{\lambda}^{\beta}\rangle \begin{pmatrix} \mathbf{C}^{\alpha} \\ \mathbf{C}^{\beta} \end{pmatrix} \equiv |\boldsymbol{\lambda}^{\mathrm{UHF}}\rangle\, \mathbf{C}^{\mathrm{UHF}} . \tag{E1}$$

These two spin components of the expansion coefficients define the spin-resolved density matrix:

$$\boldsymbol{\gamma}^{\mathrm{UHF}} = \{\boldsymbol{\gamma}^{\sigma,\sigma'} \equiv \mathbf{C}^{\sigma}\, \mathbf{C}^{\sigma'\dagger} \quad (\sigma, \sigma') = (\alpha, \beta)\} . \tag{E2}$$

After performing the relevant spin integrations one obtains the energy expression in the form of the RHF energy function of Eq. 507:

$$\begin{aligned} E^{\mathrm{UHF}}(\boldsymbol{\gamma}^{\mathrm{UHF}}) &= \mathrm{tr}\left\{\boldsymbol{\gamma}^{\mathrm{UHF}}\left[\mathbf{h}^{\mathrm{UHF}} + \frac{1}{2}\mathbf{G}^{\mathrm{UHF}}(\boldsymbol{\gamma}^{\mathrm{UHF}})\right]\right\} \\ &= \frac{1}{2}\,\mathrm{tr}\{\boldsymbol{\gamma}^{\mathrm{UHF}}[\mathbf{h}^{\mathrm{UHF}} + \mathbf{F}^{\mathrm{UHF}}(\boldsymbol{\gamma}^{\mathrm{UHF}})]\} , \end{aligned} \tag{E3}$$

where $\mathbf{h}^{\mathrm{UHF}} = \{\mathbf{h}\,\delta_{\sigma,\sigma'}\ (\sigma,\sigma') = (\alpha,\beta)\}$, $\mathbf{F}^{\mathrm{UHF}} = \mathbf{h}^{\mathrm{UHF}} + \mathbf{G}^{\mathrm{UHF}} = \{\mathbf{h}\,\delta_{\sigma,\sigma'} + \mathbf{G}_{\sigma,\sigma'} \equiv \mathbf{F}^{\sigma,\sigma'}\}$, and the four blocks of the GUHF $\mathbf{G}$ matrix are:

$$\begin{aligned} &\mathbf{G}_{\alpha,\alpha} = \mathbf{J}(\boldsymbol{\gamma}^{\alpha,\alpha} + \boldsymbol{\gamma}^{\beta,\beta}) - \mathbf{K}(\boldsymbol{\gamma}^{\alpha,\alpha}) , \quad &&\mathbf{G}_{\beta,\beta} = \mathbf{J}(\boldsymbol{\gamma}^{\alpha,\alpha} + \boldsymbol{\gamma}^{\beta,\beta}) - \mathbf{K}(\boldsymbol{\gamma}^{\beta,\beta}) , \\ &\mathbf{G}_{\alpha,\beta} = -\mathbf{K}(\boldsymbol{\gamma}^{\alpha,\beta}) , \quad \text{and} &&\mathbf{G}_{\beta,\alpha} = -\mathbf{K}(\boldsymbol{\gamma}^{\beta,\alpha}) ; \end{aligned} \tag{E4}$$

here the Coulomb and exchange matrices $\mathbf{J}$ and $\mathbf{K}$ are defined in the usual way:

$$[\mathbf{J}(\boldsymbol{\gamma})]_{i,j} = \sum\nolimits_{k,l} \gamma_{k,l}\,(k\,l\,|\,i\,j) , \qquad [\mathbf{K}(\boldsymbol{\gamma})]_{i,j} = \sum\nolimits_{k,l} \gamma_{k,l}\,(k\,j\,|\,i\,l) . \tag{E5}$$

In the simplified (UHF) treatment the off-diagonal components of the above GUHF Fock matrix are neglected so that the coupling between α and β is only indirect, through the diagonal blocks of $\mathbf{F}^{\mathrm{UHF}}$ ($\mathbf{G}^{\mathrm{UHF}}$), $\mathbf{F}^{\sigma,\sigma} \equiv \mathbf{F}^{\sigma}$, which depend on both α and β density matrices: $\boldsymbol{\gamma}^{\alpha,\alpha} \equiv \boldsymbol{\gamma}^{\alpha}$ and $\boldsymbol{\gamma}^{\beta,\beta} \equiv \boldsymbol{\gamma}^{\beta}$. Expanding the matrix expression of Eq. E3 gives:

$$E^{GUHF} = \sum_{k,l}\left[\gamma^{\alpha}_{k,l}\left(h_{l,k} + \frac{1}{2}G^{\alpha,\alpha}_{l,k}\right) + \gamma^{\beta}_{k,l}\left(h_{l,k} + \frac{1}{2}G^{\beta,\beta}_{l,k}\right) + \frac{1}{2}\gamma^{\alpha,\beta}_{k,l}G^{\beta,\alpha}_{l,k} + \frac{1}{2}\gamma^{\beta,\alpha}_{k,l}G^{\alpha,\beta}_{l,k}\right], \tag{E6}$$

$$E^{UHF} = \sum_{k,l}\left[\gamma^{\alpha}_{k,l}\left(h_{l,k} + \frac{1}{2}G^{\alpha,\alpha}_{l,k}\right) + \gamma^{\beta}_{k,l}\left(h_{l,k} + \frac{1}{2}G^{\beta,\beta}_{l,k}\right)\right]. \tag{E7}$$

Differentiations of these energy functions with respect to the relevant density matrix elements gives the following expressions for the unconstrained first and second partial derivatives of the system energy:

$$\partial E^{GUHF}/\partial\gamma^{\sigma,\sigma'}_{i,j} = F^{\sigma',\sigma}_{j,i}, \quad \partial E^{UHF}/\partial\gamma^{\sigma}_{i,j} = F^{\sigma}_{j,i}, \quad (\sigma,\sigma') = (\alpha,\beta), \tag{E8}$$

$$\partial^2 E^{(G)UHF}/\partial\gamma^{\alpha}_{i,j}\,\partial\gamma^{\alpha}_{k,l} = \partial F^{\alpha}_{j,i}/\partial\gamma^{\alpha}_{k,l} = \partial F^{\alpha}_{l,k}/\partial\gamma^{\alpha}_{i,j} = (ij||kl), \tag{E9}$$

$$\partial^2 E^{(G)UHF}/\partial\gamma^{\alpha}_{i,j}\,\partial\gamma^{\beta}_{k,l} = \partial F^{\alpha}_{j,i}/\partial\gamma^{\beta}_{k,l} = \partial F^{\beta}_{l,k}/\partial\gamma^{\alpha}_{i,j} = (ij||kl), \tag{E10}$$

$$\partial^2 E^{GUHF}/\partial\gamma^{\alpha,\beta}_{i,j}\,\partial\gamma^{\alpha,\beta}_{k,l} = \partial^2 E^{GUHF}/\partial\gamma^{\beta,\alpha}_{i,j}\,\partial\gamma^{\beta,\alpha}_{k,l} = -(kj|il), \tag{E11}$$

$$\partial^2 E^{GUHF}/\partial\gamma^{\alpha}_{i,j}\,\partial\gamma^{\alpha,\beta}_{k,l} = \partial^2 E^{GUHF}/\partial\gamma^{\alpha}_{i,j}\,\partial\gamma^{\beta,\alpha}_{k,l} = \partial^2 E^{GUHF}/\partial\gamma^{\alpha,\beta}_{i,j}\,\partial\gamma^{\beta,\alpha}_{k,l} = 0. \tag{E12}$$

REFERENCES

1. R. S. Mulliken, *J. Chem. Phys.* **23** (1955) 1835.
2. P. - O. Löwdin, *ibid.* **18** (1950) 365; *Adv. Physics* **5** (1956) 111.
3. K. R. Roby, *Mol. Phys.* **27** (1974) 81; *ibid.* **28** (1974) 1441; R. Hainzmann and R. Ahlrichs, *Theor. Chim. Acta* (Berlin) **42** (1976) 33; C. Ehrhardt and R. Ahlrichs, *Theoret. Chim. Acta* (Berlin) **68** (1985) 231.
4. K. Wiberg, *Tetrahedron* **24** (1968) 1093.
5. M. S. Gopinathan and K. Jug, *Theoret. Chim. Acta* (Berlin) **63** (1983) 497, 511.
6. R. F. Nalewajski and J. Mrozek, *Int. J. Quantum Chem.* **51** (1994) 187; R. F. Nalewajski, J. Mrozek and A. Michalak, *ibid.*, in press; R. F. Nalewajski, J. Mrozek and G. Mazur, *Can. J. Chem.* **100** (1996) 1121.
7. R. F. W. Bader, *Atoms in Molecules: a Quantum Theory* (Clarendon ,Oxford, 1990).
8. J. Ciosłowski and S. T. Mixon, *J. Am. Chem. Soc.* **113** (1991) 4142.
9. (a)R. F. Nalewajski, J. Korchowiec, and A. Michalak, in *Topics in Current Chemistry*, vol. **183**: *Density Functional Theory - Theory of Chemical Reactivity*, R. F. Nalewajski Ed. (Springer-Verlag, Heidelberg, 1996), p. 25, and references therein; (b) R. F. Nalewajski, in *Trends in Physical Chemistry: The Activated Complex in Heterogeneous Catalysis*, W Mortier and R. Schoonheydt, Eds. (Research Signpost), in press.
10. B. G. Baekelandt, W. J. Mortier, and R. A. Schoonheydt, in *Structure and Bonding: Chemical Hardness*, K. Sen Ed. (Springer-Verlag, Heidelberg, 1993), Vol. **80** p. 187, and references therein.
11. R. G. Parr, R. A. Donnelly, M. Levy, and W. E. Palke, *J. Chem. Phys.* **68** (1978) 801; R. A. Donnelly and R. G. Parr, *ibid.* **69** (1978) 4431.
12. R. G. Parr and W. Yang, *Density-Functional Theory of Atoms and Molecules* (Oxford, New York, 1989).
13. R. G. Parr and R. G. Pearson, *J. Am. Chem. Soc.* **105** (1983) 7512.
14. P. Hohenberg and W. Kohn, *Phys. Rev.* B**136** (1964) 864.
15. W. Kohn and L. J. Sham, *Phys. Rev.* A**140** (1965) 1133.
16. R. M. Dreizler and E. K. U. Gross, *Density Functional Theory: an Approach to the Quantum Many-Body Problem* (Springer-Verlag, Heidelberg, 1990).
17. R. F. Nalewajski Ed., *Topics in Current Chemistry: Density Functional Theory*, Vols. **180-183** (Springer-Verlag, Heidelberg, 1996).
18. R. T. Sanderson, *J. Am. Chem. Soc.* **74** (1952) 272; *Science* **114** (1951) 670; *Chemical Bonding and Bond Energy* (Academic, New York, 1976).
19. P. Politzer and H. Weinstein, *J. Chem. Phys.* **68** (1978) 3801.
20. R. F. Nalewajski, *Int. J. Quantum Chem.* **40** (1991) 265.
21. R. F. Nalewajski and R. G. Parr, *J. Chem. Phys.* **77** (1982) 399.
22. M. Levy, *Proc. Natl. Acad. Sci. USA* **76** (1979) 6062.
23. J. E. Harriman, *Phys. Rev. A* **24** (1980) 680.
24. G. Zumbach and K. Maschke, *Phys. Rev. A* **28** (1983) 544; **29** (1983) 1585.

25. R. F. Nalewajski, *J. Phys. Chem.* **89** (1985) 2831.
26. M. Berkowitz, S. K. Ghosh, and R. G. Parr, *J. Am. Chem. Soc.* **107** (1985) 6811.
27. M. Berkowitz and R. G. Parr, *J. Chem. Phys.* **88** (1988) 2554.
28. R. G. Parr and P. J. Chattaraj, *J. Am. Chem. Soc.* **113** (1991) 1854.
29. P. K. Chattaraj and R. G. Parr in *Structure and Bonding: Chemical Hardness,* K. Sen Ed. (Springer-Verlag, Heidelberg, 1993), vol. **80** p. 11.
30. R. F. Nalewajski and J. Mrozek. *Int. J. Quantum Chem.* **43** (1992) 353.
31. R. F. Nalewajski, *Int. J. Quantum Chem.* **44** (1992) 67.
32. R. F. Nalewajski, in *Proceedings of a NATO ASI on Density Functional Theory,* Il Ciocco, August 16-27, 1993, R. M. Dreizler and E. K. U. Gross Eds. (Plenum, New York, 1995).
33. R. F. Nalewajski, J. Korchowiec, and Z. Zhou, *Int. J. Quantum Chem. Symp.* **22** (1988) 349.
34. R. F. Nalewajski, *Acta Phys. Polon.* **A77** (1990) 817.
35. L. Komorowski and J. Lipiński, *Chem. Phys.* **157** (1991) 45.
36. R. Pariser, *J. Chem Phys.* **21** (1953) 568.
37. K. Nishimoto and N. Mataga, *Z. Physik. Chem.* **12** (1957) 335; N. Mataga and K. Nishimoto, *Ibid.* **13** (1957) 140.
38. K. Ohno, *Adv. Quantum Chem.* **3** (1967) 239; *Theoret. Chim. Acta* **10** (1968) 111.
39. R. G. Parr and W. Yang, *J. Am. Chem. Soc.* **106** (1984) 4049.
40. W. Yang and R. G. Parr, *Proc. Natl. Acad. Sci. USA* **82** (1985) 6723.
41. R. F. Nalewajski, *J. Phys. Chem.* **93** (1989) 2658.
42. J. L. Gázquez, A. Vela, and M. Galván, *Structure and Bonding* **66** (1987) 79.
43. R. F. Nalewajski, *J. Am. Chem. Soc.* **106** (1984) 944.
44. R. F. Nalewajski and M. Koniński, *J. Phys. Chem.* **88** (1984) 6234.
45. R. F. Nalewajski and M. Koniński, *Acta Phys. Polon.* **A74** (1988) 255.
46. R. F. Nalewajski, *J. Chem. Phys.* **78** (1983) 6112.
47. R. F. Nalewajski, *Z. Naturforsch.* **43a** (1988) 65.
48. J. P. Perdew, R. G. Parr, M. Levy and J. L. Balduz, *Phys. Rev. Lett.* **49** (1982) 1691; see also: J. P. Perdew, in *Proceedings of a NATO ASI on Density Functional Methods in Physics*, Alcabideche, September 5-16, 1983, R. M. Dreizler and J. da Providência Eds. (Plenum, New York, 1985), p. 265.
49. E. H. Lieb, *Int. J. Quantum Chem.* **24** (1983) 243; see also: E. H. Lieb, in *Proceedings of a NATO ASI on Density Functional Methods in Physics*, Alcabideche, September 5-16, 1983, R. M. Dreizler and J. da Providência Eds. (Plenum, New York, 1985), p. 31.
50. R. F. Nalewajski and P. M. Kozłowski, *Acta Phys. Polon.* **A70** (1986) 457.
51. H. Eschrig, in *Proceedings of the 21st Symposium on Electronic Structure* (Academic -Verlag, Berlin, 1991).
52. S. Fraga, J. Karwowski and K. M. S. Saxena, *Physical Sciences Data 4: Atomic Energy Levels - Data for Parametric Calculations* (Elsevier, Amsterdam, 1979).
53. H. B. Callen, *Thermodynamics: An Introduction to the Physical Theories of Equilibrium Thermostatics and Irreversible Thermodynamics* (Wiley, New York, 1960).

54. R. S. Mulliken, *J. Chem. Phys.* **2** (1934) 782.
55. R. F. Nalewajski, J. F. Capitani, *J. Chem. Phys.* **77** (1982) 2514.
56. R. F. Nalewajski, in *Proceedings of the International Symposium on the Dynamic Subsystems with Chemical Reaction*, Świdno, June 6-10, 1988, J. Popielawski Ed. (World Scientific, Singapore, 1989), p. 325.
57. R. F. Nalewajski, *Int. J. Quantum Chem.* **56** (1995) 453.
58. L. Tisza, *Generalized Thermodynamics* (MIT, Cambridge Mass., 1977).
59. R. F. Nalewajski, in *Structure and Bonding: Chemical Hardness,* K. Sen Ed. (Springer-Verlag, Heidelberg, 1993), vol. **80** p. 115.
60. J. Korchowiec, H. Gerwens, and K. Jug, *Chem. Phys. Lett.* **222** (1994) 58.
61. R. F. Nalewajski, *Int. J. Quantum Chem.* **49** (1994) 675.
62. T. Antonova, N. Neshev, E. I. Proinov, and R. F. Nalewajski, *Acta Phys. Polon.* **A79** (1991) 805.
63. C. J. Ballhausen and H. B. Gray, *Molecular Orbital Theory* (Benjamin, New York, 1965), p. 120; A. Gołębiewski, *Chemia Kwantowa Związków Nieorganicznych* (PWN, Warszawa, 1969), p. 457 (in Polish).
64. R. D. Levine and R. B. Bernstein, *Molecular Reaction Dynamics* (Clarendon, Oxford, 1974), p. 86; see also: P. Davidovits and D. L. McFadden, *The Alkali Halide Vapors* (Academic, New York, 1979).
65. M. Polanyi, *Atomic Reactions* (Williams and Norgate, London, 1932); J. L. Magee, *J. Chem. Phys.* **8** (1940) 687.
66. R. F. Nalewajski, *Int. J. Quantum Chem.* **42** (1992) 243.
67. K. P. Huber and G. Herzberg, *Molecular Spectra and Molecular Structure*, Vol. 4: *Constants of Diatomic Molecules* (Van Nostrand, New York, 1979).
68. R. G. Parr and L. J. Bartolotti, *J. Am. Chem. Soc.* **104** (1982) 3801.
69. R. G. Pearson, *J. Am. Chem. Soc.* **85** (1963) 3533; *Sciences* **151** (1966) 172.
70. R. G. Pearson, Ed., *Hard and Soft Acids and Bases* (Dowden, Hatchinson, and Ross, Stroudsburg, 1973).
71. C. K. Jørgensen, *Inorg. Chem.* **3** (1964) 1201.
72. G. Klopman, *J. Am. Chem. Soc.* **90** (1968) 223.
73. C. Lee, W. Yang, and R. G. Parr, *J. Mol. Struct. (Theochem)* **163** (1988) 305.
74. R. G. Pearson, *J. Chem. Ed.* **64** (1987) 561.
75. E. M. Shusterovich, in *Theory-Structure Properties of Complex Compounds*, B. Jeżowska-Trzebiatowska et al., Eds. (Polish Scientific Publishers, Warsaw, 1979), p. 209, and references therein.
76. J. Chatt, L. A. Duncanson, and L. M. Venanzi, *J. Chem. Soc.* (1955) 4455; see also: F. A. Cotton and G. Wilkinson, *Advanced Inorganic Chemistry*, (Interscience, New York, 1960).
77. F. Basolo and R. G. Pearson, *Mechanisms of Inorganic Reactions* (Wiley, New York, 1958).
78. J. Korchowiec and R. F. Nalewajski, *Int. J. Quantum Chem.* **44** (1992) 353.
79. R. F. Nalewajski, *J. Chem. Phys.* **81** (1984) 2088.
80. A. Cedillo, *Int. J. Quantum Chem. Symp.* **28** (1994) 231.
81. M. Levy, *Phys. Rev.* A**26** (1982) 1200.

82. R. G. Parr and L. J. Bartolotti, *J. Phys. Chem.* **87** (1983) 2810.
83. R. F. Nalewajski and J. Korchowiec, *Acta Phys. Polon.* **A76** (1989) 747.
84. J. L. Gázquez, in *Structure and Bonding: Chemical Hardness*, K. Sen Ed (Springer-Verlag, Heidelberg, 1993), Vol. **80**, p. 27.
85. K. Morokuma, *Acc. Chem. Res.* **10** (1977) 294.
86. B. G. Baekelandt, G. O. A. Janssens, H. Toufar, W. J. Mortier, R. A. Schoonheydt, and R. F. Nalewajski, *J. Phys. Chem.*, **99** (1995) 9784.
87. R. F. Nalewajski and J. Korchowiec, *Computers Chem.*, **19** (1995) 217.
88. R. F. Nalewajski, J. Korchowiec, and A. Michalak, *Proc. Indian Acad. Sci. (Chemical Sci.)*, S. Gadre Ed., **106** (1994) 353.
89. J. C. Decius, *J. Chem. Phys.* **38** (1963) 241; L. H. Jones and R. R. Ryan, *Ibid.* **52** (1970) 2003.
90. B. I. Swanson, *J. Am. Chem. Soc.* **98** (1976) 3067; B. I. Swanson and S. K. Satija, *Ibid.* **99** (1977) 987.
91. R. F. Nalewajski and A. Michalak, *Int. J. Quantum Chem.* **56** (1995) 603.
92. R. F. Nalewajski, *Int. J. Quantum Chem.*, in press.
93. R. F. Nalewajski and J. Korchowiec, *J. Mol. Catal.* **82** (1993) 383.
94. M. J. S. Dewar and W. Thiel, *J. Am. Chem. Soc.* **107** (1977) 4899.
95. H. Nakatsuji, *J. Am. Chem. Soc.* **96** (1974) 24, 30.
96. L. P. Hammett, *Physical Organic Chemistry* (Mc Graw-Hill, New York, 1940).
97. A. Streitwieser, *Molecular Orbital Theory for Organic Chemistry* (Willey, New York, 1961).
98. C. A. Coulson and H. C. Longuet-Higgins, *Proc. Roy. Soc. (London)* **A192** (1947) 16.
99. S. Nagakura and J. Tanaka, *J. Chem. Soc. Japan, Pure Chem. Soc.* **75** (1954) 993.
100. K. Fukui, *Theory of Orientation and Stereoselection* (Springer-Verlag, Heidelberg, 1982); *Science* **218** (1982) 747.
101. K. Fukui. and H. Fujimoto, *Frontier Orbitals and Reaction Paths* (World Scientific, Singapore, 1995).
102. K. Fukui, T. Yonezawa, and C. Nagata, *J. Chem. Phys.* **26** (1957) 831; *Bul. Chem. Soc. Japan* **27** (1954) 423; T. Fueno, *Ann. Rev. Phys. Chem.* **12** (1961) 303.
103. T. Nakajima, *J. Chem. Phys.* **23** (1955) 587.
104. G. W. Wheland, *J. Am. Chem. Soc.* **62** (1942) 900.
105. R. F. Nalewajski and J. Korchowiec, *J. Mol. Catal.* **54** (1989) 324.
106. R. F. Nalewajski and J. Korchowiec, *J. Mol. Catal.* A, in press.
107. R. F. Nalewajski, J. Mrozek, and A. Michalak, *Int. J. Quantum Chem.*, in press.
108. R. F. Nalewajski and A. Michalak, *J. Phys. Chem.*, in press.
109. M. H. Cohen, in *Topics in Current Chemistry*, vol. **183**: *Density Functional Theory - Theory of Chemical Reactivity*, R. F. Nalewajski Ed. (Springer-Verlag, Heidelberg, 1996), p. 142, and references therein.
110. L. M. Falicov and G. A. Somorjai, *Proc. Natl. Acad. Sci. USA* **82** (1985) 2207.
111. A. Gołębiewski, *Trans. Faraday Soc.* **57** (1961) 1849.
112. V. Gutmann, *The Donor-Acceptor Approach to Molecular Interactions* (Plenum, New York, 1978).

113. R. G. Pearson, *Symmetry Rules for Chemical Reactions: Orbital Topology and Elementary Processes* (Wiley, New York, 1976), and references therein.
114. R. F. W. Bader, *Mol. Phys.* **3** (1960) 137; R. F. W. Bader and A. D. Bandrauk, *J. Chem. Phys.* **49** (1968) 1666.
115. G. A. O. Janssens, B. G. Baekelandt, H. Toufar, W. J. Mortier, and R. A. Schoonheydt, *Int. J. Quantum Chem.*, **56** (1995) 317.
116. R. G. Parr and R. F. Borkman, *J. Chem. Phys.* **49** (1968) 1055.
117. R. F. Nalewajski, *J. Mol. Catal.* **82** (1993) 371.
118. M. Witko, R. Tokarz and J. Haber, , *J. Mol. Catal.* **66** (1991) 205.; M. Witko, J. Haber, and R. Tokarz, *J. Mol. Catal.* **82** (1993) 457; J. Haber and M. Witko, *Catalysis Today* **23** (1995) 311.
119. A. Gołębiewski, R. F. Nalewajski, and M. Witko, *Acta Phys. Polon.* **A51** (1977) 617; A. Gołębiewski, in *Theory-Structure Properties of Complex Compounds* (Polish Scientific Publiers, Warsaw, 1979), p. 117, and references therein.
120. R. A. Marcus, *J. Chem. Phys.* **24** (1956) 266.
121. V. G. Levich, R. Dagonadze, A. Kuznetsov, *Electrochim. Acta* **13** (1968) 1025.
122. S. J. Formosinho, in *Theoretical and Computational Models for Organic Chemistry*, S. J. Formosinho Ed. (Kluver, Amsterdam, 1991), p. 159, and references therein.
123. G. S. Hammond, *J. Am. Chem. Soc.* **77** (1955) 334.
124. R. F. Nalewajski, S. J. Formosinho, A. J. C. Varandas, and J. Mrozek, *Int. J. Quantum Chem.* **52** (1994) 1153.
125. R. F. Nalewajski and J. Mrozek, *Int. J. Quantum Chem.* **57** (1996) 377.
126. J. Mrozek, *Quantum-Mechanical Investigations of Molecular Structure and Reactivity*, Habilitation Thesis, Jagiellonian University, 1996.
127. R. F. Nalewajski, J. Korchowiec, R. Tokarz, E. Brocławik, and M. Witko, *J. Mol. Catal.* **77** (1992) 165.
128. A. Michalak and K. Hermann, unpublished results.
129. R. F. Nalewajski, A. M. Köster, T. Bredow, and K. Jug, *J. Mol. Catal.* **82** (1993) 407.
130. R. L. Kurtz, R. Stockbauer, T. E. Madey, E. Román, and J. L. de Segovia, *Surf. Sci.* **218** (1989) 178.
131. K. Jug, G. Geudtner, and T. Bredow, *J. Mol. Catal.*, **82** (1993) 171.
132. W. J. Lo, Y. W. Chung, and G. A. Somorjai, *Surf. Sci.*, **71** (1973) 199.
133. M. Grzybowska, J. Haber, and J. Janas, *J. Catal.* **49** (1977) 150.
134. K. Brückman, J. Haber, and T. Wiltowski, *J. Catal.* **106** (1987) 188.
135. K. Brückman, R. Grabowski, J. Haber, A. Mazurkiewicz, J. Słoczyński, and T. Wiltowski, *J. Catal.* **104** (1987) 71.
136. A. Bielański, and J. Haber, *Oxygen in Catalysis* (Marcell Dekker, New York, 1991).
137. J. Haber, in *Molybdenum: An Outline of its Chemistry and Uses*; E. R. Braithwaite, and J. Haber Eds. (Elsevier, Amsterdam, 1994).
138. A. Michalak, K. Hermann, M. Witko, *Surf. Sci.*, in press.
139. K. Hermann, A. Michalak, M. Witko, *Catalysis Today*, in press.

140. A. D. Bacon, M. C. Zerner, *Theor. Chim. Acta* **53** (1979) 21; M. C. Zerner, G. H. Loew, R. F. Kirchner, and U. T. Mueller-Westerhoff, *J. Am. Chem. Soc.* **102** (1980) 589; J. D. Head, M. C. Zerner, *Chem. Phys. Lett.* **122** (1985) 264; *Ibid.* **131** (1986) 359; W. P. Anderson, W. D. Edwards, M. C. Zerner, *Inorg. Chem.* **25** (1986) 2728.
141. L. Kihlborg, *Arkiv Kemi* **21** (1963) 357.
142. N. D. Mermin, *Phys. Rev.* A**137** (1965) 1441.
143. R. F. Nalewajski and M. Koniński, *Z. Naturforsch.* **42a** (1987) 451.
144. J. F. Capitani, R. F. Nalewajski, and R. G. Parr, *J. Chem. Phys.* **76** (1982) 568.
145. (a) R. McWeeny, *Methods of Molecular Quantum Mechanics* (Academic, London, 1989); (b) M. Klessinger and R. McWeeny, *J. Chem. Phys.* **42** (1965) 3343.
146. M. Teter, *Phys. Rev.* B**48** (1993) 5031.
147. M. Grigorov, J. Weber, H. Chermette, and J. M. J. Tronchet, *Int. J. Quantum Chem.*, in press; private communication.
148. J. F. Janak, *Phys. Rev.* B**18** (1978) 7165.
149. L. C. Allen, *J. Am. Chem. Soc.* **111** (1989) 9003; *Int. J. Quantum Chem.* **49** (1994) 253.
150. J. C. Slater, *Adv. Quantum Chem.* **6** (1972) 1.
151. J. Ciosłowski and B. B. Stefanov, *J. Chem. Phys.* **99** (1993) 5151.
152. P. K. Chattaraj, A. Cedillo, and R.G. Parr, *J. Chem. Phys.* **103** (1995) 7645.

SUBJECT INDEX

acid-base interactions (see donor-acceptor systems and electronic energy), 28-45, 48-50, 87, 88, 92, 149-152
acidic/basic character: 37-45, 91-160, 168, 173-175, 192
  global acid/base, 92
  local acid/base, 92
acrolein product, 197, 217-219
adsorbate (see chemisorption and catalytic systems):
  activation, 101, 197, 199, 212-214, 216, 276
adsorption molecular/dissociative, 166-175
alkali halides (see harpoon mechanism), 28, 29, 34-37, 42
  amount of CT, 34, 35
  dissociation energy of, 34, 35
alkali metals, 26-29, 34-37
allyl adsorbate, 163, 197-219
allyl oxidation (catalytic) to acrolein: 197, 199, 205, 211, 213, 215, 218, 219
  transition-states, 211, 219
atomic charges (see populational analysis schemes): 153, 161, 199, 200, 212
  effective nuclear charges, 11, 12, 30-34
atomic orbitals (AO, basis set), 8-10, 274-276
atoms-in-molecules (AIM), 1
basis set superposition error, 260
bond activation, 141, 144-146, 148, 156, 162, 166, 168, 174, 175, 189-197, 201, 212-219, 276
bond competition effect, 216-219
bond-forming-bond-breaking mechanism, 141, 159, 162, 175, 192-196, 202, 215-219
bond ionic/covalent components, 162, 173, 175, 192, 193, 274-276
bond multiplicity: 162, 163, 202-205, 210, 214-219, 274-276
  one-determinantal (finite-difference) description, 274-276
  relative changes of, 275, 276
  total/overall of reactants, 275, 276
bond polarity, 134-141, 168, 170, 173, 189-196, 212-214, 276
Born-Oppenheimer approximation, 4, 50, 131, 140, 220, 225
canonical derivatives: 4, 8, 14, 15, 24, 25, 162, 188
  atomic nuclear attraction derivative, 12-33
  chemical potential, 1
  external potential, 1
  hardness matrix, 1, 93
catalyst (substrate, see chemisorption systems): 88
  active sites, 88, 146, 160, 166, 168, 175, 176, 190, 197, 203, 212, 215, 216
  cluster representation of, 88, 200
  global reduction/oxidation of, 163, 173, 184, 185, 196, 197
  substrate softening effect, 143, 144
catalytic (heterogeneous, see chemisorption): 140, 141, 161-219
  activity, 92
  reactions, 57, 88, 215
  systems: 88, 95, 102, 130, 143, 144, 161-219
    bond-activation / dissociation, 144-146, 148, 170-175, 192-195
    charge instabilities in, 143
charge displacements: 161-219
  AIM, 64, 65
  charge transfer (CT), 1, 30-45, 48-50, 58, 71, 78, 82, 91, 97, 99, 149-152, 157-160, 197
  finite-difference (SCF MO/DFT) charge flows between reactants, 258-261
  in situ (isoelectronic), 1, 30, 32, 67-69, 71, 95, 96, 104, 105, 107, 131-141, 149, 150, 157-160, 167, 184, 185, 196, 197, 272, 273
  P/CT dominance of, 194-196, 199-219
  P/CT resolution of, 71, 74-77, 91-93
    AIM, 74, 77, 96, 97
    AIM + PNM, 74, 75, 77, 78, 92
    PNM, 74, 76
  polarization (P), 1, 31, 48-50, 58, 71, 79, 82, 91, 97, 99, 149-157, 197
charge sensitivities (see Fukui function, hardness, and softness): 2, 51-69
  atomic: 11-50
    equipotential/horizontal, 24-28
    orbitally relaxed/rigid, 24
  cluster size dependence of, 166
  computational algorithms for, 64-67
  condensed (see: hardness, softness, Fukui function), 220, 232, 233, 235, 236
  derivative identities, 14, 15
  global/resultant, 2, 6, 53
  hardness kernel (see hardness)
  inequalities between, 20-28

internal (isoelectronic, *in situ*): 60-64, 67-69
combination formulas, 67-69
linear response kernel matrix: 230-232
sum rules of, 233-235, 237
local/nonlocal, 6, 220
modelling of, 2, 8-10, 247
polarization (linear response) kernel, 6, 7, 58, 62
polarization matrix kernel, 49
polarization (linear response) matrix: 7, 62-64, 66, 112, 131, 134, 152-156, 237
eigenvectors of (see INM), 112
regional/group, 3
relaxed, 2, 23, 24
rigid, 2, 23, 24
softnesses (see softness), 1, 8
charge sensitivity analysis (CSA): 1, 51-90, 131, 134, 140, 141, 152, 153, 157, 220, 243, 276
AIM-resolved, 161-219
diagnosing of reactivity trends, 161-219
geometric concepts in, 162
modelling of, 8-10
mode-resolved, 162
orbitally-resolved [spin-orbital (SO) representation]: 243-267
frontier SO hardness in (external, HOMO-LUMO energy gap), 256
global/local SO chemical potentials in, 248,, 249
multicomponent description: 249
integral SO occupations in, 262-267
internal formulation of, 261-267
Lagrange multipliers in (occupation/population constraints), 261-264
SO average external hardness in, 257
SO average hardness in, 257
SO average orbital energy in, 256
SO chemical potential/electronegativity equalization in, 253
SO constrained equilibrium in, 253
SO effective occupations in, 260
SO Euler equations in, 253, 254
SO external FF in, 254, 255
SO external hardness in, 254-256
SO FF matrix in, 252
SO global charge sensitivities in, 253-255
SO global equilibrium criteria in, 253-255
SO global hardness matrix in, 251
SO global linear response matrix in, 254, 255
SO global softness matrix in, 250, 251
SO hardness kernel matrix in, 250, 251
SO intensive/extensive variables in, 249
SO internal electronegativity equalization (IEE) in, 266
SO internal relaxation matrix in, 255
SO linear/quadratic occupational constraints in, 261-267
SO linear response kernel matrix in, 252
SO linear response matrix in, 257
SO local energies in, 253
SO local FF in, 249
SO local orbital hardness equalization in, 251
SO local softness matrix in, 250, 251
SO occupational degrees-of-freedom in, 261
SO occupation probabilities in, 256
SO softness kernel matrix, 250, 251
chemical potentials: 4, 5, 52, 223
AIM: 7, 44, 52, 236, 272, 273
relaxed, 69
rigid, 69
atomic, 12
complementary combination rules for, 37-45, 53-59, 66, 67
discontinuity of, 161, 188, 223
displacements of, 48-50, 58, 65, 66
equalization of: 3-5, 33-40, 44, 48-50, 52, 61, 64, 65, 67, 72, 158-162, 220, 221, 230, 270
intra-reactant, 229, 230, 235
geometric mean principle, 40
global: 2, 3, 40, 52-54, 268
harmonic mean law, 40
internal, 61
in situ, 31-33, 35-37, 41, 42, 49-51, 67, 152
local, 5, 52
normal, 69-72
of reactants/fragments: 34-39, 52, 55, 67-69, 198, 199
cluster size/stoichiometry dependence of, 198, 199
regional, 2, 3, 52
relaxational corrections to, 51, 69, 188
chemical reactivity: 91-160
concepts, 91-160
indices: 2, 91, 92, 141, 162, 197
nuclear, 220
sensitivity criteria, 91, 92, 162, 220
single-reactant criteria, 10, 91, 92, 163
two-reactant criteria, 10, 91, 92, 161-219

drivers of, 224
electrophilic/nucleophilic/radical: 10, 78, 79, 91, 92, 222
extremum FF criteria, 91, 267-269
kernels, 220
single/two-reactant descriptions, 91-160, 229-241
symmetry rules of, 140
theory of, 1, 140, 152, 212, 229-241
chemical valence (quantum-mechanical, see bond multiplicity): 210, 215-218, 274-276
absolute valences, 274
atomic/diatomic components of, 274-276
relative displacements, 210, 215-218, 275, 276
chemisorption (see catalytic): 88
adsorbate/substrate subsystems, 153-156
adsorption/desorption phenomena, 88, 101, 143-146, 196
cluster approximation to: 88, 161-219
saturation of dangling valences, 199-219
forward/backward donation in, 192-195
physisorption stage of, 88, 152-156, 173-175, 178-185, 187-191, 212-214
stages, 88, 212-214
systems: 88, 97-107, 144-146, 161-219, 275, 276
allyl-MoO_3: 163, 197-219
adsorption arrangements (vertical/horizontal), 205-219
cluster reconstruction of, 166 - 175, 197, 212-216
condensed hardness matrix of subsystems, 144-146, 232, 235, 236
oxidation/reduction of, 101, 148, 159
toluene-V_2O_5 (see toluene and vanadium pentoxide): 148, 153-157, 163, 175-197
adsorption arrangements of (perpendicular/parallel), 175-197
amount / direction of CT in, 178 - 183, 187, 188, 213
charge reconstruction patterns in, 178-194
cluster size dependence in, 175, 176
global FF patterns in, 184, 185
water-rutile: 163-175
molecular adsorption in, 166-170
transition-state (NPS) in, 170-175
cluster charge dependence, 199
collective modes of charge displacements: 63, 69-90, 92, 93, 132, 134, 135, 138, 161, 162
intermediate hardness decoupling modes of reactants/fragments: 51, 70, 92, 93
externally decoupled modes (EDM): 70, 105, 108-122
algorithm for, 108, 113
characteristics of, 110, 111, 116-119
EDM (→ IDM), 109-111, 113
EDM (→ MEC), 117-120
external CT - active modes, 110, 111, 116
for (toluene|V_2O_5), 109-122
localization transformation in, 108, 109, 117
maximum overall projection criterion in, 108
maximum overlap criterion (MOC) in, 108, 109, 117
mode hardnesses of, 110, 111, 116
transformation, 108
unstable modes, 116
internal normal modes (INM), 112, 113, 132, 135, 138-141
internally decoupled modes (IDM), 70, 89, 96-107, 124, 273
for (toluene|V_2O_5), 96-107
relaxed, 96, 105-107
rigid, 96, 103, 104
inter - reactant modes (IRM): 124, 130, 162, 168-173, 199, 206, 207, 214, 215
characteristics of, 125-130, 169
complementary pairs of, 125-127, 168, 169, 206, 207, 214, 215
charge displacement hybrids, 125
delocalized (reactive) modes, 125-130
for water-rutile system, 168-170
localized (environmental) modes, 125 - 130, 168, 169
relaxational modes (REM): 270
populations/chemical potentials/hardness matrix, 270
mode-to-mode interpretation, 109, 141, 159, 170, 219
minimum energy coordinates (MEC, see also MEC/REC): 69, 79-90, 118-120, 132, 133, 138, 139, 145-146, 162
diagnosing of surface clusters, 148, 163-166
external, 80-84, 90, 117-119, 163-166
FF indices of, 82
for rutile, 163-166
for toluene, 83, 85

internal modes, 80-84, 89, 90, 133
mode hardness, 81-83, 85, 118-120, 148, 163-166
mode instability, 145, 146, 148
mode topology, 148
natural reaction coordinate, 120, 124
of molecular fragments/reactants, 87-90
FF indices, 94, 95
populational normal modes (PNM): 51, 69-82, 109, 112, 132, 140, 141, 273
acceptor/donor PNM, 78
antisymmetric (purely P) modes, 72, 74, 75, 78
chemical potential / electronegativity equalization of, 72, 73
chemical potentials of, 70, 72, 78
combination formulas in, 70, 74
CT-active modes, 76-79
CT-projected chemical potentials of, 73, 74, 78, 151, 152
CT-projected FF of, 71, 73, 74
CT-projected hardness of, 79
CT-projected hardness/softness tensors, 74
CT-projected mode populations, 71-73, 75, 151, 152
electron populations of, 70-73
for toluene, 75-79
Fukui functions of, 70, 71, 74, 77, 78
hardest/softest modes/eigenfunctions, 78, 79, 81, 82, 84, 102, 109, 112, 116, 117, 138-140, 163, 166, 226, 227
hardness equalization, 73
of reactants, 87-89, 92, 93, 96
perturbational (matrix partitioning) approach to, 93
phase convention in, 70, 71
principal hardness/softness, 70, 71, 78, 79
P-CT resolution in, 71-77
relaxational coordinates (REC, see also MEC/REC): 80, 83-90
for toluene, 83-85
mode hardness of (see hardness/relaxed), 83-85
collective modes of nuclear position displacements: 63, 132, 134,
bond-stretch MEC, 133
electron following/preceding perspectives, 91
MEC/REC, 132
minimum energy coordinates,
vibrational normal modes (VNM), 69, 135, 136, 138-141
compliance formalism:
constants (see MEC mode hardness), 81, 82, 85
interaction constants, 79, 81
minimum energy criterion, 81, 87, 89
nuclear, 79, 133
populational, 79
restoring forces, 80
compliance matrix, 8, 16, 52, 69, 80, 133
coordination bond, 42-45, 92, 101, 159, 166-168, 173-175, 189-196, 276
CT-induced polarization, 71, 74-79, 163
decoupling schemes of hardness matrix: 96-130
inter-reactant, 105-130
intra-reactant, 96-105
density functional theory (DFT, see also orbital description/*ab initio* methods): 1, 4, 10, 45-48, 57, 131, 140, 161, 162, 199, 212, 215, 217-221, 260
Kohn-Sham (KS) approximation [local density approximation (LDA), local spin density approximation (LSDA)], 45, 46, 163, 213, 215, 220-228, 239-244, 274, 275
bare-Coulomb interaction in, 224, 246
canonical orbitals in, 244
density of states in, 223-225
effective interaction in, 224, 247
energy density functional in, 246, 249
energy function of orbital occupations in, 244
ensemble formulation of, 223
FF and softness quantities in, 224
Green's function in, 223
Hartree-KS static dielectric function in, 224
internal polarization propagator in, 227
internal potential response kernel in, 227
internal static dielectric function kernel in, 227
isoelectronic (internal) kernels in: 227
spectral resolution of, 227
Janak theorem in, 244, 246-248
linear response kernel in: 227
perturbational expansion of, 226
M(S)O hardness matrix in, 248
potential response function in, 224
screened Coulomb force in, 225

softness kernel (screened static polarization propagator) in, 226
static dielectric function kernel in, 227
static polarization propagator in, 224, 226
multicomponent KS formulation of: 220, 239-241
density functionals in, 239, 249
effective potential in, 239
electron group energies in, 239
electron group wavefunctions in, 239, 240
iterative scheme of, 240, 243
KS equations in, 240
reactant densities / spin-densities in, 239, 240
strong-orthogonality condition in, 240
derivatives: 2
chain rule transformations of, 4, 39, 53-64, 66, 68, 69, 87, 94, 131, 224, 227, 232, 237, 245-248, 254
principal/canonical, 5-8
direct / indirect charge responses, 137, 138, 140, 161-219
donor/acceptor atoms, 137
donor/acceptor subsystems, 137, 257, 272, 273
donor-acceptor systems: 28 - 45, 67 - 69, 87, 88, 174
acceptor/donor behaviour in, 44, 92
alkali metal-halogen (see harpoon mechanism), 34-37
amount/direction of CT in, 34, 35, 49 - 51, 152, 157-160, 178-183, 187, 188
biased/unbiased description of, 29, 30, 34-37
CS of reactants in, 54
diatomic, 28-45
forward CT in, 34-37
interaction energy (see electronic energy) in, 48-50, 92, 93, 186, 187, 216-218
base charge withdrawal/acid charge inflow energies, 150
partitioning of, 149-152
isoelectronic derivatives of, 67-69
K + Br, 36
partitioning of FF in, 93-96
reverse CT in, 34-37
electron-cloud following/preceding perspectives (see collective modes of nuclear positions), 140, 141
electronegativity (see chemical potential):
atomic, 14, 15
combination rules for, 40, 53-59, 66
Mulliken finite-difference estimate of, 15
electronegativity equalization (see chemical potential equalization): 53, 61
in SO resolution, 250, 251
electronegativity equalization method (EEM), 1, 51, 131, 134, 135, 152, 153, 157, 161, 270
electron affinities, 2, 9, 35, 36, 46, 47
electron density: 199, 209, 215, 244, 245
difference contour diagram of, 199, 209, 215
fluctuations, 92
of polarized reactants, 229, 230
probability function, 47, 48, 60
spin densities, 222, 241, 242, 244, 245, 249, 255
electron orbital configuration: 261
virtual/occupied orbitals in, 261-264
electron population space (EPS), 140, 141
electron relaxation (see relaxational corrections), 55, 56, 162
electron repulsion energies: 5, 6, 8, 247, 248
Coulomb integrals, 9, 24
integrals: 8
interpolation formulas for: 9
electron reservoirs, 2, 3, 60, 80, 84, 90, 94, 100, 101, 116, 120, 150, 161, 173, 184, 185, 196, 197, 226, 248, 253, 254, 257, 258, 261
electronic energy: 4-7, 48-50, 149-152
atomic interpolation formulas for, 13
atomic surface, 11-28
average energy of a configuration, 244
CBO derivatives of, 245
charge transfer energy, 30, 32, 34, 41, 48-50, 149-152, 186, 187
derivatives of: 4-8, 67-69, 243-247
electrostatic energy, 30, 32, 41, 48-50, 149-152, 177, 186, 187, 211
in AIM resolution, 7, 61, 63, 149-153, 236
in MO resolution, 8, 257
in terms of CBO matrices, 245
Legendre transforms of, 6, 12-28, 45-48, 52
Morokuma partitioning of, 48-50
partitioning of, 239
polarization energy, 30, 32, 48-50, 149-152, 186, 187
quadratic Taylor expansion of: 6, 7, 11, 50, 56, 85, 157-160, 243, 264, 267
for global equilibrium state, 54
for non-equilibrium state, 54
interaction energy in donor-acceptor systems, 31-35, 41
internal (isoelectronic), 61, 62
second-order change of, 6, 7, 70, 71, 80

slope discontinuity of the g.s. energy, 46, 47, 223
variational principles for, 220-222
electrostatic (Hartree) potential, 48, 153 - 156, 177, 201, 211, 221
energetical hierarchies of adsorption arrangements, 186, 187, 193, 195, 196, 213
energy density functionals: 4-6, 45-48, 229
Coulomb-exchange-correlation interaction, 246-248
electronic energy, 4, 5
exchange-correlation energy, 221
kinetic energy of non-interacting electrons, 221
Levy construction of, 6, 222
universal HK, 5, 57
universal Levy, 6, 46
ensemble approach, 92
environmental effects, 148, 159, 163, 184, 185, 196, 197, 255, 256
environments (nucleophilic / electrophilic / isoelectronic), 255, 256, 258
equilibria: 161
constrained/regional, 2-7, 23, 48-50, 52, 57, 64-67, 220, 229-241, 243
global/internal, 2-6, 23, 48-50, 52, 53, 57, 79, 80, 87, 184, 185, 196, 197, 220, 231, 235, 243
equipotential/ horizontal atomic sensitivities, 24-28
Euler equations (see electronegativity / chemical potential equalization), 57, 61, 220, 229, 236
external potential: 7, 226
displacements, 161-219
local, 4-7
of AIM, 7, 152
point charge approximation of, 153
relative, 230
Fermi level, 223
finite-difference approach: 9, 14, 15, 29
forces on nuclei (see Hellmann - Feynman theorem), 225
frontier MO (HOMO/LUMO): 10, 91, 162, 255
densities, 222- 224
perturbational approach to interaction of, 260
representation of FF, 223
rescaled densities, 225
frontier orbital perspective on:
condensed hardness matrix of reactants, 258, 259
global chemical potential, 258, 259
global FF, 258, 259
global hardness, 258, 259
Fues potential, 131
Fukui function: 1, 6, 7, 10, 93, 225
AIM: 10, 44, 45, 53, 63, 64, 77, 91-96, 163, 164, 166, 184, 185, 196, 197, 267-269
diagonal, 56, 67-69, 98-101, 156, 199-201, 212, 219
off-diagonal, 56, 67-69, 98-101, 156, 199-201, 212, 219
in situ, of reactants, 67-69, 98 - 100, 152, 156, 166, 202, 212-214
combination formulas for, 55, 56, 66
condensed of subsystems/reactants, 94, 143
extremum criterion of, 91, 267-269
finite-difference modelling of, 10, 241, 242, 260, 261
for molybdenum-allyl system, 197-219
for water-rutile system, 167, 168, 170-174
isoelectronic (*in situ*): 30, 34, 50, 67-69, 95, 96, 98-100, 162, 165
diagnosing of surface clusters, 166
local: 10, 58, 242, 267-269
discontinuity, 223
for atoms, 242
matrix, 49-50, 230-232
spin-resolved, 222
mode resolution of:
global FF pattern in IRM framework, 125, 127, 128
global FF pattern in EDM (→ IDM) framework, 109-111, 113, 114, 116, 117
global FF pattern in EDM (→MEC) framework, 117-119
global FF pattern in PNM framework, 109, 113-115, 117, 120, 163, 164
global FF pattern in rigid IDM framework, 102, 103
global FF pattern in relaxed IDM framework, 105, 106
in situ FF pattern in EDM (→IDM) framework, 110, 111, 116, 117, 120, 122
in situ FF pattern in IRM framework, 127, 129, 130, 168-170, 208, 214, 215
in situ FF pattern in PNM framework, 117, 119, 123
in situ FF pattern in relaxed IDM framework, 105, 107
in situ FF pattern in rigid IDM framework, 104, 105
normal, 70, 71, 77

normalization of: 49, 54, 233-235, 237, 238, 267
diagonal parts, 49, 54, 59, 94
off-diagonal parts, 30, 49, 54, 67
of molecular fragments, 37-39, 53, 98 - 100, 267-269
partitioning of interaction energy: 150-152
A- / B- / and (A, B) - resolved divisions, 150-152
partitioning in reactive systems, 93-96
P/CT resolution of, 97, 99, 101
relaxational corrections to, 55
relaxed: 55, 56, 66, 87, 98-100, 102
for (toluene|V_2O_5), 96, 97
rigid, 55, 56, 89, 98-100
summation rules, 94, 95
geometric / electronic structure relations (see mapping relations), 131-141, 225
geometry relaxation: 91, 131-141, 148, 186
geometrical-electronic coupling, 91, 93
Gutmann (bond length variation) rules, 137, 138, 141
halogens, 26-28, 34-37
Hammond postulate, 157, 159
hard/soft acids and bases (HSAB): 40-43
HSAB principles, 40-43, 158, 159
symbiosis rule, 41-43
hardness:
atomic, 12, 15, 30, 51
combination rules, 37-45, 53-57, 66
condensed matrix: 37-39, 52, 55, 66, 93, 143, 144, 146, 232, 235-237
rigid, 56, 57, 143, 144, 146
equalization principle: 5, 7, 54, 59, 60, 72
in AIM resolution, 3, 237, 268, 269
in local description, 5, 59, 60, 267-269
in reactant partitioning, 235
extremum principles of, 41, 267-269
finite-difference estimate of, 9, 15
generalized kernel, 59
generalized kernel matrix, 232, 233
generalized matrix, 64, 65
global: 6, 51, 53, 54, 267-269
functional (of FF) for, 268
harmonic average laws (see chemical potential and hardness), 38, 40
internal (isoelectronic, *in situ*) quantities, 31-33, 41, 42, 51, 60-64, 67-69, 82, 83
kernel, 7, 51, 52, 57, 226, 268
kernel matrix, 230, 232, 241
local, 42, 51, 54, 57, 58, 60, 67
matrix/tensor: 7-9, 23, 24, 29, 30, 51-53, 63, 64, 80, 236, 272, 273
EDM representation of, 113, 117, 121
ellipse, 71, 81, 82
intra- and inter-reactant blocks of: 124
eigenvalue problem of (see spectral resolution of, 124
IRM representation of, 127, 130
modelling of, 8, 9, 257
reactant block structure, 92, 93, 105, 124
relaxed, 23, 56, 57, 93, 270, 271
relaxed fragment transformation of, 270, 271
rigid, 23, 270
MO representation of, 247, 248
principal (normal), 70, 77, 78
relaxational corrections, 39, 43, 44, 51, 56, 57, 146, 270
relaxed (see MEC mode hardness and relaxational corrections), 83, 85, 87
spectral analysis of tensor/kernel, 220, 226, 227, 268
hardness (interaction) representation, 52
harpoon mechanism (see alkali halides): 28 - 30, 34 - 37
K + Br, 36
chemical potential equalization perspective of, 34-37
critical distance in, 29, 30, 34-37
energy balance of, 29
ionic and covalent states in, 29
Hartree (see electrostatic) potential, 48, 221, 222
Hartree-Fock (HF) theory (see orbital descriptions/*ab initio* methods): 243-267, 274
canonical orbitals in, 243, 244
electronic energy functions/functionals: 244
equations, 263
Fock matrix in, 8, 248, 277, 278
GUHF/UHF CBO derivatives of the energy: 248, 261, 274, 275, 277, 278
CBO derivatives in, 277, 278
density matrices in, 277, 278
energy expressions in, 277, 278
Fock matrix in, 277, 278
RHF theory, 274, 277
SCF iteration process, 257, 263, 264, 266
Hellmann-Feynman theorem, 5, 12, 45, 225
hyperconjugation bond component, 189, 192
in situ quantities, 31-37, 67-69
intersecting-state model (ISM) of CT processes: 157-160, 272, 273

endothermic/exothermic activation barriers, 159, 216-219
AIM/PNM-resolved model, 157-160, 272, 273
charge stabilization / withdrawal energy curves of, 157-159
CT bond-order, 157-160, 272
CT energy profile / reaction coordinate: 157-160
N-constrained, 157-160
N-unconstrained, 158, 159
in EPS, 157-160
in NPS, 159
minimum energy path in, 273
reactant energy surfaces in, 157 - 160, 272, 273
transition-state (TS) in EPS, 157-159, 273
inductive effect, 162
ionization potentials: 2, 9, 24, 35, 36, 46, 47
VSIE, 24
Jacobians (see reduction of derivatives), 24, 25
Koopmans theorem, 10
Legendre transformed representations, 13-28, 45-48, 52
Le Châtelier principle (see stability criteria and Le Châtelier-Braun principle), 20, 22, 23, 42, 44, 226, 227
Le Châtelier-Braun principle (see stability criteria and Le Châtelier principle), 23, 39, 138, 140, 175
Lewis acids and bases (see HSAB), 41, 158
mapping relations: 63, 93, 131-141, 168, 225
bond-stretch Hessian, 131, 134
bonding/non-bonding distances, 131
constraint transformation, 131
force field, 131
for methanol, 134-141
geometric Hessian, 131
geometrical transformation, 131
in frozen charges approximation, 131
local, 132
normal, 141
Mataga-Nishimoto formula, 9
maximum chemical potential response criterion, 92
maximum spin polarization criterion, 92
Maxwell relations, 14, 15, 20, 21, 47, 58, 59
MEC/REC: 85
chemical potentials, 86, 87
CT-projected mode chemical potentials, 86, 87
CT-projected FF, 86
CT-projected hardness matrix, 86
CT-projected mode populations, 86, 87
FF indices, 86
hardness matrix, 86
metric tensor, 85
of subsystems/reactants, 87-90
softness quantities, 87
methanol, 134-141
minimum orbital energy (MOE) criterion, 263
mode stability/instability, 78
molybdenum oxide clusters, 163, 197-219
natural nuclear vibrations: 141
EPS conjugates of, 141
non-selective oxidation of hydrocarbons, 193-196
nuclear position space (NPS), 140, 141, 157, 159
nuclear reactivity indices: 225
FF/softness quantities, 225, 227
nuclear repulsion energy, 48
Ohno formula, 9, 34
one-electron density matrices, 248, 262, 263, 277, 278
orbital description: 8-10, 24
ab initio methods (SCF, SCF CI, DFT), 161, 162, 199, 203, 204, 209, 210, 243-248, 260, 261
alternative representations of, 243, 244
basis sets in (see AO), 8, 277, 278
canonical MO representation of, 243, 245, 263
CBO/density operators and matrices in, 274, 275
CBO derivatives of electronic energy in, 8-10, 245, 247, 248
charge-and-bond-order (CBO) matrix in, 8-10, 245
electron exchange effects in, 222
electron correlation effects in, 222
electronic energy functionals in, 243
equilibrium criteria and charge sensitivities in SO resolution, 250
Fock matrix in, 8
hardness derivatives of M(S)O occupations in, 247, 248
Hartree approximation of, 221, 222
Hartree-Fock theory, 221, 222
hybrid orbitals in, 125
internal (constrained) energy function in: 248, 264-267
derivatives of, 264-267

Kohn-Sham (KS) theory: 220-228, 239-243
effective potential in, 222
exchange-correlation potential in, 222
KS (LDA/LSDA) equations, 221, 222, 261
local derivatives in, 220
occupation numbers of KS orbitals, 222
LCAO MO coefficients in, 8, 10, 274, 277
local/global state variables in, 248, 249
localized orbitals in, 125
molecular orbitals (MO) and spin-orbitals (SO), 8, 10, 79, 193, 244, 274-276
MO Coulomb-exchange interaction in, 247
MO orthogonality/CBO idempotency constraints in (see Pauli principle): 248, 257, 258, 261-264
MO relaxation in, 10, 222, 223, 243, 257, 262
MO transition densities in, 140
one-electron operator in, 8
Pauli principle/Fermi statistics implications: 248, 254, 261
occupational angle variables, 257, 261
perturbation theory in, 227
self consistent field (SCF) theories, 8-10, 125, 131, 146, 153
SO Euler equation in, 250
spin-position configurational space in, 221
valence-shell ("frozen" core) approximation in, 9
Pariser formula, 9, 24, 34
partitioning of charge distribution (see populational analysis schemes): 162
AIM, 1, 261
in a molecule, 2, 3, 23, 57, 64, 65, 90
in a reactive system, 30, 34, 40-45, 48-50, 67-69, 87-90, 243
in Hilbert space of AO, 1, 8-10, 161
in physical space, 1, 161, 261
perturbation-response matrix transformations: 220
for AIM in constrained equilibrium, 64, 65, 237, 238
for AIM in global equilibrium, 59, 64
local in constrained equilibrium, 232, 233
SO-resolved, 252, 253
polarization of reactants (see reaction stages): 152, 162, 178-183, 187-191, 243
direct/CT-induced, 152, 153
populational analysis schemes, 1, 9, 153, 161, 213, 260
principal axes of hardness tensor (see collective modes of charge displacements/PNM), 70, 71
principal hardness hierarchy, 76-79, 97, 99, 112, 114, 115
promotion of reactants, 168, 174, 175, 190, 194-196
propylene adsorbate, 197
reactants:
acidic/basic (see HSAB), 149-152, 156-160
hardness/softness matching of, 159
polarization of, 152-157
reaction:
activation barrier of, 167
coordinate/mechanism, 140, 148, 166, 167, 170-173, 197, 199, 205, 211, 213-219
NPS conjugate of, 140
minimum energy path, 158, 159
progress variable, 157-160
rules, 91
stages, 1, 48-50, 87, 149-157, 177-183
stimuli, 91, 152
reactive systems: 161-219, 229-241
charge displacements in: 175
CT patterns of, 162, 175, 178-194, 197, 201, 258-261
diagonal/off-diagonal components of, 177-179, 187, 188
direct-P patterns of, 162, 173 - 175, 178-194, 197, 201
in situ FF of reactants, 260
overall (P+CT) patterns of, 178-194, 197, 201, 205-219
reactive cross-sections, 28, 29
reactive charge distribution, 143
reactive channels, 102, 109, 141, 170, 199, 215
reduction of derivatives (see Jacobians), 24, 25
reduction/oxidation (test, local), 162-166
relative external potential:
AIM indices, 63, 236
local, 46, 57, 230
relaxational corrections (see chemical potential, Fukui function, and hardness)
relaxation effects, 95, 96, 141, 270, 271
relaxation promotion, 96, 98-100, 102
relaxational hierarchy, 24
relaxational matrix (see relaxed FF), 55, 56, 66, 69, 84, 88, 90, 270, 271
resolutions/discretizations/descriptions: 161, 162, 220
AIM, 4, 7-10, 23, 51-57, 60-161, 236-239
global, 4, 51, 52, 53

group (coarse-grained), 7, 51, 52, 64, 65
local (fine-grained), 4, 7, 45-48, 51, 52, 57-60
MO: 7-10, 51, 52, 243-267
multicomponent formalism of, 249
weighted external potentials in, 249
normal (PNM), 69-79
of electronic energy, 51, 52
P/CT, 71, 74-77
reactant, 51, 52, 96, 229-241
rutile clusters, 163-175
Sanderson's principle (see chemical potential equalization), 5, 40, 162
selective oxidation of hydrocarbons, 156, 189-197, 199, 205, 211, 213, 215, 216, 218, 219
semi-empirical SCF MO theories: 9, 161, 196
SCCC, 24
ZDO (CNDO, INDO, MINDO, MNDO, ZINDO, SINDO1, Scaled-INDO), 9, 75, 97, 118, 134, 143-145, 147, 176, 177, 186, 187, 199, 200
separated atoms/ions limit (SAL), 30, 35, 42, 49, 50, 162, 202, 274
separated reactants limit (SRL), 87, 96, 97-100, 102, 188, 203, 209, 216, 218, 229, 230, 275
simple-bond-charge model, 131
Slater transition-state approach, 257
softness: 93, 225
condensed matrix, 38, 52, 232, 233, 237
generalized kernel, 59
generalized kernel matrix, 232, 233
generalized matrix, 64, 65
global, 10, 52, 53, 57, 58
discontinuity of, 223
internal (isoelectronic, *in situ*) quantities, 60-64, 67-69
kernel, 52, 57, 226, 227
kernel matrix, 230, 232
local/fragment, 38, 42, 52, 53, 57-59
discontinuity of, 223
local matrix for subsystems, 231
matrix, 8, 52, 63, 64, 80, 236
quantities in Kohn-Sham theory, 222-228
relaxed (external) of subsystems, 90
relaxed (internal) of subsystems, 89
spectral analysis of tensor/kernel, 220, 226, 227
softness (response) representation, 52
spherical AIM: 11-45
effective nuclear charge of, 30-34
non-equilibrium Legendre transforms of, 17, 19, 20, 26
periodic behaviour of, 17, 18
polarized AIM sensitivities of, 32, 33
sources of charge sensitivities for, 17, 18
states, potentials, and second-order derivatives of, 13-45
stability considerations (see Le Châtelier and Le Châtelier-Braun principles): 20, 22, 23, 93, 141-148, 220, 258, 268
atomic, 20, 22, 23
CSA internal (I)/external (E) stability criteria /regimes, 143-146, 186, 227
MEC mode instability, 145, 146
state variables: 4-6, 45-48, 51
atomic/AIM, 7, 11-52
conjugate, 13, 15, 16, 19, 20, 22, 23, 45, 52
extensive, 13, 16, 21, 22, 45, 225
global, 4-6, 51-53
ground-state, 4, 51, 52
intensive, 13, 16, 21, 22, 45, 225
local, 5, 6, 51, 52
MO, 8-10, 51, 52, 243-248
reactant, 51, 52
stiffness moduli (see hardness): 16
subsystems: 2, 3, 87-90
closed/opened: 2-5, 81
externally, 2, 3
mutually, 2, 3
charge sensitivities of, 37-45, 54, 56, 64-66
combination formulas for, 64-66
complementary, 23, 37-45, 55, 67, 84, 87, 88, 137, 143, 144, 146, 157, 258, 268, 270-273
condensed quantities, 64-66
displacement, 2
frozen, 2, 3, 55
relaxing, 2, 3, 55
response, 2
supermolecule / closed subsystem calculations, 260, 261
surface processes (see chemisorption and catalytic), 92, 101, 161-219
theory of chemical reactivity (see chemical reactivity theory), 1, 91
thermodynamic analogy (see Legendre transformed representations and electronic energy /Legendre transforms of): 13, 15, 4548, 161
non-equilibrium potentials, 17, 19, 20
thermodynamic potentials, 13, 15-17, 45-48
variational principles, 19, 20, 45-48

toluene, 75-79, 83, 85, 96-107, 109-123, 125-130, 143-146, 148, 153-157, 163, 175-197
transition metal complexes (see chemisorption systems): 41, 42
  direct influence of ligands in, 42, 43
  directing influence of ligands in (*trans-*, *cis-*effects), 43-45
  indirect influence of ligands in, 42, 43
  kinetic *trans/cis* effects in, 44, 45
  ligand substitution reactions of, 43-45
  symbiosis rule for, 41-43
two-electron density matrices: 163,202, 274, 275
  Löwdin normalization of, 274
water adsorbate, 163-175
vanadium pentoxide clusters: 96-107, 109-123, 125-130, 143-148, 153-157, 163, 175-197
  bipyramidal, 97-100, 145, 186, 187
  charge stability of, 146-148
  two-layer, 146-148, 163, 175-197
variational principles:
  CSA-AIM, 63
  for local/AIM FF, 267-269
  for optimum SO, 262-264
  in HK DFT, 4, 46, 48, 57, 220, 221, 261
  in terms of CBO matrix elements, 261
  in wave-function theory, 4, 45, 220, 221, 261
  local energy equalization, 221
  minimum principle of global hardness, 267-269
  of non-equilibrium Legendre transforms, 19, 20, 45, 46
variational theories for subsystems (see KS theory), 260
vector space:
  of AIM populations, 69-71, 74, 77, 81-83, 85, 86, 89, 90, 96, 105, 108, 117, 133, 270, 272, 273
  of complementary subsystems, 143
  of nuclear bond-stretch MEC, 132-134
    metric tensor of, 133
  of populational MEC, 133, 134
    metric tensor of, 133
  of REM, 270
vibrational normal modes (VNM, see collective modes of nuclear position displacements)
virial theorem, 131